IRON VALLEY

The Transformation of the Iron Industry in Ohio's Mahoning Valley, 1802–1913

CLAYTON J. RUMINSKI

Trillium, an imprint of
The Ohio State University Press
Columbus

Copyright © 2017 by The Ohio State University.
Trillium, an imprint of The Ohio State University Press.

Library of Congress Cataloging-in-Publication Data
Names: Ruminski, Clayton J., author.
Title: Iron valley : the transformation of the iron industry in Ohio's Mahoning Valley,
    1802–1913 / Clayton J. Ruminski.
Description: Columbus : Trillium, an imprint of The Ohio State University Press, [2017] |
    Includes bibliographical references and index.
Identifiers: LCCN 2016050468 | ISBN 9780814213216 (cloth ; alk. paper) | ISBN 0814213219
    (cloth ; alk. paper)
Subjects: LCSH: Iron industry and trade—Mahoning River Valley (Ohio and Pa.)—
    History—19th century. | Steel industry and trade—Mahoning River Valley (Ohio and
    Pa.)—History—19th century. | Iron industry and trade—Mahoning River Valley (Ohio
    and Pa.)—History—20th century. | Steel industry and trade—Mahoning River Valley
    (Ohio and Pa.)—History—20th century.
Classification: LCC HD9517.O3 R86 2017 | DDC 338.4/966910977139—dc23
LC record available at https://lccn.loc.gov/2016050468

Cover design by Andrew Brozyna
Text design by Juliet Williams
Type set in Palatino and Bell Gothic

9 8 7 6 5 4 3 2 1

CONTENTS

ILLUSTRATIONS

FIGURES

TABLES

WRITING A COMPREHENSIVE HISTORY of the iron industry in the Mahoning Valley proved an exciting, yet daunting, task. Primary resources on the subject, including company correspondence, personal letters, and ledgers, are extremely limited and, in most cases, nonexistent. Because nearly all Youngstown area iron companies were absorbed by larger entities such as Youngstown Sheet & Tube, Republic Iron and Steel Co., and U.S. Steel in the early twentieth century, records from what were some of the largest iron enterprises in the state of Ohio were destroyed or lost. Perhaps for this reason, many historians have neglected Youngstown iron and steel in the nineteenth century and have largely focused research efforts on deindustrialization and the fall of the steel industry in the late twentieth century. Thus, traditional use of manuscript collections were not applicable for this study. However, printed and published sources proved absolutely imperative in piecing together the Mahoning Valley's previously underrepresented and unexplored early industrial development.

Perhaps one of the more unorthodox and extensive primary resources used in this study are photographs. Along with contributing important visual aids that help interpret and decipher the many furnaces, rolling mills, foundries, ironmasters, and industrial leaders in the Youngstown area, the photographs provide clues to the technological and geographical development, business decisions, and changing industrial landscape of the iron and steel industry that we lack in written records. In addition to the photographs, industry trade journals and newspapers played an equally important role in interpreting the market trends, innovations, and responses to economic fluctuation by Mahoning Valley ironmasters.

Of course, searching for the photographs used in this book was a lengthy and exciting venture. Each photograph, whether discovered at the onset of this project or after the manuscript was completed, helped unravel and interpret the region's industry with new clues and information. Along the way, I met numerous good people who assisted me in my search, from archivists in large collections, to devoted curators of small historical societies, to iron and steel industry enthusiasts, to private collectors. When I first began my research, I was fortunate to collaborate with Rick Rowlands and Rich Rees of the Youngstown Steel Heritage Foundation, caretakers of the historic Tod Engine and many other wonderful steel industry artifacts. Their knowledge of iron- and steelmaking processes helped influence the direction of this study and gave me a much better understanding of the subject that historic manuals and books alone could not provide. I give my deepest gratitude to the Youngstown Historical Center of Industry and Labor archivist and librarian, Martha Bishop. Her help in locating collections and allowing me to browse the stacks resulted in some marvelous discoveries. I also owe a debt of gratitude to the faculty at the Youngstown State University history department, including Thomas Leary, Donna DeBlasio, Martha Pallante, and David Simonelli. Their support of this project from its humble beginnings has made this book possible.

I could not have completed this book without the help of Pamela Speis, archivist at the Mahoning Valley Historical Society. Her gracious devotion to my seemingly endless requests for photographs and manuscript material are, and always will be, greatly appreciated. I also want to thank Mahoning Valley Historical Society director Bill Lawson for his support of my project. Indeed, numerous historical societies across the Mahoning Valley gave their time and effort to help me locate rare photographs. I want to thank Ralph and Cecilia Cooper and the Hubbard Historical Society; Marian Kutlesa and the Struthers Historical Society; Audrey John and the Niles Historical Society; Roberta Lawrentz and the Girard Historical Society; Larry Baughman and the Poland Historical Society; and the Ohio History Connection. I also want to thank the National McKinley Birthplace Museum for allowing me to use items from their collections and Bill Lewis, who gave me access to the priceless photograph collection in *The Vindicator's* archives. Of course, not all photographs come from archives. A big thank you goes out to Shelley Richards, Tom Molocea, Joe Tucciarone, and Dave Madeline for sharing their photographs and family history. It was indeed a pleasure to speak with you all. I also want to thank my editors and production team at The Ohio State University Press, as well as the anonymous reviewers chosen by the Press. Their critique and suggestions were extremely helpful, and I am indebted for their hard work and support.

Lastly, this project could not have been completed without the aid and support of my parents, family, and friends. My interest in the iron and steel industry began when my father told me stories of growing up in Struthers, Ohio, in the shadow of Youngstown Sheet & Tube's Campbell Works. Sharing a small house on the hillside with his eight brothers and sisters, he sat on the roof watching the blast furnaces and coke plant operate while the glow of the red-hot iron lit up the night sky. Being too young to personally witness the Mahoning Valley's steel mills in operation, his stories encouraged me to seek out photographs of what once made Youngstown the "Steel Valley." The impressive size and the surreal nature of the industry peaked my unending curiosity, and I never looked back.

IN DECEMBER 1910, fifty of industrialist Joseph G. Butler Jr.'s associates and closest friends honored the veteran manufacturer's seventieth birthday with an extravagant dinner at the Union Club in Cleveland. Some of the most famous "Captains of Industry" attended the dinner. These included high-profile names such as Charles Schwab, Carnegie Steel president Alva Dinkey, William G. Mather and Samuel Mather of Cleveland, James A. Campbell, and the young Henry H. Stambaugh, Butler's longtime business associate, friend, and president of Youngstown's Brier Hill Iron and Coal Company. Butler also received congratulatory telegrams from Andrew Carnegie, former Ohio governor Myron T. Herrick, and U.S. Steel chairman Elbert H. Gary. "I don't like to talk about myself," said Butler, "but I'm certainly proud to receive so many congratulatory telegrams and dispatches from my friends in this country and abroad."[1] Perhaps best known for founding the Butler Institute of American Art in Youngstown, Butler was one of the brightest minds in iron and steel and highly respected among his contemporaries. That evening in Cleveland, he reminisced about his experiences in the old iron industry, from his childhood working for James Ward & Co. in Niles, Ohio, to his first associations with the Brier Hill Iron and Coal Co. in the late 1870s. During Butler's long career, the economic, geographic, and technological structure of the iron and steel industry changed dramatically. He witnessed the shifts from small-scale, independent enterprises to immense industrial combines and from the use of localized raw materials to the discovery and exploitation of huge deposits of iron ore near Lake Superior, as well as the transition from iron to steelmaking in the late nineteenth century, to name a few. When Butler first started in the iron business in the 1860s, the United States rolled less than 17,000 tons of steel per year; in 1910, that number reached nearly 20 million tons.[2] Nonetheless, after fifty years in the business, Butler's Brier Hill Iron and Coal Company remained strictly a merchant pig iron manufacturer and never engaged in the production of steel (figure I.1).

Ultimately, merchant pig iron firms like Brier Hill—defined as independent companies that produced only pig iron, the base product of the iron and steel industry, for outside consumption—did not require extreme product diversification.[3] Throughout the nineteenth century, the Mahoning Valley was one of the principal merchant pig iron centers in the Midwest, shipping its product to major iron and steel firms in Pittsburgh, Cleveland, and other dominant industrial centers rather than consuming it locally. Even as independents fell to the corporate monopolies forged by Schwab, Carnegie, J. P. Morgan, and other prominent financiers at the turn of the twentieth century, many small merchant furnace companies in the Mahoning Valley still had ample markets for their pig iron. Accordingly, there was still a strong demand for their product after the establishment of the United States Steel Corporation in 1901 that continued for another three years. However, demand fluctuated considerably until the start of the First World War, and Valley industrialists took notice.[4]

By the eve of the Great War, the Mahoning Valley's time-honored and once great iron rolling mills vanished in the face of Big Steel. The rolling mills' value, however, was limited, unlike the region's merchant iron furnaces that provided these mills with pig iron to make their wrought-iron products. The large steel combines had successfully exonerated themselves of these often strike-ridden and costly plants, which resulted in higher profits and productivity, but precipitated a gradual loss for merchant furnaces and smaller, independent iron companies that remained. In addition, companies like Republic Iron and Steel and Carnegie Steel sought to fully integrate operations to depend less on outside sources of raw materials and pig iron. However, the conversion from iron to steelmaking in the Mahoning Valley occurred so swiftly that every steel company in the Valley relied on backward integration, giving the region's vestigial merchant furnace companies a strong but increasingly temporary local market. Steel companies constructed large-scale Bessemer plants and open-hearth steelworks with such rapidity that most still purchased pig iron from outside markets in order to feed their steel furnaces. Acquisition and construction of blast furnaces proved incredibly successful for major corporations like U.S. Steel, but the integration of iron and steel production came slower for other companies.

FIGURE I.1. Many companies like Brier Hill Iron and Coal Co. that remained in Youngstown and the Mahoning Valley after the turn of the twentieth century first began operations in the Civil War era. Rather than slowly dying out, merchant pig iron firms in the region prospered after the introduction of steelmaking in the United States. This c. 1905 image shows laborers loading pig iron into rail cars for shipment at Brier Hill's Grace furnace. Courtesy of Tom Molocea.

Between 1905 and 1907, merchant furnace companies in the Mahoning Valley were both prosperous and profitable.[5] As a result, several firms in the region streamlined their furnaces in order to remain competitive and increase their output; between 1908 and 1915, furnace operators rebuilt and modernized four of the Mahoning Valley's seven merchant stacks. However, the immanency of vertical and horizontally integrated steel firms loomed over the Mahoning Valley's merchant pig iron operators. The constant fluctuation in the market often prompted impulsive responses from the Valley's pig iron manufacturers. In 1910, an editorial in *Iron Age,* an important industry trade journal, stated that "for 20 years each interval of depression, or even of slackness in the iron trade, has brought forward the proposal that Mahoning . . . Valley blast furnaces making iron to be sold to steel companies should join in the formation of a steel company and become the consumers of their own pig iron."[6] After the turn of the twentieth century, the business of providing steel companies pig iron for converting to steel was more profitable than catering to rapidly vanishing rolling mills and foundries alone. For years, companies

FIGURE I.2. This c. 1912 photo of Youngstown Sheet & Tube's East Youngstown Works (later renamed the Campbell Works after Sheet & Tube's president) shows the company's incredible expansion from a small-scale iron mill to one of Youngstown's largest integrated steel producers within a period of eight years. The company's new blast furnaces appear in the back right, while a Bessemer converter can be seen "making a blow" in the left foreground. Courtesy of the Youngstown Historical Center of Industry and Labor (MSS 140).

like Brier Hill Iron and Coal Co. sold their output to U.S. Steel and other emerging steel companies in the Youngstown region. However, these same companies began to construct their own blast furnaces, nearly eliminating their reliance on the open market (figure I.2).[7] Likewise, U.S. Steel not only made sure their Youngstown plants were self-sufficient in iron production but also invested in blast-furnace construction for the majority of their ill-equipped steelworks around the Pittsburgh, New Castle, and Cleveland areas. Thus, the writing was on the wall for the Mahoning Valley's remaining merchant furnace companies: conform to steel production or accept increasing insignificance or bankruptcy in the iron and steel industry. These blast furnace firms' efforts to adapt to a new market exemplified their tenuous foothold in the industry as independent merchant pig iron firms declined and unchallenged steel companies grew to dominate America's industrial economy.

Like many in the United States, remaining Mahoning Valley iron companies struggled for survival in the early twentieth century, but most in the region enjoyed incredible affluence in the nineteenth century, when iron dominated over the novelty and curiosity that was steel. Despite its well-documented supremacy in the iron and steel trade, Pittsburgh's manufacturers relied on the Mahoning Valley's furnaces for supplies of crude iron to convert into stronger, finished products well into the 1880s. Even steel magnate Andrew Carnegie looked to the Valley's high-capacity iron furnaces as inspiration to build his own such mills in Pittsburgh in the early 1870s.[8] However, as with other major manufacturing centers, the Valley's industry had modest, and mostly insignificant, beginnings. In 1802, James and Daniel Heaton constructed Ohio's first iron furnace along Yellow Creek, in what became the city of Struthers. Despite naming the furnace the "Hopewell," the small enterprise proved more historically symbolic than financially prosperous for the two brothers. Today, the remains of the Heatons' iron furnace still rest along Yellow Creek, and the city of Struthers embraces its history as the "Cradle of Steel," representing the Valley's rise into an industrial power.[9]

Ironically, however, the Hopewell furnace itself did nothing to "cradle" the Mahoning Valley's iron and steel industry. After selling the Hopewell furnace, James Heaton moved to what became the city of Niles, Ohio. There, he built another small iron furnace and established the town that, along with Youngstown, became one of Ohio's leading Civil War–era iron producers. Nevertheless, such large-scale iron production was not as common in the years leading up to the Civil War. The localized distribution of raw materials and lack of efficient transportation methods throughout areas in Ohio served to fragment the iron industry and isolate the state's existing industrial complexes. Thus, before the 1840s, agriculture, not industry, dominated the Mahoning Valley and most other Ohio towns. The discovery of a local source of high-grade raw materials such as iron ore and coal in the 1840s and 1850s, combined with the opening of the Pennsylvania and Ohio Canal, launched the Mahoning Valley into industrialization. This critical period in the region's history rightly served to cradle the Valley's iron and steel industry by pushing the construction of iron mills and attracting capitalists and ironmasters alike from western Pennsylvania and the Western Reserve. Several prominent families thus established themselves with wealth attained from coal mining and iron making. Among the most important were the Tod family, namely David Tod, Ohio's Civil War governor who capitalized on the region's vast coal reserves and later developed extensive iron mills with his four sons. The Stambaugh family also joined alongside the Tods in their coal mining and iron enterprises, while the Wicks became a dominant force in the local consolidation of the Valley's iron mills in the late nineteenth century. The region's industrial elite became a tight-knit group, often adjoining industrial ventures rather than opposing each other in direct competition. The result was a small but dominant group of industrialists and mills that had little diversification at a time when expanding operations meant survival in the industry. Nonetheless, these older social elite laid the foundations of near-exponential economic and industrial growth in the Mahoning Valley. By the early to mid-twentieth century, steelworkers and their families fittingly christened the area the "Steel Valley." This was not the region's only industrial moniker. During the Second World War, the Mahoning Valley was nationally known as "America's Little Ruhr" after the largest steel-manufacturing center in Germany.[10] But Youngstown and the Mahoning Valley were not built on steel, nor was steel manufacture a factor in the region until the turn of the twentieth century.

A number of circumstances led to the Mahoning Valley's longtime reliance on iron over steel production. These include factors ranging from technological innovation, to product niches in the industry, to competition among nearby industrial districts, to the decision among Valley iron manufacturers to remain mostly small-scale, market-driven producers. Thus, this book uses a comprehensive chronological narrative to highlight the major people and events that transformed the Mahoning Valley from an agricultural backwater into an industrial powerhouse by the beginning of the twentieth century. However, the primary theme of this study is not so much the birth of the steel industry in the Mahoning Valley, but the life and eventual death of the iron industry in the face of big steel production. For much of the nineteenth century, Mahoning Valley iron manufacturers failed to see the prestige of steel—generally considered the stronger metal—especially for rails and structural material. Several companies in the United States began mass-producing steel after the Civil War using the pneumatic process pioneered by American William Kelly and Englishman Henry Bessemer. Nonetheless, wrought iron, which, unlike steel, could not be mass-produced and relied on the intense labor of highly skilled ironworkers called puddlers, still had numerous applications in the 1880s. Furthermore, steel plants required massive capital. Notwithstanding a failed attempt at producing steel during the Civil War, many industrialists in the Mahoning Valley lacked the money and initiative to bring together a successful steel company. In addition, steel production was not a requirement in the region until the 1890s, when the pressure to produce the metal from other markets caught up to Valley iron manufacturers. Even after the Mahoning Valley produced its first steel in the 1890s, iron remained the dominant metal in the Valley until the consolidation of most of the country's iron and steel firms after 1899. In Pittsburgh, many small iron and steel manufacturers remained independent after the turn of the twentieth century and chose not to compete with large-scale steel production by successfully making specialized products.[11] In the Mahoning Valley, iron manufacturers had little choice in the matter because their products were not specialized enough to maintain independent operations, nor were their operations as modernized as other prominent districts'. Mahoning Valley industrialists' reliance on iron manufacture thus gives an interesting perspective on reluctant adaptation amidst an ever-modernizing industry.

In the late nineteenth century and throughout the twentieth century, Youngstown and the Mahoning Valley had strict competition from nearby industrial giants Pittsburgh and Cleveland, which loom larger in both historical studies and the public consciousness.[12] As one nineteenth-century journalist put it, Pittsburgh had long "sat upon the iron throne of America with a never fading wreath of smoke streaming from her proud head."[13] However, western Pennsylvania's Shenango Valley also served as one of the most important merchant pig iron centers between Cleveland and Pittsburgh.[14] The region sits directly across the Ohio and Pennsylvania border, its major waterway being the Shenango River, and it includes the towns of Sharpsville, Sharon, Farrell, West Middlesex, and New Castle. Often, historians include the Mahoning Valley with the Shenango because of their geographic proximity. Kenneth Warren collectively refers to the Mahoning and Shenango Valleys as the "Valleys" district.[15] This book, however, treats them as largely separate regions and does not include the Shenango Valley in extreme detail. There are several reasons for this distinction and partial omission. In nearly all of the nineteenth-century iron and steel industry trade publications, editors separated the two valleys, referencing them individually. For example, when giving statistics for iron and steel production, the American Iron and Steel Association always separated the two regions and never combined their annual iron or steel production numbers. In many circumstances, the two valleys competed for higher pig iron production totals.[16] Second, a detailed study of the two valleys is beyond the scope of a single book of this nature. Although it is true that several Mahoning Valley industrialists held interests in iron firms within both valleys, most of these business connections occurred before and during the Civil War. The separation of industrial interests among iron manufacturers in both valleys became more distinct after the 1870s. By the 1890s, very few, if any, iron manufacturers had interests in both valleys. A final reason for the separation of the two valleys is that the American Iron and Steel Institute did not form the "Youngstown District" until 1916.[17] Under the institute's definition, the Youngstown District included all iron and steel mills in both the Mahoning and Shenango Valleys. The date of the new district's formation is beyond the time frame of this book. Nevertheless, as a rival of Mahoning Valley iron manufacturers, the Shenango Valley does play certain roles within the context of this study.

Throughout *Iron Valley*, there is an overarching theme of technological innovation and stagnation among the Valley's iron companies. In Wollman and Inman's *Portraits of Steel*, the authors emphasize that "iron and steelmaking are industries driven by technology."[18] In the Youngstown area, as with all other iron- and steel-producing districts, the machinery and technology of the industry went hand in hand with labor, production outputs, and many other facets of the trade. Iron and steel companies tried to mechanize the industry as much as possible, which both limited the leverage of labor unions and increased profits and productivity. Thus, technological development played a major role as the Mahoning Valley transformed from an iron to a steel center, often leading to hostile labor relations that thwarted skilled labor and helped dismantle the iron industry. Such themes are prevalent throughout the book.

Chapter 1, "Development and Struggle, 1802–1840," discusses the early attempts at marginal iron making in the Mahoning Valley. Youngstown was slow to embrace manufacturing, and it was poised to become an agricultural region. This chapter briefly introduces the geographical disposition of the Valley and documents the Heaton family's attempts at small-scale, charcoal pig iron manufacturing in Struthers and Niles, Ohio. Charcoal iron production—a success in other regions of Ohio—proved a failure in the Valley, and further iron manufacture seemed unfeasible in the Youngstown area until the discovery of a better fuel source and the construction of practical transportation.

Chapter 2, "Brier Hill Coal and 'Merchantable' Pig Iron, 1840–1856," discusses the Mahoning Valley's entrance into economic prosperity. The opening of the Pennsylvania and Ohio Canal in 1839, and the subsequent opening of coal mines in Youngstown, spurred the region's transformation into an industrial center. The Valley's local coal deposits, known as Brier Hill coal, had special characteristics that allowed ironmasters to use it as a fuel directly in the blast furnace, therefore foregoing use of charcoal. By the early 1850s, ironmasters from western Pennsylvania and coal speculators from Cleveland began building blast furnaces and opening coal mines in the Valley.

Chapter 3, "Railroads, Coal, Iron, and War, 1856–1865," covers the Valley's period of railroad growth and its response to the Civil War. As pig iron production developed in Youngstown, other areas in the Mahoning Valley boomed after the 1856 opening of the Cleveland and Mahoning Railroad, which connected Cleveland to Youngstown. The consistent and steady growth of coal mining and pig iron production made the Mahoning Valley a major supplier of iron to the Union during the Civil War. Throughout the war years, the need for iron spurred additional manufacturing in Youngstown, which put the region, like Cleveland and Pittsburgh, on a course for exponential industrial growth in the postwar era.

Chapter 4, "Expansion and Depression, 1865–1879," looks at the extreme periods of boom and bust after the Civil War. During the

Reconstruction Era, the building of railroads and the country's continued westward expansion fueled the demand for Youngstown iron. Not only did the Mahoning Valley increase its pig iron production, but other industrialists began building rolling mills for the manufacture of wrought-iron products for farming, railroad construction, and shipbuilding. However, Valley ironmasters did not embrace the production of steel, and the Panic of 1873 bankrupted many of the region's small iron companies. This led to a large reorganization of most of the Valley's iron mills under Youngstown capital in the mid- to late 1870s.

Chapter 5, "The Pressure of Steel, 1879–1894," examines the Mahoning Valley's response to decreasing markets and labor strife. As the region failed to embrace steel production—even as rival industrial centers such as Cleveland and Pittsburgh began producing the metal—Youngstown ironmasters had to develop a niche market. They did this by first providing pig iron to foundries amidst rapid construction of new steelworks in the United States, and later Bessemer-grade pig iron to steelworks in Pittsburgh and Chicago. Many of the iron mills in the Valley became obsolete, while owners of Youngstown's major iron companies felt the pressure to make steel products over iron. As a result, Youngstown iron manufacturers organized the region's first Bessemer steel company in 1892, just before the great Panic began.

Chapter 6, "Steel, Consolidation, and the Fall of Iron, 1894–1913," covers the Valley's entrance into steel production. Although Youngstown produced its first steel in 1895, the steelworks would be the only one of its kind organized by local iron companies. In order to weather economic turmoil and reduce competition, the country's leading industrialists began to consolidate independent iron and steel mills. This led to the creation of immense national trust companies, the largest being the United States Steel Corporation. The formation of these trusts and the merger of many of the Valley's iron mills swiftly modernized and centralized steel production in Youngstown, leading to the gradual downfall of both the wrought-iron and merchant blast furnace industries in the Mahoning Valley.

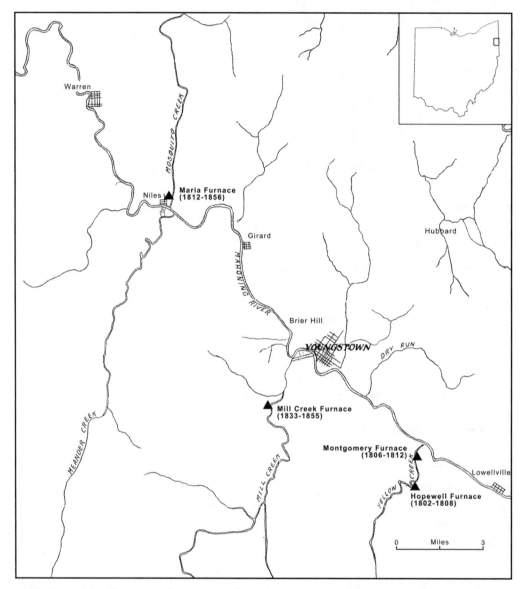

FIGURE 1.1. Map illustrating the meager industrial advances in the Mahoning Valley between 1802 and 1835, with triangles representing blast furnaces, all of which used surrounding forests for fuel during this period. Despite efforts from a few ironmasters, early iron production stagnated in the face of a small yet dominant agricultural trade and fuel and transportation issues.

DEVELOPMENT AND STRUGGLE, 1802–1840

> As early as 1803 a start was made in an industry that was destined to become the very backbone of the growth and prosperity of the Mahoning Valley, although Mahoning Valley residents, who leaned towards agriculture did not realize this.
> —Joseph G. Butler Jr., *History of Youngstown and the Mahoning Valley, Ohio*, vol. 1 (Chicago: American Historical Society, 1921), 174.

BEFORE SETTLERS in the Mahoning Valley began to utilize the area's vast resources for industrial enterprises, and before Ohio became a state in 1803, the Valley was a lush, relatively untamed wilderness. Once part of the Connecticut Western Reserve, the land that now hosts the cities of Youngstown, Warren, Niles, Hubbard, Struthers, and other towns and villages is a narrow river valley that stretches twenty-three miles in a northwest-southeast orientation, with its major waterway being the Mahoning River. The name *Mahoning* comes from the Native American word *Mohoning* or *Mahonik*, meaning "salt lick."[1] The river itself varies in width from 100 to 300 feet and stretches 108 miles southeast. It begins in Stark County and flows through Portage County and the southwest portion of Trumbull County, then continues its southeasterly flow through Mahoning County and into Lawrence County, Pennsylvania, where it joins the Shenango River in New Castle. In total, the river flows through eight different counties in northeast Ohio and the extreme western portion of Pennsylvania.

A number of important tributaries feed the Mahoning River. These include Mill Creek in Youngstown, Mosquito Creek in Niles, Yellow Creek in Struthers, Eagle Creek in Leavittsburg, and Meander Creek in Austintown and Niles—all of which supplied early industry with vital sources of waterpower. The Valley as a whole appears as a rolling plain that slopes to the north and has a maximum elevation of about 500 feet above the water level of the Mahoning River. Very little of the surface lands in either Trumbull or Mahoning County is level; they alternate between rounded hills and broad valleys, and Mahoning County has a significantly higher level of erosion and a well-defined valley. The valley averages about a half mile in width from Warren to Lowellville, where it becomes progressively narrower, with steep bluffs, as it nears the Pennsylvania border. In Trumbull County, the river flows through rolling country with relatively flat or gently sloping banks on both of its sides until it reaches Mahoning County, where it encounters heavily bedded sandstone that overlies seams of coal. Underneath the valley sit rocks belonging to the carboniferous system that includes a conglomerate of sandstone, shale, limestone, fire clay, coal, and iron ore.[2]

The Mahoning River and its various tributaries were the lifeblood of the Valley. In 1806, the river was declared a navigable stream as far west as Newton Township in Trumbull County, but in 1829, navigability was limited to Warren because the clearing of timber along the riverbed greatly altered the river's flow by generating sporadic periods of exceptionally high and low water levels, a common concern in industrializing areas throughout nineteenth-century America.[3] To settlers of the Western Reserve, however, these waterways provided a means of early transportation and power sources for grist and flouring mills built along their banks, the first being Lanterman's Mill along Mill Creek in Youngstown. Like many regions in the United States at the time, the Mahoning Valley was rich with undeveloped wilderness, ample waterpower, and hardwood forests that allowed early settlers to fashion homesteads and tools. The same forests held the potential to provide an abundance of charcoal fuel for the manufacture of pig iron.

Early speculators, however, did not immediately recognize the Valley's rich ferrous mineral resources. Instead, they focused their early efforts on salt production. The saline waters after which the Mahoning River was named proved vital to early inhabitants of the region and attracted others who wished to make a profit. Native Americans knew about the great salt springs, located on a tract of land between Weathersfield and Austintown, south of the Mahoning River, long before the springs caught the attention of Pennsylvania colonists in 1755. When General Samuel H. Parsons, a lawyer and Revolutionary War soldier from Middletown, Connecticut, purchased the 24,000-acre tract of land containing the salt springs from the state of Connecticut in 1788, he surveyed the site himself and established a temporary saltworks from which he intended to make a fortune. Parsons, however, died in 1789 while attempting to cross Beaver Falls in a canoe, leaving the saltworks abandoned. Other early settlers attempted salt production at the site but with tragic results. A storekeeper for traders Duncan & Wilson was killed by Native Americans in 1786, while the same fate befell another salt maker in 1804. The salt springs were also the site of the only known threat of a Native American uprising in the Mahoning Valley.[4] Despite the sacrifices made in attempts to reap its financial benefits, the salt content in the springs was not high enough to yield any profit, and the only value the land provided was as a gathering place for deer. Consequently, for much of the nineteenth century, the swampy marsh of the salt springs supplied ample hunting grounds for settlers and was eventually abolished by a road leading from Youngstown and, later, the Baltimore & Ohio Railroad's line. In 1798, the state of Connecticut reacquired the springs, and they became a portion of the roughly 3.3 million acres included in the Connecticut Western Reserve.[5]

Though the salt springs functioned as little more than a hunting grounds for early inhabitants, the rest of the Valley offered other exploitable natural resources. The bustling town of Youngstown was still a distant dream to pioneers, one of whom was land speculator John Young. Born in New Hampshire in 1763 to a father of English descent, Young emigrated to Whitestown, New York, around 1780. In the spring of 1796, Young left Whitestown for the Western Reserve, where he agreed to buy Township Number 2, Range 2, pending close inspection of the land. Young did not go alone, but brought with him surveyor Alfred Wolcott and his assistants Isaac Powers, a woodcutter from Western Pennsylvania, and Phineas Hill of North Beaver, Pennsylvania. In their travels, the group met Daniel Shehy, an outspoken opponent of the British government who had $2,000 in gold he planned to invest in land. Young convinced Shehy to join their group, and in June 1796, they arrived at a point near Spring Common, the present-day site of Youngstown.[6] In August of the same year, Young agreed to sell Shehy 1,000 acres of land in the Township for $2,000. Shortly thereafter, the small company of land speculators built a cabin on the north bank of the Mahoning River, just to the west of Spring Common, where Young established a real estate office for potential settlers. In February 1797, Young temporarily returned to Connecticut and purchased over 15,000 acres of land from the Connecticut Land Company for $16,085, thus establishing the township and village that would later bear his name.[7]

Colonists from the East Coast began moving westward through the dense wilderness of Pennsylvania, crossing the state line into the Northwest Territory and the Connecticut Western Reserve, and eventually arriving in the Mahoning Valley. John Young founded his new village as an agricultural community, and much of the surrounding land outside of what would become downtown Youngstown was immediately mapped into farmland. Young divided the central portions of the land into lots suitable for houses and other buildings, such as stores, blacksmith shops, furniture shops, and hotels. Indeed, new settlers to the Valley envisioned "fertile and prosperous farms surrounding a village containing only such small factories and mechanics as could produce the simple necessities of life."[8] These settlers came equipped with primitive cast-iron cutting tools and iron plows, but supplies were extremely limited. Some arrived with sickles and scythes for reaping grain, and a select few brought a single cow from their old farmsteads.

JAMES AND DANIEL HEATON

For settlers to the Valley, farming became the primary means of livelihood for several decades, though some did try their hand at capitalizing on the region's mineral resources. Around the turn of the eighteenth century, brothers James, Daniel, Isaac, Rees, and Jacob Bowen Heaton arrived in Trumbull County, the seventh such county in the Northwest Territory, established in July 1800. The brothers' father was Isaac Heaton, a merchant, manufacturer, and disputed Revolutionary War veteran born in Connecticut in 1730. Isaac moved to Morris County, New Jersey, when he was about ten years of age and married Mary Booth around 1760. Soon thereafter, Isaac and Mary moved to Berkeley County, Virginia, where they established a successful sawmill operation and acquired several parcels of land in Greene County, Pennsylvania. However, around 1759, Mary and their only child died of unknown causes; Isaac married Hannah Bowen a year later. In 1785, Isaac and Hannah relocated permanently to Greene County, Pennsylvania, where the family developed extensive farming, milling, and iron smelting operations.[9] Isaac established two charcoal-fired blast furnaces near Clarksville, though sources vary and some contend that Isaac constructed only a single furnace.[10] The Heatons' farms provided raw materials for the iron furnace and gristmills, as well as other water-powered mills. Throughout the 1790s, Isaac's sons helped work the furnace and obtained the skills needed to operate a successful ironworks.[11]

After the Heaton brothers left Greene County for the Mahoning Valley, two of the brothers, James and his younger brother Daniel—both in their early thirties at this point—remained in the Youngstown area. The two brothers are now mythical figures in the Mahoning Valley's history, forever immortalized in the song "Youngstown," rock-legend Bruce Springsteen's solemn ode to the city and its iron and steel heritage. Born in 1773, Daniel—described as a tall, stout, and muscular man who possessed a vigorous constitution—was a manufacturer, politician, and representative of his people. Though he was not formally educated as a child, Daniel was an avid reader with strong social graces. An ardent Christian, he published a book, *The Christian's Manual*, through which he promoted religious piety and advocated for universal brotherhood. Daniel was also vehemently antislavery, and in 1811, he instituted Ohio's first Temperance Society. Beyond his status as an iron manufacturer,

FIGURE 1.2. Portrait of James Heaton (1771–1856), c. 1845, a pioneer ironmaster in the Mahoning Valley. Courtesy of Niles Historical Society.

Daniel served as a senator from Trumbull County and later as a representative in the state legislature. He married Amy Hill, daughter of Robert Hill and Prescilla Bowen, and together they had eleven children.[12]

Daniel's older brother James (figure 1.2) had a more direct impact on the development of the Mahoning Valley. Born in Berkeley County, Virginia, in 1771, James moved with his family to Greene County, Pennsylvania, in 1785, where he worked in his father's iron furnace. Around 1793, he married Margaret Williams in Greene County, where they had their first five children before moving to the Mahoning Valley. James was less eccentric than his younger brother, who had gone so far as to change his last name to Eaton by act of state legislature, claiming that the "H" was superfluous. Politically, James was faithful to the Whig Party and was a devout reader of the Baltimore-based national weekly news magazine, *Niles' Weekly Register,* published by editor Hezekiah Niles.[13] In fact, James was such an ardent follower of the Baltimore paper that he later named the Mahoning Valley town of Niles in Hezekiah's honor. After relocating to the Valley with his brothers, James and his wife had five more children, four of whom died very young. Born in 1806, James's only daughter, Maria, was also the only one of his children born in the Valley who lived past the age of twelve.[14]

With knowledge obtained from working their father's iron furnace, the Heaton brothers took advantage of the Valley's resources and established a blast furnace along Yellow Creek, in what became Struthers, Ohio. The exact dates of the construction and blowing in of the furnace are unknown, as several local history sources contain conflicting and contradictory information.[15] According to evidence gathered through archaeological excavation by the late professor John White of Youngstown State University, the Heatons began construction on the furnace in 1802, and historians generally recognize it as Ohio's first blast furnace and the first blast furnace built west of the Allegheny River.[16] Other small charcoal furnaces, such as William Turnbull's Alliance Ironworks and Isaac Meason's Union furnace, constructed between 1789 and 1791 in Fayette County, Pennsylvania, were the first west of the Allegheny Mountains.[17] These stacks, however, were not situated westward enough to cross the Allegheny River's junction with the Ohio River in Pittsburgh. The Heatons named their new furnace "Hopewell" after Mark Bird's successful ironworks in Berks County, Pennsylvania, and they no doubt hoped for the same prosperity in their new venture. On August 31, 1803, Daniel Heaton secured a contract with Lodwick Ripple for the right to mine iron ore on Ripple's tract of land at a rate of twelve and a half cents per ton. He also gained the right to use all of the land's wood for charcoal production to fuel the furnace.[18] By October 1804, Heaton had secured additional contracts for contiguous woodlands. Although the exact date is unknown, these written contracts for raw materials suggest that the Heatons first started making iron at the furnace sometime in 1803.

Thus, the product of the Hopewell furnace—pig iron—became the Mahoning Valley's first heavy industrial commodity. The name "pig" iron originates from most ironmasters' rural and agricultural roots. Workers cast the molten iron into sand molds arranged with individual ingots at a right angle to a central runner, which extended from the opening at the bottom of the furnace, or its tap hole. The resultant formation resembled a litter of suckling pigs. Pig iron, unlike steel or wrought iron, was a brittle product that had little use other than direct casting, serving as a starting point for various refining processes and the base product of the iron and steel industry. The characterizing feature of these metals was largely their carbon content, which determined the strength of iron or steel. Pig iron fresh from the blast furnace had a comparatively high carbon content, ranging from 2% to 5%. Iron artisans required cold pig iron to work it into stronger wrought iron, while in the latter half of the nineteenth century, steel companies needed pig iron in either a cold or a molten state to convert it to steel. Wrought iron is the oldest form of iron made by man and the name literally means "worked iron," or iron forged by hand. Unlike pig iron, it was a commercially pure form of iron with a low carbon content (0.00%–0.03%) that was highly ductile, could withstand corrosion, and was strong in both tension strength and compression. By contrast, steel is a malleable alloy of iron and carbon, with a carbon content that does not usually exceed 1.5%.[19] Steel combines the hardness of pig iron with the versatility and malleability of wrought iron.[20] Each metal held varying degrees of significance and application throughout the industrial revolution, while wrought and pig iron largely dominated in the nineteenth century.

The blast furnace itself was and still is the primary unit for producing pig iron. These metallurgical structures are tall, cylindrical furnaces that reduce iron ore by using a "blast" of air along with a carbon fuel (coal, coke, or charcoal), which provides carbon for the reduction of iron oxide and also administers the necessary amount of heat to carry out the smelting process. In the eighteenth and early nineteenth centuries, blast furnaces like the Hopewell were usually built of quarried stone in heavily forested areas. The size and shape of a furnace largely dictated its capacity. Traditionally, ironmasters measured their furnace by its height and diameter of its bosh—the lower section of the blast furnace shaft where the walls taper inward to meet the crucible, or hearth, where the molten iron accumulated. As technology advanced and other building materials became available, ironmasters constructed furnaces with an outer shell made of sheet iron or steel with a heavy firebrick lining on the inside to withstand the intense heat. Such construction allowed for larger furnaces and higher outputs. Workers called top fillers "fed" the furnace by filling it with fuel and igniting it at its top. As the flame reached the bottom, the furnace crew charged alternating layers of carbon fuel, limestone, and locally mined iron ore into the furnace via a wooden bridge that spanned from the top of the furnace to an adjoining bluff (figure 1.3). The limestone functioned as a fluxing agent that separated the impurities of the carbon fuel and iron ore called slag, which is lighter than iron and floated to the top of the molten mass inside the furnace.

Along with technology, fuel sources changed throughout the nineteenth century. Early American ironmasters like the Heaton brothers used charcoal to reduce iron ore almost exclusively. Produced by coaling adjacent hardwoods in the absence of oxygen, charcoal was initially an abundant fuel in the vast wilderness, but it significantly restricted the life-span of a blast furnace complex. Over the next few decades, ironmasters began to experiment with other, more efficient fuels like bituminous and anthracite coal. However, raw coal required the use of preheated blast air for combustion purposes, a technology patented by Scottish ironmaster James Neilson in 1828. Neilson's contemporary claimed the hot blast was "one of the grandest epochs in the history of the manufacture of iron."[21] Indeed, it spurred incredible industrial growth in both Great Britain and antebellum America. Early hot-blast technology consisted of cast- or wrought-iron pipes encased within a brick stove heated by the open flame of the furnace. This was a makeshift and rudimentary method, as the extreme heat from the furnace flame often damaged the cast-iron pipes, while the air within the stove only reached a maximum of about 800 degrees Fahrenheit. After the Civil War, however, English engineer Thomas Whitwell developed large, cylindrical firebrick stoves using the regenerative heating process that superheated air up to 1,500 degrees Fahrenheit.[22] Blowing engines, sometimes known as the "lungs"

FIGURE 1.3. Reconstructive drawing depicting James and Daniel Heaton's Hopewell furnace along Yellow Creek. Casting iron required a shed, or casting house, that protected both the workers and the molten iron from the elements. There, laborers painstakingly molded the sand beds in preparation for casting. Sketch by Chet Hunt from Poland Historical Society.

of the blast furnace, used water and later steam power to compress the air and force it into the bottom of the furnace through nozzles called tuyeres.

As Robert Gordon put it, making iron required "considerable intelligence" and "courage."[23] Blast furnace workers were subject to great physical labor and often incredibly dangerous and fatal situations. Many workers and ironmasters alike believed that the blast furnace had a mind of its own. Popular historian John Struthers Stewart referred to the furnaces in the Mahoning Valley as having "an almost lifelike character. . . . We thought of the furnaces as a sort of living entity."[24] In 1918, engineer and metallurgist Joseph E. Johnson Jr. stated, "The blast furnace . . . in operation is often so uncertain, incomprehensible and capricious as to have earned universally the human pronoun 'she.'"[25] Blast furnaces—like most heavy machinery—acted unexpectedly, and workers often could not escape the horrendous and deadly dangers associated with these fiery, mechanical beasts. In his 1938 book *Blast Furnace Practice*, Ralph Sweetser remembered his own experiences with the dangers posed by nineteenth-century blast furnace operations:

The author's first year at a blast furnace plant of four furnaces was so filled with narrow escapes from instant death by explosions, breakouts, gas and falling objects that he made up his mind to do all he could do to make blast furnacing safe. . . . Many men were killed or maimed by blast furnace accidents, accidents that were terrific and horrible.[26]

The small, primitive iron furnaces like those built by the Heatons were relatively tame compared to those constructed later in the nineteenth century. New technology and increasing size often presented a perfect combination for hazardous conditions. In 1916, an unwary group from Chicago's Crane Co. visited a blast furnace along the Calumet River. After "signing away" their lives to gain access to the furnace, they rode a steam-powered elevator hoist 100 feet to the top of the stack, reaching the charging floor that extended "across the chasm to the top of the furnace."[27] The visitors later wrote that the "incessant quiver of the iron plates beneath our feet with the rumbling and groaning from the inside of this monster are disquieting and the thought constantly recurs: 'What if this powerful creature should just now rebel?'"[28] This unsettling thought was typical, not only for visitors to these massive mills but also to those who worked the furnace eight to twelve hours a day. Death by explosions and other terrible accidents were relatively commonplace at blast furnace plants in the Mahoning Valley and around the country. For example, in the early hours of the morning on October 28, 1871, the Haselton furnace just east of Youngstown "burst like a steam boiler or an immense cannon."[29] The *Western Reserve Chronicle* of Warren, Ohio, reported, "The huge stack of iron and brick, filled with upwards of one hundred tons of stock, was lifted from its foundations into the air, then burst asunder, pouring a fiery torrent of seething metal over the stock house, consuming it, and making a general wreck of the concern."[30] Local newspapers described the hellish account of two workers trapped at the top of the furnace:

No one who was in the casting house or in the vicinity of the stack escaped without injury. Patrick Dormany and Joseph Hester were in the hoisting house when the explosion occurred, and the flames, which burst out, seemed to cut off their way to the ground. Hester, however, ran down the stairs through the flame, and got out with a severe burning. . . . The man [Dormany] employed on the top to fill in ore, coal, etc., attempted to save himself from being swallowed by the opening stack, by jumping a distance of fifty feet, alighting on the stock-house, the leap resulting in breaking his back.[31]

Four men were killed and five severely injured during the explosion. In one of the first comprehensive reports on safety conditions in the iron and steel industry, U.S. Labor Commissioner Charles P. Neill found that between June 1908 and June 1910, there were 129 fatal accidents at blast furnace plants throughout the country. Bottom fillers, who worked in the furnace stockyards manually loading thousands of pounds of raw materials into barrows (figure 1.4), were exposed to the greatest danger:

> These men take the material such as ore, coke, and limestone from the bins and wheel it in barrows to the hoist, by which it is carried to the top of the furnace. The fatalities among them [between 1908 and 1910] were 1 due to burns, 3 due to crushing injury, and 1 due to falling objects.[32]

Most crushing injuries occurred when bottom fillers manually shoveled raw material from the base of stockpiles, thus loosening a mass of iron ore, coke, or limestone, which then fell on the workers and crushed them to death. Neill reported that among deaths to top fillers in the same period, one was killed by crushing injury, one by falling objects, one fell from the top of the furnace, and one was asphyxiated by furnace gases.[33] Unsurprisingly, burns caused 24.8% of deaths at blast furnaces, while crushing injuries accounted for 22.5%, asphyxia 14%, falling objects 11.6%, and workers falling off platforms 10.9%.[34]

Early in the nineteenth century, however, the Heaton brothers had little to worry about in terms of dangerous operations, but making the iron—and making it profitably—proved difficult. Their small Hopewell furnace was constructed of stone slabs carved out from the hillside, and it utilized iron ore that yielded about 30% iron. This meant that in order to produce a ton of iron, about three tons of iron ore was charged into the furnace. The stack was first blown with a cold air blast produced by a trompe, described as a primitive blowing device "in which the air is sucked through holes in the upper end of a wooden tube and is led to the furnace by a steady stream of falling water."[35] The apparatus produced a continuous blast pressure of one and a half to two pounds per square inch. But this system was not efficient enough to maintain the blast pressure needed to run even a small blast furnace, and in 1805 the Heatons replaced the blasting apparatus with a small overshot water-wheel.[36] The wheel worked the bellows and produced a pulsing air blast that traveled from the bellows through a pipe and up the hillside into the furnace's tuyere. Although this system provided a stronger blast of air relatively free of moisture, it was still considered weak and production barely increased, if at all.

While the Heatons struggled to turn a profit, other new settlers attempted making iron at Yellow Creek. Baltimore-area native David Clendennin and Pennsylvanian Robert Montgomery erected a similar furnace just three quarters of a mile downstream from the Heatons' Hopewell furnace in 1806.[37] Montgomery became a partner in the new

FIGURE 1.4. Bottom fillers had the grueling task of loading cast-iron barrows with 500 to 700 pounds of iron ore and other raw materials in preparation for dumping the contents into the top of the furnace. This unknown bottom filler, posing with his ore barrow c. 1916, had worked at the furnaces in Hubbard, Ohio, for over forty-seven years when this photograph was taken. Courtesy of the Youngstown Historical Center of Industry and Labor (MSS 140).

enterprise after he negotiated a contract for land and timber with John Struthers, who owned a farmstead on the west side of Yellow Creek's junction with the Mahoning River. According to John Struthers's son Thomas, both furnaces were of about equal capacity and produced about two and a half or three tons each per day. The metal was primarily run into molds for kettles, bake ovens, flatirons, stoves, andirons, and other articles needed by new settlers, while any surplus was sent in pig form to the Pittsburgh market.[38] In Pittsburgh, only one foundry existed, built by Joseph McClurg in 1804, which he later converted into a cannon foundry in 1812.[39] By 1810, Pittsburgh had four nail factories that produced about 200 tons of cut and hammered nails annually.[40]

FAILURE OF THE YELLOW CREEK FURNACES

In the Mahoning Valley, both Yellow Creek furnaces struggled to produce pig iron efficiently with such primitive machinery, and the Heatons' Hopewell furnace achieved little success. Only a few years after the furnace commenced operation, Daniel bought his brother James's interest and attempted to improve operations amidst a fuel shortage, as trees for charcoal became increasingly scarce. After about six years of operation, all of the timber within nearly a half-mile radius of the furnace was cleared.[41] In an effort to delay permanent closure of the furnace, Daniel switched from using only charcoal to a mixture of charcoal and bituminous coal found in outcroppings along the creek bed. Although this was a distinct innovation well ahead of its time, the technology of the period did not allow for efficient smelting of iron ore with bituminous coal. The absence of hot blast resulted in an incomplete combustion of the coal, which formed a tarry substance that choked the furnace. Despite the unsuccessful outcome, Daniel's experiment is the earliest known instance of using raw coal as a blast furnace fuel, something that ironmasters would not revisit for another thirty years.[42]

After Daniel failed to revive the Hopewell, profit loss forced him to sell the furnace and its accoutrements to Robert Montgomery and John Struthers (who later sold his share to David Clendennin) in 1808 for $5,600.[43] Montgomery purchased the furnace for its waterpower, iron ore, and timberland, as any resurrection of the Hopewell furnace required a virtual overhaul and complete reconstruction. Montgomery never operated the Hopewell furnace again, with fuel concerns being the primary culprit. Montgomery's new furnace was more efficient, thus consuming about 250 acres of hardwood each year. Had both furnaces operated simultaneously within less than a mile of each other, neither would have made any type of profit.[44] A charcoal-fired ironworks operating in New Jersey in 1833 faced similar fuel consumption problems. The *Pittsburgh Gazette* reported on the situation:

> A gentleman who has two furnaces and a forge in New Jersey states that his annual consumption of fuel is about 16,000 cords. 20,000 acres of woodland such as is to be found in the iron regions of New Jersey is deemed hardly sufficient to ensure a permanent supply of fuel for a single furnace.[45]

In general, one acre of timber provided about thirty cords of wood, while each cord yielded about forty bushels of charcoal. Thus, an acre of timber provided 1,200 bushels of charcoal. Producing a ton of pig iron required 180 bushels of charcoal, and one acre of timberland supplied enough fuel to make six and two-thirds tons of pig iron.[46]

The fuel question was even more drastic in the Mahoning Valley because its furnaces were closer in proximity, resulting in even more charcoal consumption. However, the Montgomery furnace remained in operation until 1812, when the U.S. army drafted all of its available laborers to fight the British. The stack was never operated again.[47] Although it is unclear if Montgomery remained in the Valley, a contract dated July 8, 1814, between Montgomery and other notable Valley pioneers, including Colonel William Rayen, Judge George Tod, and Homer Hine, called for the construction of a dam and furnace just north of the Mahoning River, along Dry Run in Youngstown, but the furnace was never built.[48] Unsurprisingly, before the start of the war in 1812, iron production in the Youngstown region was minimal. The American Iron and Steel Association estimated a total of 53,908 tons of pig iron were produced in the United States in 1810.[49] Although no official numbers for the Valley's overall iron manufacture exist for this period, it is plausible to estimate annual production values from the known average output of the Yellow Creek furnaces. At an average of two tons per day operating six days a week (taking into account that the furnace was usually banked on Sundays), the Hopewell furnace probably produced about 400 to 450 tons of iron per year from 1803 to 1808. Following the abandonment of the Hopewell, Montgomery's stack likely produced a slightly larger tonnage of about 500 to 550 tons per year until 1812.

Joseph Butler notes that the Montgomery furnace was an improvement upon the Heaton stack and therefore produced, on average, more iron per day. Unlike the Hopewell furnace, Montgomery's stack was blown by a waterwheel and walking beam. Through this operated a crude air compressor of very low volume and pressure, but, unlike the Hopewell, the cold blast did not contain any moisture, which decreased production and the quality of the iron.[50] The annual production numbers proposed here for the two furnaces are most likely higher than the actual average because of imperfect operations and primitive technology, which undoubtedly caused significant downtime for repairs.

Beyond the Mahoning Valley, inefficient and primitive iron production defined most of the existing iron enterprises throughout the country. Thanks to minimalized, high-cost iron production and American ironmasters' reliance on old technology, the United States was a large net importer of British pig iron.[51] For decades, Britain had been incorporating advanced iron-making technology and giving preference to new fuel such as coke over time-honored charcoal. Coke is the residue produced when high amounts of heat are applied to coal without direct contact to air that drives out impurities, thus creating a solid, porous fuel that is nearly pure carbon.[52] In the eighteenth and early nineteenth centuries, English and later American ironmasters made coke by partly burning coal in open banks. After 1843, coke was usually made in rows of bee-hive coke ovens, dome-shaped firebrick chambers with a hole in the roof used to charge coal. In the 1750s, British ironmasters began to adopt coke as a regular blast furnace fuel, and by 1796, they had abandoned nearly all of the country's charcoal furnaces.[53] Along with being limited in supply, charcoal was very soft and not able to handle heavier burdens, consequently preventing ironmasters from building larger furnaces or increasing their output.[54] Removing any reliance on Britain's rapidly depleting forests enabled ironmasters to exploit the island's extensive coal deposits and increase their overall production rates. Therefore, Britain became a net exporter of pig iron, making it difficult for American ironmasters to compete, even under the imposition of tariffs.

Although the Mahoning Valley witnessed the construction and operation of Ohio's first iron blast furnace, the War of 1812, combined with foreign competition and the continuous extraction of timber-lands that once plagued Britain's ironmasters, left the Mahoning Valley's iron industry in a precarious situation. Agriculture continued to dominate the region economically, and iron production proved only a small supplement to the former. Meanwhile, the Valley's largest village, Youngstown, developed slowly. Between 1805 and 1810, the village contained two hotels and two stores, one started by Henry Wick, a merchant and landholder of English descent who came to the area around 1795, and the other by Colonel William Rayen, a man of numerous professions including merchant, postmaster, colonel in the War of 1812, and judge of the court of common pleas. The rest of the village encompassed a hat maker, a blacksmith shop, a boot maker, one lawyer, and one doctor.[55] Turhand Kirtland, who helped John Young lay out the village, described Youngstown in 1810 as "a sparsely settled village of one street, the houses mostly log structures, a few frame buildings excepted; of the latter character was the dwelling house and store of Colonel Rayen."[56]

A NEED FOR IRON

With minimal ferrous industries in the Valley and none in Youngstown, settlers now relied on outside markets for iron, but the lack of good transportation in and out of the area made the metal hard to obtain. In addition, little, if any, refined iron was available, and settlers continued to use comparatively weak cast-iron tools. After James Heaton dissolved his partnership with his brother in the Hopewell furnace, he and his family moved to Howland and shortly thereafter to a plot of land at the junction of the Mahoning River and Mosquito Creek in Weathersfield Township. In 1806 and 1807, James constructed a gristmill and sawmill along the north side of Mosquito Creek and erected a dam to utilize waterpower. In 1809, he built a finery forge at the mouth of the creek, where he produced Ohio's first malleable bar iron from pig iron shipped up the Mahoning River from Montgomery's furnace along Yellow Creek.[57] In the process of refining pig iron in his finery, Heaton remelted the pig and cast the metal into plates roughly two feet square and one inch thick. The red-hot metal was then chilled with cold water, broken into small pieces, and melted in the charcoal-fired finery with the aid of bellows, transforming the iron into a spongy mass similar to that produced in a puddling furnace. Lacking steam or machine power such as the squeezer or trip hammer to mechanically form the refined metal, Heaton and a select few laborers physically hammered the red-hot iron. This instantaneously expelled the slag and shaped the metal into a bar of wrought iron. According to one contemporary observer, it was impossible to hammer small-sized iron bars, and country blacksmiths were tasked with fashioning whatever items pioneers needed from the iron. Wrought-iron articles for household use were very crude, as most bore the marks of the blacksmith's hammer.[58]

Following the abandonment of the Montgomery furnace, Heaton sought another source of pig iron and, with no other iron enterprises in the immediate area, decided to construct his own iron furnace on the west side of Mosquito Creek adjacent to his finery forge. It is not clear whether James enlisted the expertise of his brother Daniel in the construction and working of the new iron smelter, but Daniel did follow his brother and settled near Mosquito Creek after selling the Hopewell furnace.[59] Once finished, James's son Warren, following an old English custom, carried his six-year-old sister Maria "to light the first fire in the new charcoal blast furnace," named Maria in her honor.[60] Appropriately, the small village became known as Heaton's Furnace.

The Maria furnace improved on the technological imperfections of the Yellow Creek furnaces and produced up to three tons of iron a day. Heaton constructed his new thirty-five-foot by eight-foot furnace with native flagstone, including for the inner lining, as more viable firebrick was not yet available. Consequently, the furnace could not operate over six months without maintenance to the inner walls. The blowing engine was of the primitive, water-powered type, though it greatly improved upon the design used at the Hopewell furnace:

The blast cylinder was operated by a slow moving water wheel connected by a walking beam, and at each half revolution emitted a

FIGURE 1.5. Remains of Gideon Hughes's Rebecca furnace in Lisbon, Ohio, c. 1890. Portions of the furnace are still visible at the site along Little Beaver Creek. Courtesy of Niles Historical Society.

terrifying groan, and to equalize the pressure the air passed from this cylinder into a square wooden receiver with a bellows top weighted with water, so that when the cylinder was idle at the end of its stroke, the pressure of the water forced the air into the furnace.[61]

Like most charcoal furnaces, the Maria sat up against a bluff along the creek, where a wooden bridge connected the top of the furnace with the level ground at the summit of the small bluff. The complex employed about twenty men, two of whom ran the actual furnace while the others produced charcoal and gathered and transported ore. Heaton paid his employees in provisions and supplies obtained from country merchants in exchange for cast-iron products manufactured at the furnace, such as stoves containing the brand "James Heaton Maria Furnace." Heaton sometimes paid his laborers meager wages, and they only received one dollar in cash on holidays.[62] Excess iron not consumed by Heaton's forge was sent in pig form down the Mahoning and Beaver Rivers on flat boats, where it was tugged to the then-nominal Pittsburgh market.

Despite having the distinction of hosting the first iron furnace and the first bar iron produced in the new state of Ohio, industry in the Mahoning Valley came to a relative standstill with the exception of Heaton's small ironworks. Elsewhere in the Valley, only grist and planing mills lined portions of creeks and the Mahoning River, providing essential flour and building materials for the necessities of pioneer life. About thirty miles south of Youngstown in Lisbon, Columbiana County, Gideon Hughes, a Quaker, constructed a charcoal furnace around 1807 or 1808, the third such furnace built in Ohio (figure 1.5). Named the Rebecca furnace after Hughes's wife, the small stone stack was built on the bank of Little Beaver Creek, where it utilized waterpower and abundant nearby forests for charcoal production.[63] Similar to the Yellow Creek and Maria furnaces in the Mahoning Valley, the Rebecca stack primarily served pioneer life by producing cast-iron products over pig iron for other markets. Other small charcoal furnaces began to crop up in Summit County near Akron, Ohio, but most were abandoned by the 1820s or 1830s.[64] Despite their fleeting existence, William B. McCord, editor of the 1905 *History of Columbiana County, Ohio,* emphasized the important role that the Rebecca stack and other similar furnaces played in serving Ohio's early settlers:

> To build his log cabin without a nail, a spike, a hinge—that was the problem that confronted the early settler. The weight of manufactured iron products rendered their transportation over the early roads across the mountains and into the wilderness next to impossible. . . . Some pig-iron was made, but the ultimate production of the furnace consisted chiefly of plow-shares, dog-irons, flat-irons, pots, kettles, Dutch-ovens and other household utensils used in that day.[65]

Because of their short lifespan, these small, isolated furnaces had little effect on the region's long-term expansion of industry. Nevertheless, well-known industrial giants like Cleveland and Pittsburgh did not see any pig-iron production for another half-century because their local carbonate ores were buried too deep to merit any economical use.[66] They therefore relied on the output of furnaces scattered about northeast Ohio and western Pennsylvania, particularly those in Fayette County, which had a higher density of charcoal-fueled blast furnaces than any other county in Pennsylvania by 1820.[67]

As settlers continued to move westward, small villages and towns such as Youngstown, Heaton's Furnace, and Pittsburgh required more industry and goods to serve their increasing population. Iron was still at a premium as James Heaton's Maria furnace and finery remained the only source of iron production in the Youngstown region for over twenty years following their construction. Moreover, thanks to minimal transportation services, importing iron was expensive.

The passage of the Tariff of 1823 worked in the Valley's favor. The tariff's heavy tax on imports aided in the development of Northern industry, and the Heatons found themselves in need of an additional supply of pig iron to meet market demands for both pig and bar iron, particularly as the Pittsburgh market developed. Christopher Cowen built the Pittsburgh Rolling Mill, the first of its kind in the town, in 1812. It produced only sheet iron, nail, and spike rods from bar iron procured from other establishments. In 1819, Baldwin, Robinson, McNickle & Beltzhoover constructed the city's second rolling mill, which contained four puddling furnaces, the first in Pittsburgh. By 1829, Pittsburgh contained eight rolling mills that consumed 6,000 tons of refined iron, largely purchased from fineries and bloomeries of Pennsylvania's Juniata Valley, as well as 1,500 tons of pig iron from small furnaces outside of Pittsburgh. The city's nine foundries consumed an additional 3,500 tons of pig iron.[68] The tremendous growth of not only the city's iron production but also machinery, glass, river steamboats, breweries, tanneries, salt works, and other manufactories warranted its distinction as the "Birmingham of America."[69]

The production of wrought iron in the puddling furnace served to transform Pittsburgh into the iron manufacturing center of the United States. Many ironmasters in the Mid-Atlantic used fineries to directly convert pig iron into wrought iron. However, these small-scale fineries required charcoal fuel and a consistent blast of air, which proved inefficient for the needs of a growing and industrializing country. The puddling process, patented by Englishman Henry Cort in 1784, diminished much of the labor involved in making wrought iron in a refinery. More importantly, the process eliminated the need for blowing machinery by using a natural draft, thus giving ironmasters the ability to use coal instead of charcoal, a much cheaper and more plentiful fuel.[70] Puddling itself was almost analogous to washing dirt from clothes, but rather than soap and water, ironworkers used fire and air. Those who worked the puddling furnace were highly skilled workers who often took two years to personally learn the process. It was labor intensive and required patience, dexterity, and strength, as each heat was manufactured by hand and the scale of production depended on the physical strength of the worker.[71] Pittsburgh laborers in John Fitch's *The Steel Workers* described puddling as "very hard, hot work"; few other positions in an iron or steel mill were as physically demanding.[72] Former puddler and secretary of labor James Davis recounted, "It is no job for weaklings. . . . I grew into manhood with muscled arms big as a bookkeeper's legs."[73] In his 1922 book *The Iron Puddler,* Davis illustrates work in the puddling mill as a hellish scene of men, muscle, and machine:

> Patches of white heat glare from the opened furnace doors like the teeth of some great dark, dingy devil grinning across the smoky vapors of the Pit. Half-naked, smoot-smeared fellows fight the furnace hearths with hooks, rabbles and paddles. Their scowling faces are lit with fire . . . the sweat runs down their backs and arms and glistens in the changing lights.[74]

The chemical process of puddling was important for the manufacture of usable products such as tools, structural pieces, rails, and other

FIGURE 1.6. The puddler's helper lifts the fire door while the puddler uses long tongs called telegraphs to remove the near-molten ball of wrought iron from the hearth at Youngstown Sheet & Tube's puddling mill in the 1920s. The Mahoning Valley Historical Society, Youngstown, Ohio.

commodities. In the puddling furnace (figure 1.6), puddlers remelted about 600 to 700 pounds of pig iron from the blast furnace in a bath of its own fusible compounds, or slag. The puddling furnace itself was a type of reverberatory furnace, where the material being refined did not come into contact with the fuel, but rather with the hot air and gasses from that fuel. It consisted of two main sections: the fireplace where the puddler's helper stoked the fire with coal, and the hearth, where the pig iron was charged. In the early stage of puddling, the carbon and silicon became oxidized by the oxygen and other materials, such as mill scale and ores, and the iron solidified into small particles. Each particle was enclosed in fusible slag released from the melted iron that contained sulphide or phosphide. The puddler used a long iron prod called a rabble to include slag and violently reduce the amount of phosphorous and carbon to create wrought iron. This slag gave wrought iron a "grainy" look that resembled wood and thus became malleable, bendable, and easily welded and worked. After this two-hour process, the puddler formed about four 100-pound balls of semi-molten wrought iron, removed them from the furnace, and transferred them to the rolling department, where they were shaped and worked into usable products. Once shaped, the bloom of refined iron was transferred to muck rolls with grooves of

FIGURE 1.7. Rollers at Youngstown Sheet & Tube shape a bloom of wrought iron from the squeezer (which shaped the puddle ball into a bloom, seen behind the roller on the right) into muck bars by continuously passing it back and forth through the rolls. ©*The Vindicator*.

varying sizes, where a roller inserted the iron and created muck, or the rough, unfinished bar iron (figure 1.7). This initial process of rolling also served to remove additional slag in the iron and to shape the puddled iron into bars that usually measured ¾ inch thick by 2.5 to 8 inches wide and 15 to 30 feet long.[75] After the iron was rolled into muck bars, it was cut into short lengths and stacked in rectangular piles. The pile was heated in a furnace to a welding temperature and rolled again into muck bars. This process was then repeated a second time. The reheating process further treated the iron, and each heating added 7,000 to 9,000 pounds per square inch of tenacity. After the rolling and heating was complete, the homogenized iron was further rolled into finished products on special trains of rolls in a fashion similar to rolling muck bar. Although ironmasters first introduced wrought-iron production to the Mahoning Valley in the early 1840s, it would not become a major factor in the region until after the Civil War.

YOUNGSTOWN ENTERS THE IRON INDUSTRY

By the 1820s and 1830s, the market for pig and bar iron increased enough in the Pittsburgh region to merit additional pig iron production in the Mahoning Valley. James Heaton retired from active business in 1830, and four years later, he laid out the village plat of Heaton's Furnace, simultaneously changing the name to Nilestown, later shortened to Niles in 1843. Though James had largely left the iron business, his sons Warren and Isaac continued the trade; they were the last in the

FIGURE 1.8. Ruins of Isaac Heaton's Trumbull furnace along Mill Creek in Youngstown, c. 1890. All of the furnace's equipment was stripped from the site, leaving only the stone stack. Notice the man standing on the furnace's second tier of stone in the upper center of the photograph. The Mahoning Valley Historical Society, Youngstown, Ohio.

Heaton family line to be active in iron manufacture in the Mahoning Valley. In their youth, Warren and Isaac helped their father work the Maria furnace and developed both skills in iron production and acumen for business. After his father's retirement, Warren, who married Elizabeth McConnell in Niles in 1830, continued iron production at the Maria furnace with his business partner Josiah Robbins, albeit rather unsuccessfully. The economic Panic of 1837 drastically reduced demand for pig iron, leaving the market extremely limited and the proprietors unable to provide compensation to their workforce, with the exception of some provisions.[76] However, this venture was relatively short-lived, as Warren died at an early age in 1842. The furnace was then leased to a variety of firms that included McKinley, Reep, and Dempsey, of which William McKinley Sr., father of the twenty-fifth president of the United States, was a partner. James Heaton's finery forge along Mosquito Creek faded from existence in the early 1830s, and only the Maria furnace thereafter manufactured castings and pig iron sold to the Pittsburgh market. With relatively little modernization, the Maria furnace could produce no more than four tons of iron per day, but the exhaustion of nearby forests required fuel to be transported farther and farther as bog and local ores became increasingly scarce.

Isaac, on the other hand, left his father's town for Youngstown in the mid-1820s, where, at the age of thirty-nine, he constructed the Mahoning Valley's last charcoal-fired blast furnace—named Trumbull—along Mill Creek. Isaac probably began construction on Youngstown's first furnace in early 1833 (figure 1.8). By September of the same year, he took an ad out in Warren's *Western Reserve Chronicle* calling for a workforce

FIGURE 1.9. James Heaton's former woolen mill and Isaac Heaton's furnace store-room as it appeared in the late 1860s. Notice the barren hillside. Many of the trees provided charcoal for the Trumbull furnace, which sat at the foot of the hillside just outside of the image to the right. Some ironmasters sustained production with second-growth timber, although this did not occur in the Mahoning Valley. Author's collection.

of 100 that included wood choppers, colliers, ore diggers, and molders "to whom liberal wages will be given—one half in store goods, the other in cash."[77] Several yards from the furnace site sat an abandoned woolen mill constructed by Isaac's father James in 1821 and 1822, an unsuccessful venture that operated only until 1830. Isaac utilized the sandstone building as a storeroom for furnace materials such as sand, ore, tools, and possibly finished cast-iron products and pig iron (figure 1.9). Like the Maria and Yellow Creek stacks, the Trumbull furnace produced iron for stove and kettle castings as well as axes. By necessity, the stack was constructed in an isolated, wooded area about three miles from the center of Youngstown for a source of fuel, which later proved a great disadvantage. Sources of iron ore came from outcrops of kidney ore that lined nearby creek beds. In the late 1970s, an iron ore mine was also discovered on the east side of man-made Lake Cohasset in Mill Creek Park, only a short distance from the furnace site.[78]

Like other charcoal furnaces, the Trumbull stack sat positioned against a hillside and measured about thirty-five feet in height. At the top of the furnace, the trunnel-head into which top fillers charged charcoal, iron ore, and limestone measured a mere three feet in diameter and doused uncareful laborers with flames and toxic gases.[79] Transportation of both raw materials and finished products to and from the furnace was still primitive and exceedingly difficult. Horses and oxen pulled heavy carts of charcoal from pits located on the acres of woodland surrounding the furnace complex and also carted iron ore from mines and outcroppings. Even worse was the task of transferring finished cast-iron products and heavy pig iron over the sloping hills and primitive paths that led to Youngstown. Despite its location, however, Isaac's furnace achieved a more or less successful run throughout the relative prosperity of the early 1830s.[80]

After ten years of operation, Isaac sold the Trumbull furnace, along with all of his landholdings, and subsequently purchased a small furnace in Venango County, Pennsylvania, but he never made a profit from the operation. He quickly sold his new furnace, and he and his wife, Elizabeth, moved to Kinmundy, Illinois, where they purchased a quarter section of land and established a successful farming business.[81] Isaac never returned to the Mahoning Valley. He died in Illinois in 1872.

Following Warren Heaton's unexpected death and Isaac's permanent departure from the area, the Heaton name was no longer directly associated with any of the Valley's remaining iron enterprises.[82] By the early 1840s, the two Yellow Creek furnaces sat abandoned and deteriorating, covered with weeds and briers (figure 1.10). Only two blast furnaces quietly roared throughout the Youngstown region, but both suffered unstable ownership, saw only sporadic operation, and had limited success throughout the decade. Iron production continued to be minimal as both furnaces produced a maximum of four tons on a successful day, and it seemed that the Valley's iron industry had reached a rather unremarkable zenith.

As early as 1817, Valley residents discussed the appeal of a canal through the region to encourage the development of local industry and aid in the growth of those that already existed. Despite their pleas, no canal materialized over the next two decades, which stifled industrial progress and stunted economic expansion. In 1830, Ohio produced a total of about 5,400 tons of pig iron with seven blast furnaces that produced 250 tons of castings, ranking it second among the country's iron-producing states. Pennsylvania ranked first with forty-five furnaces and 31,056 tons of pig iron produced in 1830, though it is possible that not all iron companies reported their annual production totals.[83] By 1840, however, iron production in Ohio rose to an average of about 20,000 to 22,000 tons of pig iron per year, with the Mahoning Valley contributing only a small fraction of the state's total; overall production in the United States reached about 330,000 tons.[84] In 1840, the firm of Spencer and Co. made a largely unsuccessful attempt to start a new iron enterprise in the Mahoning Valley. The firm's steam forge was a mostly forgotten enterprise that antedated all rolling mills in the Valley. According to Butler, the forge, after operating for a short period, failed to garner any profit for its owners and was moved several miles north of Youngstown. Its new owners never operated the mill.[85] The rather quick failure of Spencer & Co.'s

FIGURE 1.10. The long-abandoned Hopewell furnace, shown here about 1890, was a popular photographic spot among Valley residents. Photo courtesy of the Struthers Historical Society and Museum; Marian Kutlesa, curator.

forge was symbolic to the Youngstown region's struggle for industrial growth. The lack of any known and widespread supply of high-quality raw materials in the Valley, with the exception of its low-iron-content bog ores, discouraged most other ironmasters and industrialists from coming to the region for a number of years.

IRON OUTSIDE THE VALLEY

As the Mahoning Valley's iron production stagnated, other regions in the state quickly began to increase their overall output and production, namely Lawrence, Jackson, Gallia, Vinton, and Scioto Counties in the Hanging Rock Iron Region of southern Ohio. This region, taking its name from a cliff that projected over the north side of the Ohio River, rapidly became one of the most important pig iron producers in the state and country. Between 1826 and 1880, ironmasters constructed sixty charcoal furnaces in an area of about 1,000 square miles.[86] The region was rich in iron ore and contained vast, dense forests for charcoal production, as most Hanging Rock furnace companies owned between 5,000 and 10,000 acres of additional land at about $1.25 per acre.[87] These iron plantations were far more extensive in operation than the few charcoal furnace complexes in the Mahoning Valley. Each furnace spawned its own village, and furnace laborers and their families relied on the company for nearly all aspects of their lives. Ironmasters provided housing near the furnace for their employees, stores for merchandise, stables for oxen and horses, and even schoolhouses for the employees' children. Often six to eight teams of oxen were seen hauling ore and charcoal several times a day while smoke from charcoal pits constantly seeped from the barren and treeless hillsides.

Unlike the Mahoning Valley, Hanging Rock ironmasters had a reliable outlet—the Ohio River—to ship their product. Because finery forges and rolling mills were not constructed in the Hanging Rock region on a noteworthy scale until the 1850s, ironmasters sent their pig iron via the river to Pittsburgh and Cincinnati for use in these cities' abundant foundries and rolling mills.[88] In the first half of the nineteenth century, foundries used much of the region's charcoal iron for making strong, high-quality cast-iron rolls for use in rolling mills. Such rolls made from Hanging Rock iron had a reputation for being durable and withstanding breakage under the harsh conditions of shaping and rolling red-hot iron.[89] Foundries also relied on the charcoal iron for making pieces of machinery needed to withstand large amounts of tension and weight. Hanging Rock ironmasters were also more innovative than those in the Mahoning Valley. By 1836, they began implementing new technology such as the hot blast, only eight years after James Nielson introduced it in Scotland.[90]

For much of the first half of the nineteenth century, the Hanging Rock district greatly surpassed the Mahoning Valley as a center of iron production. So too did other Ohio enterprises such as the extensive charcoal ironworks in Lake County, owned by Judge Samuel Wilkeson, the fifth mayor of Buffalo, New York. In 1802, Wilkeson married Jean Oram, and soon after they moved to the Mahoning Valley, where they established a farm and built a gristmill in Poland, Ohio. In 1806, they gave birth to their eldest son, John, one of seven children. Not long after the arrival of their son, the family moved to the shores of Lake Erie, where Wilkeson honed his craft as a master shipbuilder. During the War of 1812, he was asked to furnish a fleet of ships for the U.S. Army at Buffalo, where he eventually settled and concentrated his efforts as a merchant. In 1820, he took charge of the construction of the city's first harbor; though he knew little of engineering, he had a commanding presence and voice that aided in its completion. Wilkeson was also an advocate of the Erie Canal, but he realized that without manufactured goods and boats to ship them, the canal was useless in Buffalo. As a result, he encouraged local industry around Lake Erie to support Buffalo's economic growth. In addition, Wilkeson helped establish several ironworks and foundries, which, in 1830, included the purchase and enlargement of the Erie Furnace Company in Madison, Ohio.[91]

Wilkeson undoubtedly conveyed his sense of hard work and dexterity to his eldest son, John. After moving with his family from the Mahoning Valley to Buffalo at the young age of eight, John became familiar with business and industry early in his life. In Buffalo, he learned the business of manufacturing cast-iron stoves and furnaces and became interested in oil and timberlands in Pennsylvania.[92] Samuel Wilkeson put John in charge of the old Erie furnace after purchasing the company in 1830. In 1834 and 1835, after a devastating fire two years before, twenty-five-year-old John immediately added a second furnace and reorganized and incorporated the Arcole Furnace Company with a capital of $100,000—one of the largest capitalizations of any business in Ohio at the time.[93] John Wilkeson's brother-in-law Frederick Wilkes, a native of Portsmouth, England (whose family was in the British iron business), received a job at the Arcole furnace as a clerk when he was sixteen, but he left after five years of employment because his pay consisted solely of room, board, and clothes.[94] Regardless, Wilkeson's extensive ironworks employed a workforce of nearly 2,000 and was the largest single iron manufacturer in Ohio, producing about 1,000 to 1,500 tons of iron per year while supplying stoves, potash kettles, and other castings to settlers on the shores of Lake Erie and the state of Michigan.[95]

By 1840, however, John Wilkeson's Arcole furnaces were at a crossroads. Bog ore found in swales and swamps became scarce, charcoal fuel became more expensive and distant, and sand bars had formed at the mouth of Arcole Creek, making it impossible to use as an outlet for the company's product.[96] In 1840, a desperate Wilkeson used raw coal from Greenville, Pennsylvania, in the Arcole furnace, but, as Daniel Heaton had learned thirty-two years earlier, raw coal was unsuitable for smelting iron ore without hot blast, and Wilkeson's attempt only "met with a small measure of success."[97] He had no other choice but to abandon the ironworks and, together with his brother-in-law Frederick Wilkes, cast his eyes on his childhood home, the Mahoning Valley. Perhaps fortuitously, both men dreamed of industrial rebirth just as the Youngstown region neared a substantial economic and industrial transformation.

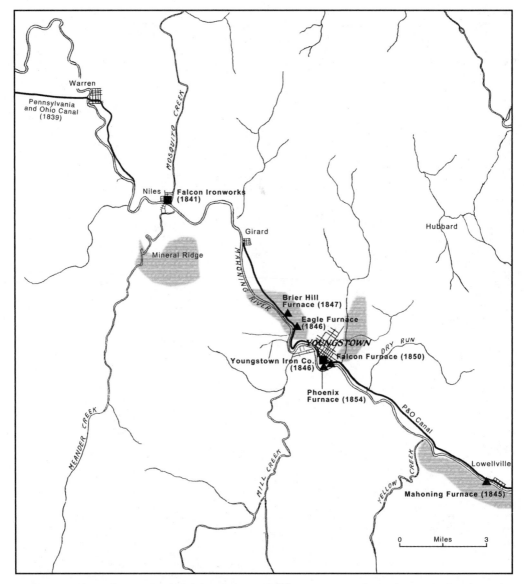

FIGURE 2.1. Map depicting industrial and transportation progress in the Mahoning Valley through 1856. Triangles represent blast furnaces, squares signify rolling mills, and shaded areas depict major coal mining areas during this period. The arrival of the canal prompted construction of blast furnaces and rolling mills along its route, primarily in Niles, the west side of Youngstown, and Lowellville on the Valley's southeasternmost edge.

gave it a "rather dingy appearance."[5] Nevertheless, the convergence of the Allegheny, Ohio, and Monongahela Rivers facilitated ample transportation and freight passages in and out of Pennsylvania's major industrial city.

Nestled between Cleveland and Pittsburgh, the still agriculture-dominated and underdeveloped Mahoning Valley nevertheless sought a way to increase its economic standing by connecting with the outside world. Unlike the cities to its northwest and southeast, the Valley had limited transportation sources, as the Mahoning River was much smaller in comparison to the Cuyahoga, Ohio, Allegheny, and Monongahela Rivers. However, in the 1820s, Valley merchants and industrialists advocated a lake-to-river canal constructed from Youngstown to Lake Erie, similar to the Ohio and Erie Canal that linked Akron to Cleveland. At the time, the Mahoning Valley's transportation consisted of a few primitive wagon roads that stretched to Lake Erie, and they provided little in the way of bulk transportation at minimum costs. In 1825, the state launched several canal projects, none of which cut through the Mahoning Valley. Finally, in January 1827, the Ohio legislature granted a charter to begin a canal project that would cut east through the Mahoning Valley from Akron and across state borders near New Castle, Pennsylvania. From there it would connect with the Erie Extension Canal. Due to political uncertainty, however, the project never got off the ground. For a brief period in 1827, a proposed lake-to-river railroad called the Ohio and Erie Railroad (which would connect Ashtabula and Columbiana Counties) overshadowed the canal project. Though a charter was passed and a capital of $1 million obtained, the ambitious railroad proposal frightened leery Trumbull County capitalists, who soon thwarted the project.[6]

In the fall of 1833, delegates from Portage, Trumbull, Columbiana, Stark, and Wayne Counties in Ohio and from Allegheny, Beaver, Mercer, and Philadelphia Counties in Pennsylvania met at a convention in Warren, Ohio, to revive discussion of the six-year-old crosscut canal scheme. Ultimately, the group decided to privately fund the project, as Ohio delegates believed that a canal linking their state to Pennsylvania was, strangely, not worthy of state funding. Two years later, in 1835, the Pennsylvania and Ohio Canal Company was formed, with a number of directors, including staunch abolitionist and Ohio state senator Leicester King. Also on board were Ohioans Eliakim Crosby and William Ryan, along with several businessmen who hailed from Philadelphia and Beaver County, Pennsylvania.[7] The Panic of 1837 delayed construction on the canal until 1838, but upon completion, the Pennsylvania and Ohio (P&O) Canal, also known as the Mahoning Canal, signaled a major engineering feat. Seasoned experts such as Colonel Sebried Dodge, a member of the Ohio Corps of Engineers and the project's chief engineer, ensured the canal's success. The canal stretched sixty-seven

FIGURE 2.2. The Pennsylvania and Ohio Canal parallels the Mahoning River at the Spring Common bridge on Youngstown's west side, c. 1869. Author's collection.

miles from Akron to the Pennsylvania state line, then ran another fifteen miles east into New Castle, Pennsylvania, for a total of eighty-two miles. The P&O Canal began at its western terminus in Akron, and from there connected to the Ohio and Erie Canal. It also extended through Akron and traversed northeast through Cuyahoga Falls and Kent until it reached its summit in Ravenna. From there, the canal entered Trumbull County, going through Newton Falls, where it crossed the East Branch over a stone masonry aqueduct before traversing southeast through the Mahoning Valley. From there the canal entered Warren, where it held to the north bank of the Mahoning River as it passed through Niles, Girard, Youngstown, Struthers, and Lowellville before crossing the state line and terminating just south of New Castle (see figure 2.1).

The official opening of the canal in August of 1840 had major economic implications for the Mahoning Valley. The canal promised "active commerce carried on between the city of Pittsburgh and the western part of Pennsylvania," and it had the ability to connect the flow of goods from the major cities of Philadelphia to Pittsburgh, up through Cleveland to Lake Erie, and south toward the Ohio River to Cincinnati.[8] Initially, Youngstown, Lowellville, and other Mahoning Valley towns and villages along the path of the canal were not expected to compete with bigger, anticipated boom cities like Philadelphia, Pittsburgh, and Columbus. Nevertheless, with the canal completed from its southern terminus to Warren in May 1839, Valley residents declared a holiday when the first boat reached its northern terminus. Once the first canal boat arrived at the foot of Main Street in Warren, "they were greeted by the Warren band, and a procession formed which marched through the square."[9] The canal handled nearly all of the freight moving in and out of the Mahoning Valley (figure 2.2). Goods shipped out of the Valley primarily consisted of agricultural products such as wheat, flour, cheese, pork, beef, and butter; property and goods coming into the Valley consisted of paper, lumber, stoves, furniture, lake fish, salt, and whiskey. Iron products shipped from the Valley were typically limited to pig iron or castings produced at the Maria furnace in Niles, while finished products such as nails were primarily imported from Pittsburgh, since they were manufactured at one of the city's many rolling mills. In 1841, for example, 82,520 pounds of pig iron, 25,662 pounds of castings, and 3,027,514 pounds of iron, nails, and glass arrived via the canal.[10]

Despite the opening of the canal, the Mahoning Valley's iron industry experienced only anemic growth for several years, but the Valley's improved transportation outlets eventually led to the unearthing and shipping of one of the region's most important resources: coal. Once established, the Valley's coal mines fueled the region's incredible industrial growth throughout the remainder of the nineteenth century. The Mahoning Valley is part of the Appalachian coal basin, one of the world's largest coal-bearing regions. In its entirety, the Appalachian basin spreads from central Alabama northeast through portions of Tennessee and Kentucky before it reaches West Virginia, Pennsylvania, and Ohio. The sulfur content in the coal was higher in the westward end of the basin, in eastern Ohio and western Pennsylvania. The east contained coal with lower sulfur content, which made up the country's primary anthracite coal reserves. In the early 1840s, ironmasters used these reserves to smelt iron ore in northeastern Pennsylvania, and they turned the region into ground zero for American pig iron production in the mid-nineteenth century.[11] Anthracite coal is a hard variety of coal that has a high carbon content and very few impurities; it was often referred to as a "natural coke."[12] Coal mined in portions of the Mahoning Valley was of the bituminous type, a soft coal that is high in volatiles and other impurities such as sulfur. In most cases, this soft coal was used in coking to expel gasses and form nearly pure carbon. It also produced heat for domestic consumption and fueled industrial steam production.[13]

According to surviving records, the first coal mine in the Mahoning Valley opened in 1826 on land owned by Mary Caldwell, along Crab Creek on Youngstown's east side.[14] Early settlers to the area, like Daniel Heaton at Yellow Creek, often found coal cropping out from hillsides and streams. Others encountered it when sinking wells and digging cellars for houses.[15] Early pioneers, however, did not recognize the coal's worth beyond a supply of domestic fuel. In 1829, Colonel William Rayen installed a coal stove that he purchased in Pittsburgh at his tavern purely as a novelty, though word of the stove's "luxuriant heat" spread quickly around the village.[16] Area blacksmiths probably used the coal, but the region's relatively heavy forest cover supplied the wood that served as the primary source of fuel for heating homes, steam boilers, and businesses for many years.

Throughout the 1830s, small mines were opened throughout Youngstown, though not for large-scale industrial or commercial use. Unknown to Valley residents before 1840, the region's coal was of a special quality and "one of the purest and best seams in the State."[17]

Identified variously as "splint," "block," or "Brier Hill" coal, it was mined mostly in large blocks, had a laminated structure, and split readily into sheets that had a soft, fibrous, and carbonaceous texture resembling charcoal.[18] The coal was difficult to break in the opposite direction to the laminate, and the resulting fracture exhibited a "splinty" structure marked by opposite layers of dull and bright-shining coal. According to an analysis done by professor E. S. Gregory of Youngstown in the early 1870s, this special coal was rich in carbon, at 60.62%, and contained 34.03% combustible matter, 2.12% ash, and 3.23% moisture.[19] Ohio state geologist J. S. Newberry commented in his *Geological Survey of Ohio*:

> The coal of the Mahoning valley, nearly all of which is first-class coal, [is] superior to that from any other coal field in the State, and not excelled by any bituminous coal mines anywhere. . . . It is a generally dry, open burning coal, its mechanical structure causing it to take fire rapidly through the centre of the largest pieces, especially adapting it to the smelting of iron.[20]

Unlike the bituminous coal later mined south of Pittsburgh in the Connellsville area—which, due to the coal's tarry nature, could not be used as a furnace fuel until coked—the Valley's block coal could be used directly as a blast furnace fuel in place of charcoal.[21] Additionally, block coal did not have the chemical properties attributed to other coking coals. It failed to swell, change form, or run together during the coking process, thus eliminating the expense of building coke ovens and the need for extra transportation of the fuel to the blast furnace.

The Ohio Geological Survey classified the block coal seam underneath the Valley as coal No. 1. Within an area of about thirty miles, the coal rested at depths ranging from sixty to three-hundred feet in veins that varied from eighteen inches to six feet in thickness.[22] In the Mahoning Valley, the No. 1 coal seam wound underneath Youngstown, Austintown, Boardman, and Poland in Mahoning County, and beneath Hubbard, Brookfield, Liberty, Vienna, and Weathersfield in Trumbull County. The same seam extended into Pennsylvania, where state geologist Henry Darwin Rogers referred to the coal as Mercer County block coal or Sharon coal.[23] Although the chemical makeup and quality of coal can vary over just a small area, Sharon coal was described as having a similar composition to Mahoning Valley block coal, being "a species of semi-cannel-coal with a slaty structure, and a dull, jet-black lustre [sic], with a thickness of from three to four feet."[24] The Sharon seam had a large industrial influence on towns flanking the Shenango River, which included West Middlesex, Wheatland, Sharon, Sharpsville, and Greenville. Most of these towns became significant pig iron manufacturing centers in the mid- to late nineteenth century. Moreover, block coal was

FIGURE 2.3. Portrait of Governor David Tod (1805–1868), Youngstown's pioneer coal operator and one of its earliest industrialists. From Joseph G. Butler Jr., *History of Youngstown and the Mahoning Valley,* vol. 3 (Chicago: American Historical Society, 1921), 596.

described as "an exceedingly irregular seam" not found in any other regions outside of the Mahoning and Shenango Valleys or in the entire Appalachian coal basin.[25] Mahoning Valley resident David Tod, Ohio's Civil War governor from 1862 to 1864, recognized the coal's potential, and he played a major role in exploiting the region's exclusive coal reserves.

Born in 1805, David Tod (figure 2.3) was the son of Judge George Tod of Suffield, Connecticut, and Sally Isaacs, daughter of Ralph and Mary Isaacs. David's father graduated from Yale College in 1795 and subsequently studied law in New Haven, Connecticut. After marrying his wife, Sally, in 1797, George moved to Youngstown in 1801, where Governor Arthur St. Clair appointed him secretary for the territory of Ohio. George Tod settled with his family on a farmstead just west of Youngstown, which he named Brier Hill after the numerous briers that dotted the local hillsides. George's son David grew up on the farm at Brier Hill and, unlike his father, did not attend a well-established university. He was self-educated but nonetheless possessed a broad knowledge of culture, politics, and law. With little more than an elementary education, David moved to Warren in 1827, where he was admitted to the bar and subsequently established a law practice at the young age of twenty-two. Thanks to his "unsurpassed ability in the examination of witnesses . . . and gaining and holding the confidence of the jury," Tod became one of the area's finest lawyers.[26]

Tod first entered politics as a state senator in 1838, but his interest in industrial activity and business also grew. That same year, he joined with Colonel Simon Perkins (the son of Akron city founder Simon Perkins, who married Tod's daughter, Grace), James R. Ford, Daniel Townsend, John Williams Jr., George B. Martin, and Arad Kent to establish the Akron Manufacturing Company. The Ohio state legislature granted the company a charter of thirty years, specializing in the manufacture of iron, castings produced from blister steel, nails, stoves, pig iron, and other castings.[27] The company included a foundry and the Etna blast furnace to provide the former with pig iron. The latter was a small, thirty-foot stone-stack charcoal furnace powered by an overshot waterwheel. Little else is known about the company, though the opening of coal mines in the Mahoning Valley influenced the physical location of the Akron-based enterprise in the future.

After the death of his father in 1841, David Tod inherited the family homestead at Brier Hill and subsequently sought to develop the vast seams of block coal that wound underneath his feet. Although historians debate the exact date, Tod probably opened the first coal mine on his property around 1841 with Clevelander Daniel P. Rhodes, a pioneer in the Cleveland coal mining business and a contributor to the development of railroads in northern Ohio. Once opened on the Mahoning River and canal, the mine—one of the most famous in the Mahoning Valley— initially produced about fifty tons of coal per week.[28]

Tod hired Welsh miner John Davis to develop and operate the coal mines on his property. The Welsh, along with the Germans, were among the first immigrants to come to Youngstown and the Mahoning Valley. Many had experience working in the mines of Wales and Great Britain, and as coal mining was only in its infancy in the Valley, Davis even set out to recruit miners from his home country to work in the Youngstown region.[29] Trumbull and Mahoning Counties subsequently became a major destination for Welsh immigrants to the United States, and several Welshmen played a major role in developing the region's iron and coal industry. After working the mines in Wales, for example, Anthony Howells came to the Mahoning Valley in 1850 and worked at Tod's Brier Hill mines until 1855. Later in life, he launched Howells Coal Company and served as Ohio state treasurer and senator before building a luxurious home in Canton, Ohio, where President William McKinley was his next-door neighbor.[30]

EXPLOITING BRIER HILL COAL

The completion of the Pennsylvania and Ohio Canal provided great impetus for the development of Brier Hill's mines, though it took some convincing on Tod's part to sway Lake Erie steamboat captains—as well as a fairly uninterested public—to switch from wood fuel to Brier Hill coal. Since 1830, local citizens had been aware of Youngstown coal's slow-burning qualities and plentiful heat. Tod attempted to market the coal's special qualities to Cleveland citizens and businessmen, though coal shipped to Cleveland from Henry Newberry's coal lands in the Western Reserve proved unpopular among the city's residents: an attempt to sell it by hauling a load around the streets of Cleveland resulted in no buyers.[31] Citizens believed the coal was "dirty, nasty, inconvenient to handle, [and] made an offensive smoke," and they "did not see the use of going long distances to procure a doubtful article of fuel, neither as clean, convenient, nor cheap as hickory or maple."[32] There was little progress in reintroducing coal as a popular fuel after that early attempt until David Tod found a method to publicize his Brier Hill coal. Tod had fifty coal stoves manufactured in Erie, Pennsylvania, and offered citizens and businessmen a free ton of coal with the purchase of one of his stoves.[33] Tod's plan succeeded, though supplying coal to homes and businesses paled in comparison to the vast wealth to be made by supplying steamships. Tod made several trips to Cleveland in an effort to persuade steamship companies to use Brier Hill coal in place of wood. Though there was "much opposition on the part of boat captains and crews," a trial run using Tod's coal "proved that coal was far superior to wood, requiring less work in stoking, as well as less room in the boats."[34] The opening of the P&O Canal and the subsequent adoption of coal in the Cleveland area dramatically increased shipments of Mahoning Valley coal to the city. In 1832, a grand total of only 1,550 tons of coal had been shipped to Cleveland since the city's founding.[35] By 1865, the city saw an average of 465,550 tons of coal imported per year, a number that nearly quadrupled by 1884.[36]

Youngstown's coal mining industry thrived following the first canal shipments of Brier Hill coal to Cleveland in 1841; however, it would be another four years before ironmasters began to use the coal as fuel for smelting iron ore. In the nearby Shenango Valley, David Himrod (figure 2.4) made a major discovery that changed the course of the Mahoning Valley's industrial history. Recognized throughout the Mahoning and Shenango Valleys as a well-respected iron manufacturer, Himrod was a native of Erie, Pennsylvania, where he became a prominent contractor of public works and improvements. Following the Panic of 1837, Himrod, along with his business partner Bethnel B. Vincent, established the firm of Himrod, Vincent & Co. This firm revived the city of Erie during a period of economic depression when they finished construction of Erie County's first blast furnace in 1841.[37] After operating the furnace for a few years, Himrod and Vincent relocated to the Shenango Valley, where they established several iron furnaces, including the first of many in Sharpsville, Pennsylvania, one of Mercer County's most important pig iron producing centers.[38]

FIGURE 2.4. Well respected ironmaster David Himrod (1806–1877), an Erie, Pennsylvania, native, built several blast furnaces in the Mahoning and Shenango Valleys. His use of raw coal as fuel in the Clay furnace east of Hermitage, Pennsylvania, helped shape the direction of the Mahoning Valley's industrial development. From *Historical Atlas of Erie County, Pennsylvania* (Philadelphia: Everts, Ensign, & Everts, 1876), 107.

In early 1845, Himrod and Vincent constructed the Clay furnace along Anderson's Run, just east of Hermitage, Pennsylvania, and only a few miles from the Ohio state line. Named after Henry Clay, the Kentucky statesman who promoted tariffs to foster American industry, Himrod put the small stone charcoal furnace into operation in July of 1845. After about three months of operation, however, a continual supply of charcoal became difficult to obtain.[39] Himrod attempted coking coal from the Brier Hill and Sharon seams, but the result was an insufficient coke that lacked density and hardness.[40] Desperate, Himrod ordered the furnace's founder, Carson Davis, to persist in coking the bituminous block coal and to mix it with charcoal in the furnace burden, which created a successful cast of iron.[41]

Not long after achieving an effective mix, the furnace's colliers and cokers struck for higher wages. This left Himrod in a precarious situation, as he, Vincent, and Davis were forced to choose between banking the furnace or using raw coal without coking. Himrod chose the latter option, a decision that caught the attention of other ironmasters:

> These experiments were watched with the greatest interest by others,
> and when, by the results of the trial, the suitability of raw coal for use
> in the furnace was established, this material was soon adopted by other
> furnaces in the Shenango and adjoining Mahoning Valley.[42]

The Clay furnace continued to work well and produced a "fair quality of metal," though it is unclear if Himrod employed the use of hot blast, a technology other ironmasters lacked during previous attempts to use raw coal to smelt iron ore.[43] According to Ohio state geologist Edward Orton, the Clay furnace was the first in the United States to use raw bituminous coal, although John White's archeological analysis of the Hopewell furnace eventually discredited this claim. Regardless of the

FIGURE 2.5. John Wilkeson (1806–1894), a pioneer in the smelting of iron ore with raw coal, was one of the Mahoning Valley's most influential antebellum ironmasters, despite his short tenure in the region. Courtesy of the Buffalo History Museum, used by permission.

"fair" quality of iron produced, the raw block coal of the Brier Hill and the Sharon No. 1 seam proved suitable for smelting iron ore, an innovation that signaled an era of change in the region.

Elsewhere in Ohio, charcoal fuel remained scarce, and many ironmasters sought new sources of raw materials. Following the abandonment of John Wilkeson's Arcole furnaces in Lake County, he and his brother-in-law, Frederick Wilkes, moved to the Mahoning Valley, where the iron industry was in its infancy, local raw materials were plentiful, and the newly opened canal promised a reliable outlet for manufactured goods.[44] Wilkeson (figure 2.5), who already had some experience in smelting with raw coal, presumably followed the experiments that took place at the Clay furnace and sought to construct his own furnace built specifically to utilize Brier Hill coal. Together, Wilkeson and Wilkes formed Wilkeson, Wilkes Co. Although Butler in his *History of Youngstown and the Mahoning Valley* states that David Tod also had stakes in the enterprise, no other source corroborates this claim.[45] The company chose a tract of land in the small borough of Lowell (later named Lowellville) on the easternmost edge of the Mahoning Valley, just northeast of Wilkeson's hometown of Poland. The land sat along a hillside on the south bank of the Mahoning River, and it lay within close proximity to a source of raw materials. Lowellville was incorporated as a village in 1890 (figure 2.6); with a population of only 1,155 as of the 2010 census, it is still one of the Mahoning Valley's smallest towns. Despite its small size, Lowellville was a good location for Wilkeson and Wilkes to establish their new furnace. The Mahoning River divided the center of the borough, and the canal followed suit along its north bank. Brier Hill and Sharon seam coal ran underneath the town, and limestone sat in the hills along the southern bank of the river and continued east into Hillsville, Pennsylvania. In addition, the hillside provided a means for charging the furnace with the nearby raw materials.

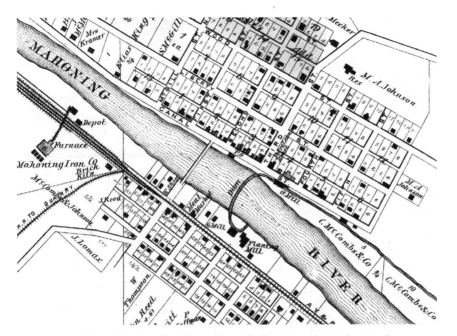

FIGURE 2.6. Map of the small village of Lowellville, 1874. The Mahoning furnace sits just west of the town center on the south side of the river (labeled as Mahoning Iron Co.). Between the furnace's construction in 1845 and the creation of this map, Lowellville made little other industrial progress apart from gristmills and limestone quarries. From D. J. Lake, *Atlas of Mahoning County, Ohio* (Philadelphia: Titus, Simmons & Titus, 1874), 59.

Between 1840 and 1845, it seems likely that ironmasters in the region exchanged ideas on the smelting of iron ore with raw coal, as was the case in eastern Pennsylvania when ironmasters first started smelting with anthracite.[46] Before working as founder under David Himrod at the Clay furnace, Carson Davis was founder at John Wilkeson's Arcole furnaces in Madison, Ohio, for a number of years, until their abandonment after 1840.[47] In August of 1845 in Lowellville, construction of Wilkeson, Wilkes Co.'s new stack, called the Mahoning furnace, began under millwright William McNair, who utilized quarried stone for the furnace's composition. Carson Davis visited the furnace while it was under construction, probably around the same time that the Clay furnace began experiments using raw coal. Swank notes,

> Mr. Davis . . . visited and inspected Mahoning furnace while work upon
> it was progressing, and that he had been an employe [*sic*] of the owners
> of Arcole furnace and was familiar with the experiments that had been
> made at it in the use of raw coal about 1840.[48]

Davis likely instructed McNair and the furnace's owners in the proper construction and configuration of the furnace.

On August 8, 1846, the Mahoning furnace was put into blast under furnace manager and founder John Crowther. A native of Shropshire, England, Crowther was an experienced blast furnace manager who operated seven furnaces in Staffordshire, England, before immigrating to the United States in the early 1840s. Before his employment at the Mahoning furnace, he operated the Brady's Bend Iron Co.'s two furnaces in Brady's Bend, Pennsylvania, and later instructed his three sons, Joshua, Joseph, and Benjamin in the working and management of raw coal furnaces in the region.[49] Upon taking management of Wilkeson's aptly named Mahoning Iron Works, Crowther successfully smelted the first iron in Ohio with raw coal, and several Valley newspapers reported the incident, likely in a tendentious manner. The September 2, 1846, edition of the Canfield, Ohio, *Mahoning Index,* for example, noted, "The Mahoning Iron Works are of the largest capacity and of the most perfect construction— They are provided with powerful machinery, and embrace all the modern improvements in use in Scotland and at the furnaces of Staffordshire."[50] On the first day of operation, the furnace produced 100 pigs of No. 1 gray foundry iron that resembled the "pig iron held in such high estimation in the Eastern cities for fine castings, being soft and close grained."[51] The Mahoning furnace's primary source of coal came from both the Sharon seam on the state borders and Mount Nebo coal mines. Both coal deposits were located in the area between Struthers and Lowellville where the Heatons quarried much of their limestone for the Hopewell furnace.

Wilkeson, Wilkes Co. discovered that smelting iron ore with raw coal proved cheaper than doing so with charcoal. Local ores, which furnace managers considered "harsh," contained siliceous matter, and they were reduced more easily with the concentrated heat and blast of the raw coal furnace than by a charcoal stack. In addition, iron manufacture was not constrained by the need for timber, which spared the region's forests from harm. The introduction of preheated blast air also made the Mahoning furnace's operation more efficient. The 1847 issue of the *Journal of Mining and Manufactures,* which covered the first half of the year, noted that the Mahoning furnace utilized hot blast in the smelting process, the first in the Valley to adopt the technology.[52] The use of hot blast was not only necessary to properly smelt the raw coal, but it also reduced fuel costs, as the furnace required three to three and a half tons of coal per ton of iron and produced about five tons of iron per day. Comparatively, the Trumbull furnace along Mill Creek, which continued to use a cold blast, required four to four and a half tons of charcoal per ton of iron while producing only three to three and a half tons of iron daily.[53]

Initially, newspaper reporters and trade journalists deemed the Mahoning furnace the first to utilize raw coal as fuel. On August 15, 1846, the Warren, Ohio, *Trumbull Democrat* stated that "to these gentlemen [Wilkeson, Wilkes Co.] belongs the honor of being the first persons in the United States who have succeeded in putting a furnace in blast with raw bituminous coal."[54] The Heatons had attempted to use raw coal in their furnace as early as 1808, and many ironmasters experimented with raw coal in the early to mid-1840s. However, most, if not all, of these attempts were acts of desperation when charcoal fuel proved scarce.[55] No other blast furnace was constructed specifically with the intent to smelt iron with raw bituminous coal before 1845. This resulted in a competition between John Wilkeson, Frederick Wilkes, and David Himrod for the honor of the first furnace operator to successfully smelt iron ore with local block coal:

> According to Mr. Wilkes, writing from Painesville, April 2, 1869, this [Clay] furnace was run with coke several months, but at what time it does not state. It is admitted that Mr. David Himrod, late of Youngstown, produced the first metal with raw coal, about the close of the year 1845, and has continued to use it ever since. The friends of Wilkinson [sic] & Co. claim that it was an accident, and a necessity, while their [Mahoning furnace] works were built and intended for raw coal.[56]

This debate notwithstanding, the 1891 issue of *The Popular Science Monthly* noted the importance of the successful smelting of the Valley's block coal in Lowellville when the publication highlighted the event as "the commencement of the use of raw bituminous coal as a blast-furnace fuel in the United States."[57]

FOLLOWING WILKESON AND HIMROD'S EXAMPLES

The Mahoning Valley was poised for an industrial renaissance of sorts, though other changes occurred simultaneously in the Valley's municipal distribution and political spectrum. In March 1846, Mahoning County was formed from portions of Trumbull and Columbiana Counties, mainly due to political divisions among the counties' residents. An argument for the county seat immediately followed, with the towns of Canfield, Ellsworth, and Youngstown all vying for the prestigious title. Canfield won out by virtue of being in the center of the county. The town had little in the way of industrial activity, its population was a paltry 300, and it was well off the path of the canal, but Canfield did sit on the main stagecoach road from Cleveland to Pittsburgh.[58] The new county had a land area of 422 square miles, a population of 22,000, and 4,000 voters, over half of whom were Democrats. Canfield's Democratic newspaper, *The Mahoning Index*, described the new county and its political importance, stating, "The democratic party cannot afford to lose any of its voters in Mahoning at the coming election, and whosoever throws firebrands amongst us, stirring up strife, dissension, and division is indiscreet, unwise, a near-sighted politician, and an enemy to the cause."[59] Apart from party politics, Mahoning County's northern separation with Trumbull County left much of the Valley's major block coal reserves in the towns and villages of Hubbard, Girard, Churchill, Mineral Ridge, and Niles, on the southernmost edge of Trumbull County, though it would be another decade before many of these were worked.

Youngstown—incorporated as a village in 1848 and located just a few miles south of the new county line—began to see additional industrial growth beyond that of David Tod's coal mining operations. By 1850, the Mahoning Valley hosted four furnaces that utilized raw coal. William Philpot constructed the first in Youngstown not long after the completion of Wilkeson's furnace in Lowellville. In the 1840s, Philpot, whose coal mining experience stretched all the way back to his childhood in Shropshire, England, became interested in mining coal near Brier Hill. As a young man, he moved to Wales, where he worked in the mines for three pence a day but nonetheless toiled his way up to the position of overseer and eventually took his own coal contracts.[60] In 1835, Philpot used his $8,000 savings to travel to the United States, which afforded greater opportunities for small businessmen and capitalists alike. He made his way to Pittsburgh, where he immediately began leasing various coal mines in the region. After dropping his interest in a Pittsburgh coal firm following difficulties among the partners, he moved to Portage County, Ohio, in 1838 with his wife and two daughters.

In 1843, Philpot developed and improved the famous Chippewa Mines in Wayne County, but by 1845, he sold half his interests to Clevelander Lemuel Crawford. A year later, Philpot moved to Youngstown, where he continued his coal mining endeavors. He leased the Manning and Wertz bank at Brier Hill, where, while banking for coal, he also discovered higher-grade iron ore.[61] In an unsuccessful attempt to form a furnace company out of Pittsburgh, Philpot returned to Youngstown and, together with Jonathan Warner, David Morris, and Harvey Sawyer, purchased land from Dr. Henry Manning and constructed a furnace along the north bank of the canal two miles northwest of downtown Youngstown and about 300 yards southeast of David Tod's Brier Hill farmstead.[62] This new enterprise, known as the Ohio Iron & Mining Co., mined the majority of the coal used at the furnace from mines located on Manning's property; the company also used charcoal in the burden mixed with block coal. Philpot, Warner, and Morris negotiated the first "coal lease" in the township with Dr. Manning, agreeing to pay royalties of one cent per bushel of coal (a bushel of coal weighed eighty pounds, and about twenty-five bushels equaled a ton of coal).[63] Like Wilkeson's Mahoning furnace, Philpot's stack probably utilized hot blast in order to smelt raw coal.

Philpot proved to other ironmasters that Youngstown's block coal and iron ore reserves held the potential to bring profits and further financial opportunity. At the old Trumbull furnace along Mill Creek, new owner David Grier of Pittsburgh converted the furnace to use block coal rather than charcoal around 1846 or 1847. The experiment initially proved effective, but it failed in the long-term due to the furnace's geographic positioning:

> The expense of transportation, in wagons, of coal from the mines in a distant part of the township, and of ore and limestone from the canal, and of pig-iron to the canal, from two to three miles, was so great that there was no profit in running it.[64]

The furnace made about 1,000 tons of iron from a mixture of charcoal and raw coal in 1853, but only 100 tons of iron in 1855, when Grier abandoned it and stripped all of the machinery from the site with the exception of the furnace itself.[65]

As it did for Grier, the discovery of the use of block coal as a furnace fuel attracted other investors from western Pennsylvania. About the same time that Grier converted the Trumbull furnace to use raw coal, Pittsburgh industrialist James Wood erected Youngstown's third blast furnace in two years, named Brier Hill. Unlike in charcoal furnaces in southern Ohio and western Pennsylvania, ironmasters could build raw coal furnaces in close proximity to others because their fuel source was concentrated in high volumes underground rather than in the nearby wilderness. Accordingly, Wood built his forty-one-foot-tall furnace 300 yards northwest of Philpot's stack along the canal.[66] Before coming to Youngstown, Wood worked as a riverboat captain until 1835, when he moved to Pittsburgh, where he first established a cotton mill later destroyed by the Great Fire of 1845 that engulfed the city.[67] The tragedy left Wood poor, though shortly after he played a major hand in establishing the Hecla Iron Works of Wood, Edwards & McKnight along Wood and Front Streets in Pittsburgh.[68] The firm produced bar iron and nails and was one of ten major iron manufacturers in Pittsburgh at the time. The need for a sustainable and reliable supply of pig iron for the company's rolling mill prompted the construction of the Brier Hill furnace.

CONSUMERS OF PIG IRON

Unlike the old stock of charcoal furnaces constructed in the Mahoning Valley between 1802 and 1833, the Mahoning furnace, Philpot's furnace, and the Brier Hill furnace manufactured pig iron specifically for consumption on the market for further refinement (table 2.1). This

TABLE 2.1. Mahoning Valley Blast Furnaces and Their Markets, 1848

Furnace Name	Year Built	Owner and Location	Product Produced	Primary Market
Maria	1812	Various lessees, Niles	Castings; pig iron	Local
Mill Creek (Trumbull)	1833	David Grier, Youngstown	Castings; pig iron	Local; Pittsburgh
Mahoning	1845	Wilkeson, Wilkes Co., Lowellville	Pig iron	New Castle
Philpot's Furnace	1846	William Philpot & Co., Youngstown	Pig iron	Local; Pittsburgh
Brier Hill	1847	James Wood & Co., Youngstown	Pig iron	Pittsburgh

was a major change, as ironmasters previously manufactured cast-iron products for local settlers rather than catering to the iron rolling mills in western Pennsylvania. Furthermore, the canal opened up access to regional markets (as opposed to just local markets), thereby expanding the number of pig iron consumers beyond the Mahoning Valley.[69] Wilkeson and Wilkes' Mahoning furnace in Lowellville provided pig iron principally to a rolling mill in New Castle, Pennsylvania, owned by prominent iron, coal, and railroad manager Alexander L. Crawford and his brother, J. M. Crawford. One of western Pennsylvania's most notable iron manufacturers, Alexander Crawford was primarily an iron speculator. He built, bought, and sold blast furnaces when the market for iron was high, and he accumulated fortunes from ventures like the Etna furnaces in New Castle, which he sold in 1872—just before one of the greatest economic depressions of the nineteenth century.[70] Early on in his career, however, A. L. Crawford helped rejuvenate New Castle's iron industry by founding the Cosalo Iron Company in 1850. Originally constructed in 1839 by James D. White, Shubal Wilder, and Joseph H. Brown, the small rolling mill lacked puddling furnaces and initially only used wrought iron blooms from the Juniata Valley. Upon White's death in 1841, A. L. Crawford and his brother purchased the mill and enlarged it by adding puddling furnaces to manufacture bar iron and nails.[71] Crawford bought the Springfield furnace in 1842 for pig iron production and in 1847 built the Tremont furnace in New Wilmington, Pennsylvania, although these small stacks alone could not wholly supply the mill after Crawford added iron rail production in the early 1850s.[72] In Lowellville, financial problems prompted Wilkeson and Wilkes to sell their Mahoning furnace to the Crawford brothers in the summer of 1853. Wilkes stated that the furnace "was a great success, but a financial failure; I there lost eight years of the best of my life and nearly $3,000."[73] Not long after, Wilkes moved with his wife to Mayville, Wisconsin, where he became manager of the North Western Iron Company's furnace.

FIGURES 2.7 AND 2.8. James Ward (1813–1864), left, and his older brother William (1806–1888), right, cofounded the iron firm of James Ward & Co. with Thomas Russell, which became the nucleus of the village of Niles after the Civil War. Courtesy of National McKinley Birthplace Museum and Niles Historical Society.

The Mahoning Valley, however, saw its first attempt at finished iron production in the 1840s, when the predecessors of two of its largest nineteenth-century iron plants began modest operations in Niles and Youngstown. In 1841, James Ward, his brother William, and Thomas Russell (James Ward & Co.) constructed the Mahoning Valley's first rolling mill in Niles on the western side of Mosquito Creek, which emptied into the Mahoning River. The mill, which became the nucleus of James Ward & Co.'s substantial Niles ironworks in the coming years, was the first of its kind in the Valley, and according to Butler, "they rolled the first iron made in the Mahoning Valley by this process, perhaps the first made in Ohio."[74]

James Ward was born in 1813, and his brother William in 1806, near Dudley, Staffordshire, England, one of the leading iron manufacturing centers in Europe at the time (figures 2.7 and 2.8). In 1817, the Ward family immigrated to Pittsburgh, where their father established a manufactory for handcrafting wrought iron nails. At the age of thirteen, James began assisting his father in the nail business and remained his assistant for another six years, at which time he began to study engineering, a profession he remained in until 1841.[75] While in Pittsburgh, James and William became associated with Thomas Russell, a devout Christian and native of Bliston, England, who came to the United States in 1822. Russell gained a great knowledge of industry while in Pittsburgh and became "thoroughly acquainted with the iron business."[76] Russell's experience, along with the youth and pragmatism of the Ward brothers, made for a dynamic trio that had the proficiency to create a successful iron enterprise. In 1841, Russell and the Ward brothers travelled to Lisbon, Ohio, where they commenced the construction of a small rolling mill. However, the three men found the conditions in Lisbon unsatisfactory after a flood wiped out much of their equipment. Shortly afterward, they removed what remained of their mill to Niles.[77] The new rolling mill—named the Falcon Iron Works and put into production in late 1842—consisted of a small frame building that measured forty-five by ninety feet. It contained only one set of muck rolls, one fifteen-inch train, a single puddling furnace, one nail machine, a trip hammer for forging blooms, and two boilers—all powered by a single seventy-five horsepower steam engine formerly used in a river steamboat. The mill was also the first in the Valley to use a puddling furnace for wrought iron production.[78]

Conditions in both the mill and the market were difficult, though the Ward brothers and Russell, described as "men of small means, but full of energy and skilled in the business," found a foothold in the industry.[79] The three men performed much of the labor in the mill themselves for the first year of operation, with only three other laborers employed for additional help. James Ward was manager and engineer, William Ward puddler and heater, and Thomas Russell the roller.[80] Teams of mules and miners hauled all of the coal used at the mill from mines in nearby Mineral Ridge, the earliest such mines developed in Trumbull County, while William Ward worked scrap iron from local farms and pig iron purchased from the proprietors of the Maria furnace to produce wrought iron.[81] Although the Maria furnace still produced charcoal iron only a few hundred yards north of the Falcon mill, James Ward & Co. had difficulty in securing a constant supply. This problem aside, James and William Ward's mill became a relative success and produced a per-week average of three tons of finished iron and thirty kegs of nails (one keg of nails usually weighed 100 pounds; twenty kegs equaled one net ton).[82] The three men turned out bar iron, horseshoe iron, sheet, and a special mixture of iron and scrap called "Dandy Tire," a product in great demand for wagon tires due to its good welding and wearing qualities. The Wards shipped "Dandy Iron" west to several markets connected by the newly opened canal.[83]

James Ward's Falcon mill remained in steady production during the 1840s, but output was limited due to the mill's small size and a period of free trade throughout the decade. In the middle of this nonrestrictive period of importing and exporting, Youngstown's first rolling mill appeared in 1846 on the north side of the Mahoning River, along the north bank of the canal. Hugh Bryson Wick and his brothers Colonel Caleb B. and Paul, along with Isaac Powers, Robert W. Taylor, and others, put the first Youngstown rolling mill into operation in response to the construction of Philpot's furnace the same year, when the group of local investors sought to produce finished products from the pig iron manufactured at the small furnace. Hugh B. Wick was the primary catalyst of what became the new Youngstown Iron Company. Wick was a prominent Youngstown financier and son of Henry and Hanna Baldwin Wick, both pioneers of the Mahoning Valley with close connections to the area's business and social interests.[84] The Wick family became one of the Mahoning Valley's most prominent names in the latter half of the nineteenth century, not only in financing, but also in coal mining and iron interests.

The Youngstown Iron Company, however, failed almost as quickly as it began. In contrast to the collective experience of the partners of James Ward & Co., only one of the Youngstown Company's original organizers had any practical experience with ironworking. The mill first began operation in the spring of 1846; it consisted of only four puddling furnaces and rolled bars, bands, and sheets refined by charcoal fire and a light blast.[85] Youngstown Iron Co. operated for only a few years and rarely produced seven tons of finished iron per day. Financial troubles resulting from Congress lowering the tariff on British iron, along with a lack of skilled workers, decreased the mill's iron production. As a result, the mill extinguished its fires for several years. This limited Youngstown's capacity to finish iron production and left Pittsburgh as the Valley's major purchaser of pig iron. From Pittsburgh, the pig iron was converted into finished products and sent back to Youngstown for consumption.[86] Consequently, finished iron production remained relatively insignificant in the Youngstown region compared to its growing capacity to produce pig iron.

A SHIFT IN INDUSTRIAL IDENTITY

The 1840s were a period of technological experimentation and financial trial and error in the Mahoning Valley. Hundreds of entrepreneurs entered the iron business in the antebellum period, yet these artisans were very different from the steel barons of the late nineteenth century in both profit margins and the scale of industry.[87] In the 1850s, ironmasters in the Valley began to develop an identity just in time for westward expansion, a period of growth and demand for iron and iron products. Most of the Valley's existing ironworks changed ownership in the beginning of the new decade, and many newcomers to the region found a foothold in Youngstown's young industry. Several established iron furnaces transformed from the familiar isolated stone structures that dotted the countryside into more modernized complexes that took advantage of new construction methods and technology. These furnaces were also built closer in proximity to urbanizing areas, such as downtown Youngstown, to reap the benefits of new transportation networks. During the first half of the nineteenth century, ironworks west of the Appalachians largely obtained their supply of raw materials locally, whereas older mills east of the Appalachians could not find ore and coal together in the same vicinity.[88] The geography of raw materials acquisition—specifically iron ore—began to shift in the 1850s, which greatly benefited ironworks west of the Appalachians, especially those in the Mahoning Valley. Nonetheless, Pittsburgh remained the largest producer of finished iron in the region: it vastly outproduced Youngstown and Cleveland, the latter of which did not see its first rolling mill until 1854 or 1855 and its first blast furnace until 1864.[89] An 1850 Pittsburgh directory listed the city as having "thirteen rolling mills, with a capital of about $5,000,000, and employing 2,500 hands. These mills consume 60,000 tons of pig iron, and produce bar iron and nails amounting to $4,000,000 annually."[90] Growth in the manufacture of finished iron products in Pittsburgh over the following two decades sparked remarkable growth in pig iron production in the Mahoning Valley. Consequently, furnaces in the Mahoning and Shenango Valleys increased their importance in providing pig iron to mills in Pittsburgh, which for years obtained the majority of their supply from furnaces in central and eastern Pennsylvania, namely the Juniata Valley.[91] Henry B. Robson of the Zaleski Furnace Company in Vinton County, Ohio, reinforced the notion that "the manufacture of merchantable iron from stone [block] coal is just beginning to receive attention in Ohio."[92] By the eve of the Civil War, Youngstown area pig iron dominated in the Pittsburgh market.

Between 1840 and 1850, Youngstown's population had increased by almost 100%, to 1,500, while village newspapers made the bold prediction that Youngstown would become a "great manufacturing city" in the coming decade.[93] The village limits now extended from Fifth Avenue on its west side to the easternmost end of Federal Street on the east side. Youngstown's north-south borders were Wood Street and the southern banks of the Mahoning River. Saloons were "found in abundance—more notable for their number than for their quality."[94] Churches grew in number and influence, from only three in 1840 to a "noticeable increase" ten years later, which represented the Valley's earliest religious organizations, including Presbyterian, Methodist Episcopal, Protestant, Roman Catholic, and Evangelical Lutheran.[95] Church numbers and denominations naturally increased when deeply religious Welsh, Scotch-Irish, and German immigrants came to the Valley during the 1830s and 1840s.

FIGURE 2.9. Portrait of Charles Templeton Howard (1822–1905), a pioneer in the use of Lake Superior ores. From Ralph D. Williams, *A Biographical Sketch of the Lake Superior Iron Country* (Cleveland: The Penton Publishing Co., 1907), 78.

Other neighboring towns and villages grew slower by comparison and remained agricultural centers, with the exception of Niles, whose iron mills continually grew in size under the management of James Ward.

Among Youngstown's more important and influential immigrant newcomers in the early 1850s was Charles Templeton Howard (figure 2.9). Born in 1822 in Belfast, Ireland, Howard came to the United States at the age of twenty and later moved to Akron, Ohio, where he accepted the position of laborer at the Etna furnace of David Tod's Akron Manufacturing Company. Howard learned his way around the blast furnace and labored for another six years at the Akron firm before coming to Youngstown, where he found employment as a clerk under James Wood at his Brier Hill furnace.[96] Howard soon proved his worth as a competent businessman when the company's head bookkeeper badly mishandled the account balance. After Howard solved the situation, the bookkeeper gave him a position in the office, which helped launch his career in the iron business.[97]

Despite his lower-class origins, Howard's accrued knowledge of both business and practical foundry and furnace work allowed him to expand his interests in the region's developing industrial infrastructure. In 1850, Howard bought a small furnace built in 1847 or 1848 by "a company of Welshmen" near Akron, Ohio.[98] He rebuilt the furnace using its old machinery along a bluff in Youngstown on the south bank of the canal, naming it the "Falcon." In a letter, Howard stated, "I had saved up $700 of my salary, and I thought I was rich enough to be an iron master on my own account. I found out later that $700 was not anything like enough money to build and run a blast furnace."[99] In 1851, he sold the small

FIGURE 2.10. Lemuel Crawford (1805–1868), one of Youngstown's earliest major coal developers and ironmasters. From W. J. Comley and W. D'Eggville, *Ohio: The Future Great State* (Cincinnati: Comley Brothers Manufacturing and Publishing Co., 1875), 160.

furnace to James Ward & Co. of Niles and moved to Massillon, Ohio, where he built and operated the Massillon furnace for some time. After a few years of prosperity, Howard promptly returned to Youngstown in 1854, where he formed a partnership with Lemuel Crawford (Crawford & Howard) for the manufacture of merchant pig iron. Crawford (figure 2.10) was a Schoharie County, New York, native and real estate holder who moved to Cleveland in 1846. He believed that David Tod's Brier Hill coal would permanently replace wood as steamship fuel and invested $40,000 in the Chippewa Mines sold to him by William Philpot. Along with Tod, Crawford became one of the Mahoning Valley's first major coal developers.[100]

Following William Philpot's death in 1851, Crawford purchased the Welshman's furnace in 1853. Collaborating with David Morris of Girard, Crawford organized the coal mining and pig iron firm of Crawford, Morris & Co.[101] After he repaired Philpot's furnace and put it into successful operation in 1853, Crawford's business grew, and he looked to further extend his reach in the Valley's pig iron industry. Upon collaborating with Howard, the firm constructed a modern furnace plant, named the Phoenix furnace, along the canal only 200 yards west of Howard's relocated Falcon furnace.

Crawford & Howard specialized in foundry iron for fire castings for stoves, mill gearing, and other types of foundry work (figure 2.11). The new enterprise was a significant advance for the Valley's young industry, marking the first substantial technological improvement in the region's iron trade as well as a focal point for others in the industry. The *Mahoning Free Democrat* heralded Crawford & Howard's furnace "to be the greatest improvement on the Blast Furnace for the last half century."[102] *The Mining Magazine*'s editor, William J. Tenney, echoed this rather substantial claim shortly after operations began at the Phoenix stack. Tenney

FIGURE 2.11. This 1855 sketch, looking east, illustrates Crawford & Howard's Phoenix furnace on the right (south) side of the canal, with the Youngstown Iron Company's rolling mill on the left. These early furnaces often dumped their slag into the Mahoning River or the canal, causing blockage that prompted lawsuits from concerned citizens and businessmen who relied on the canal to transport goods. The Mahoning Valley Historical Society, Youngstown, Ohio.

wrote, "Charles Howard, of Ohio, has changed the form of the blast furnace. He has advanced the art of smelting iron ores. He has cheapened the cost of iron, to all consumers of the most valuable of the metals."[103]

Crawford & Howard completed the Phoenix furnace in just four months and ten days, and after only six days in blast, the stack made ten tons of iron every twenty-four hours.[104] The furnace itself was forty-seven feet high with a twelve-foot bosh. It diverted slightly from traditional American furnace construction: it consisted of sandstone for the first eighteen feet. The next twenty-nine feet were made of firebrick thirteen inches long and molded to fit the circle at the hearth. At this period, the furnace's inside shape was entirely different from any stack constructed before 1854.[105] Rather than the traditional design of drawing the furnace in smaller from its widest point at the bosh just above the hearth, the lining remained the same diameter at the top as it was at the bosh. This radical new design allowed an equal blast pressure from the tuyeres as it was not hindered by the usual narrow, high hearth.

Crawford & Howard's Phoenix furnace was the first stack in the Mahoning Valley built in an open area away from a hillside in order to locate it near the Pennsylvania and Ohio Canal, a benefit that allowed for convenient transportation of iron to its purchasers. Previously, all furnaces utilized a bluff or hillside in order to connect a wooden charging bridge to the top of the stack for charging raw materials. Crawford & Howard's innovative design permitted the construction of a larger furnace that simultaneously produced higher quantities of iron than the smaller stone stacks in the area, but required the use of a vertical, water-balanced hoist for transferring raw materials in barrows to the furnace's top. Crawford's furnace may have been the first to introduce this method of charging. In Port Henry, New York, Wallace T. Foote put the new iron-shell Port Henry furnace into blast in November 1854, which utilized the first vertical hoist.[106] If the start date of the Port Henry stack is correct, Crawford's furnace began production three months before Foote's furnace, making the Phoenix the first to adopt this widely used nineteenth-century technological innovation. Between 1856 and 1860, all existing iron firms in the Valley adopted the vertical, water-balanced hoist, and every new furnace built after 1854 utilized some form of vertical hoist, whether it was water-balanced, pneumatic, or steam-driven. Thus, the old practice of ironmasters constructing furnaces along a hillside ended in the Valley.

After several months, Crawford and Howard's Phoenix furnace produced between fifteen and eighteen tons per day, a substantial output compared to the smaller stone furnaces in the region that produced an average of four to ten tons per day. Simultaneously, coal mining continued to grow in the Valley, as several new firms opened mines throughout Youngstown, including A. & W. Powers; Crawford, Morris & Co.; and Jesse Thornton's coal banks. Together with the established firms of Crawford & Murray and David Tod's coal mining interests, over 77,800 tons of coal were mined in the Mahoning Valley in 1853.[107] Edward C. Wells & Co. established a new firm for the production of castings, steam engines, and steam pumps for the region's blast furnaces and coal mines, while other new steam foundries, such as J. & C. Predmore, used the high-quality iron produced at the Phoenix furnace to manufacture castings. The region's founders and ironworkers sought out Crawford & Howard's iron after a stove plate made with sixty pounds of iron from the Phoenix furnace and forty pounds from iron scrap proved highly durable when subjected to a test over Snider & Woodruff's blacksmith fire in Salem, Ohio.[108] The Salem firm "had no such metal for general use in their foundry for years."[109]

Despite their success, Crawford and Howard's business relationship developed tension after only two years, when Lemuel Crawford's son, W. W. Crawford, became a partner in the firm, representing his father at the furnace. Howard described young Crawford as a "wild colt" whom he "could not get along with," prompting Howard to sell his interest in the Phoenix back to Crawford in April 1856, leaving Crawford and his son to continue the business alone.[110] Howard seemingly had enough money to construct and operate his own modern blast furnace, as he had intended back in 1850. In 1856, he purchased his old Falcon furnace back from James Ward & Co. and "rebuilt the stack and made improvements of different kinds," including a water-balanced hoist.[111] The furnace built upon the advancements made at the Phoenix stack and, for the most part, held the same measurements. The major difference was in its construction. As was the case with the Port Henry furnace built in 1854, Howard's new Falcon furnace utilized an all-iron shell casing "formed of circular cast iron plates ¾ inch thick in segments of three feet each way, and banded with strong wrought iron hoops."[112] Howard put the furnace into blast on July 4, 1854. His original intentions were to manufacture iron for use in foundries and castings, similar to the Phoenix furnace. However, he quickly transitioned to making rolling mill iron for mills in Pittsburgh, which was, as Howard noted, "the principal market."[113] Throughout the late 1850s, the Falcon furnace received extensive attention in the realms of the iron industry, but the majority of its innovation in blast furnace practice stemmed from the Phoenix furnace, considered an abnormal stack because of its numerous departures from standard blast furnace practice. Both furnaces' designs, particularly the Falcon, became influential not only to ironmasters in the United States but also to others across the Atlantic: "The Falcon furnace . . . became the banner stack of the United States at the time. People interested in the iron business came from all parts of the country, and some even from Europe to see and copy it in whole or in part."[114]

Although the Falcon was exceedingly successful in operation, the iron it produced was not high quality enough for its clients at the Pittsburgh mills, whose primary product was bar iron and nails. Shortly after operations began, Howard smelted local lean ores from nearby mines, but he noticed the low quality of his iron and sought another solution, particularly a different grade of iron ore that mixed well with local ores and produced a higher-quality product. In 1856, an ad in the *Cleveland Herald* announced a shipment of Lake Superior ores by the Cleveland Iron Mining Company to supply blast furnaces in Ohio and western Pennsylvania. Howard's interest in mixing local ores with another grade seemed possible, which prompted him to travel to Cleveland to discuss terms of shipment with Cleveland Iron Mining Co. directors William J. Gordon and Samuel Mather. In August 1856, Howard agreed to purchase 500 tons of iron ore at eight dollars a ton from the struggling iron mining company.[115] The shipment left Marquette, Michigan, on the steamship *Ontonagon* and arrived in Cleveland in late June 1856.[116]

The use of Lake Superior iron ore in blast furnaces was a foreign concept to all ironmasters in northeastern Ohio and northwestern Pennsylvania before 1853. Established in 1847, the Cleveland Iron Company initially sought to mine silver or copper after Michigan state geologist Douglass Houghton issued a report that suggested copper deposits in the state's Upper Peninsula could be mined commercially.[117] Houghton's discoveries were confined mainly to the coastal regions of the Upper Peninsula, but massive iron ore deposits sat further inland. The regions contained iron ore reserves that many thought were "inexhaustible"; these reserves were over 60% iron, whereas most other North American mines produced iron ore with iron content between 20% and 50%.[118] After a group searching for copper and silver north of Jackson, Michigan, found outcrops of iron ore in 1845—which seemed "potentially valuable"—a group of Cleveland associates turned from copper to iron and claimed stakes on the "mountains of iron."[119] Cleveland Iron Co.'s ores, however, would not catch on with ironmasters until the mid-1850s, as they were not yet considered sufficient for smelting in blast furnaces. In addition, transportation of the material to inland ironworks proved expensive. Between 1847 and 1850, new iron mining companies, such as the Marquette Iron Company, erected bloomeries and forges on the shores of Lake Superior to convert their iron ore directly to wrought iron blooms for shipment to other ironworks in Ohio and Pennsylvania—especially Pittsburgh. For several years, this was the extent of the use of these companies' iron ore, and transportation costs across Lake Erie to Pittsburgh and other markets in addition to financial problems that culminated in charcoal shortages for the bloomeries proved disastrous for investors. As a result, directors of the various mining companies purchased and consolidated mines and iron ore lands. They established the Cleveland Iron Mining Company in 1853, which began shipments of iron ore to ironworks in the Shenango Valley the same year.

In September 1853, the Cleveland Iron Mining Co. shipped 152 tons of Lake ores to the Sharon Iron Co.'s Clay furnace (which David Himrod sold to the Sharon company in 1851). The company used "four vessels to move the ore from Marquette to Sault Ste. Marie, where it was portaged over the falls" to the port of Erie and sent via canal to Sharon.[120] Members of the Sharon Iron Co. (formed in 1850 by coal merchant General Joel B. Curtis), along with experienced ironmasters David Agnew and Dr. Peter Shoenberger, acquired an interest in the Jackson Iron Co.'s 640-acre iron ore mine in Jackson County, Michigan. This mine served to motivate company directors to use their Lake ore successfully and profitably.[121] Although the shipment was intended for the Clay furnace, the road from the canal to the furnace was nearly impassable and partly frozen. This promoted experimental use of ore in the John and David Agnew–owned Sharpsville furnace, which sat directly along the Erie Extension Canal. According to a letter written by David Agnew, the Lake ore was mixed with local ores, taken to the Sharon Iron Works, and "there converted into bar iron and nails of very superior quality."[122]

At the Clay furnace, manager Frank Allen smelted the Lake ores alone and without mixing them with local ores, producing high-quality pig iron. However, the hardness of the ores left the inside of the furnace in a terrible state. David Agnew, writing from his office in Sharpsville, stated, "We have tried more experiments with that ore than has been profitable, and will never put another pound of it into our furnace."[123] Despite Agnew's remarks, directors of the Cleveland Iron Mining Co. were still enthusiastic over the idea of their ore used in Mahoning and Shenango Valley furnaces, but subsequent tests conducted using Lake ores in blast furnaces in 1854, 1855, and 1856 proved the ore was too hard for furnace operations and ultimately worthless.[124] On October 27, 1855, Sharon Iron Co. director M. C. Trout wrote to Cleveland Iron Mining Co. president William J. Gordon, in regard to additional tests of the ore in furnaces:

> The experiments with Lake Superior ore in blast furnaces have not been made under circumstances to enable me to state the best method of using it. The improvements now being made at Clay furnace will enable us to test that metal this fall or winter.[125]

The suggested improvements were to line the inside of the blast furnace with iron ore, similar to how a puddler prepared his furnace. This helped reinforce and protect the blast furnace's inner lining. Following the longer than expected repairs to the furnace, Frank Allen noted,

> In October 1856, we gave the Clay furnace a general overhauling, putting in new lining and hearth, and made material changes in the construction of the same, put her in blast late in the fall and in a few days were making a beautiful article of iron from Lake Superior ore alone.[126]

These experiments proved successful and convinced several blast furnace operators in northeastern Ohio and western Pennsylvania that smelting with Lake ores could produce high-quality pig iron and more tons per day, and consume less coal, despite costing three times as much per ton as native ore.[127] One handicap that potentially inhibited use of Lake ores in the furnaces of the Mahoning and Shenango Valleys was the persistent belief that smelting such ores with raw coal was impossible, and that any coal used required transformation into coke. However, this speculation proved false, and smelting with Lake ores was an immediate success that vastly improved the quality of iron Howard produced for his customers at the Falcon furnace. This process proved equally successful at the Agnew brothers' Sharpsville furnace. Howard noted that using the ore "worked nicely from the start, improving the quality of iron by giving it body and very much increasing the output of the furnace per day and making the cost per ton for labor less, and the iron more salable as well."[128] Howard's mixture of Lake and local ores was the first such occurrence in Ohio, a practice that became commonplace for decades. Nearly all Midwestern iron and steel manufacturers transitioned to using Lake ores by the turn of the twentieth century.

Howard's former business partner turned competitor also understood the advantage of using Lake ores. Following Howard's departure from their partnership in 1856, Lemuel Crawford used the money to overhaul his old Philpot furnace and reorganize his company. Not to be outdone by Howard's new Falcon Iron Works, Crawford put his new furnace into blast on September 8, 1856, following five months of repairs and an overall structural transformation. Crawford christened his new furnace the "Eagle," yet another named for a bird of prey (possibly alluding to an unspoken business rivalry between Crawford and Howard). At forty-nine feet tall, the Eagle furnace was slightly taller than the Phoenix or Falcon, making it the largest in the Mahoning Valley at the time.[129] Crawford adopted many of the principles used in the construction of his Phoenix furnace, in particular the stone and brick construction, rather than the new ironclad composition that his rival Charles Howard used for his Falcon furnace. Crawford's new Eagle furnace produced sixteen to seventeen tons of iron per day and, in one instance, as much as twenty-one tons in a single day. This was a considerable amount for a time when most small furnaces around eastern Ohio and western Pennsylvania averaged only five to ten tons of iron daily.[130] Like Howard, Crawford began using Lake ores at both his Phoenix and Eagle stacks. He mixed them with local ores and, by 1859, switched to Lake ores entirely.[131]

REBIRTH OF THE ROLLING MILLS

Charles Howard and Lemuel Crawford drastically improved Youngstown's status as an important iron manufacturing town, but the Mahoning Valley's finished iron capacity was minimal and confined to the two small rolling mills in Niles and Youngstown that produced no more than 1,000 tons annually.[132] Youngstown's rolling mill failed to remain in operation into the 1850s and sat idle until 1855. The *Mahoning Free Democrat* responded to Youngstown's indifferent outlook toward its idle works and the village's lack of an operable rolling mill:

> There is no place in this section of the State where all the facilities for making iron are more easily attainable than here, and a rolling mill in the midst of us can secure all the metal necessary to a successful run. What good policy is there in having metal sent from this to Pittsburgh, and

FIGURE 2.12. Portrait of Joseph H. Brown (1810–1886), president of Youngstown's Brown, Bonnell, Westerman & Co. (later changed to Brown, Bonnell & Co. during the Civil War). He quickly joined the ranks of the city's social elite, becoming directly involved in numerous Youngstown banks and industries. From Joseph G. Butler Jr., *History of Youngstown and the Mahoning Valley,* vol. 2 (Chicago: American Historical Society, 1921), 180.

there made into nails and bar iron, and sent back to us for consumption, thus paying two freights beside the profit to the manufacturer?[133]

Shortly after the *Democrat*'s statement, the long-idle rolling mill went up for sale by the village in September 1854. The next month, William Powers, a merchant, coal operator, and one of Youngstown's leading socialites, purchased the deteriorating mill and property for $14,006 in order to rehabilitate the machinery for resale to a competent firm in the near future.[134] Investment in the rolling mill came in the form of displaced brothers—and experienced ironworkers—Joseph Henry, Richard, and Thomas Brown. They partnered with William Bonnell and together came to the Mahoning Valley from New Castle, Pennsylvania, in search of new business prospects following a tremendous financial loss.

The arrival of the Browns and Bonnell in Youngstown was quite possibly one of the most significant events in the Valley's nineteenth-century industrial history. Of the three Brown brothers, Joseph Henry Brown (figure 2.12) was the oldest and most experienced in terms of both business and practical ironworking. He was born July 24, 1810, in Glamorganshire, Wales, where his father and grandfather, both well-respected iron manufacturers, exposed him to the industry at an early age.[135] At the age of four, he immigrated to the United States with his parents, who settled in Maryland. There, his father, John Brown, built the Ellicott iron and copper mills near Baltimore. Joseph helped work the mill with his father and, coming from a prominent background, made use of his family's extensive library. When he was twenty years old, he left home against his father's wishes to work at the iron mills in Mont Alto, Pennsylvania,

FIGURE 2.13. Richard Brown (1824–1903), younger brother of Joseph H. Brown and member of the firm of Brown, Bonnell, Westerman & Co., managed the mill itself. He was a devout Methodist and later aided in the development of Youngstown's library and the Young Men's Christian Association. From Thomas W. Sanderson, *20th Century History of Youngstown and Mahoning County* (Chicago: Biographical Publishing Co., 1907), 582.

and later became superintendent of an ironworks in Antietam and Harrisburg. Brown gained a strong reputation as a competent manager in the iron industry, so much so that James D. White, one of the wealthiest men in New Castle, Pennsylvania, with business interests in New York, Philadelphia, and Cincinnati, sent for Brown to manage and aid in the construction of New Castle's first rolling mill in 1839.[136] After White's unexpected death in 1841 and the acquisition of the mill by Alexander L. Crawford and his brother, Brown purchased an interest and remained manager. Following four years of success under the Crawford brothers, Joseph Brown partnered with Joseph Higgs and Edward Thomas and built the Orizaba Iron Works in New Castle in 1847.

Not long after the construction of the Orizaba Mill, Joseph's younger brother Richard obtained a position in his brother's mill as a roll turner. He was, however, less skilled than his older brother and did not have the same advantages in age and experience. Richard (figure 2.13) was born in 1824 near their father's Ellicott mills in Baltimore, but unlike his brother, he received a more formal education in the area's local schools. A devoted Christian, Richard went to New Castle in 1845 when he was only twenty-one years old; his only possessions consisted of a church letter, 100 dollars, and some change.[137] The bank in which he invested his 100 dollars failed shortly after he arrived, leaving Richard nearly penniless. After a bout with illness that left him near death, Richard's brother Joseph recruited him in his mill, where "he was accustomed to begin work at five in the morning and work until four in the afternoon at the rolls, and from four until six he worked at roll turning, clearing $3.00 a day."[138] Not long after, Richard met his wife, Henrietta Chenoweth,

FIGURE 2.14. William Bonnell (1810–1875), who worked under Joseph H. Brown in New Castle, managed Brown, Bonnell, Westerman & Co.'s office affairs. From *The Iron Trade Review* 44 (January 21, 1909): 185.

Brier Hill
Coal and
"Merchantable"
Pig Iron

at Sunday school, and they married in 1849. Richard's interest in the iron industry grew, and for the couple's honeymoon, they toured England and Wales, where Richard's primary purpose was to visit these countries' iron mills.[139]

Business at the Browns' Orizaba Iron Works flourished for several years, and eventually Richard and Joseph's other brother, Thomas, became a partner. Moreover, the Brown brothers took in another promising partner named William Bonnell (figure 2.14). Bonnell, the eldest of the group, came to the United States from Yorkshire, England, settling in Cincinnati. After struggling to earn a living wage, Bonnell came to New Castle in 1850, landing a full-time bookkeeping position at Joseph Brown's rolling mill.[140] Together this diverse group of practical ironworkers and businessmen became well respected throughout New Castle, though their prosperity was not without drawbacks. In the early 1850s, Joseph Brown required additional capital to expand operations at the mill, so he admitted three more partners. Brown managed the mechanical portions of the mill profitably, but "through the dishonesty of one of the partners, the financial management was a failure."[141] Not long after, a fire ravaged the Orizaba Iron Works, which left Bonnell and the Brown brothers penniless. After hearing of the sale of the former Youngstown Iron Company's mill along the canal in Youngstown, the brothers traveled twenty miles northwest to the Mahoning Valley and inspected the small mill for a potential overhaul in operations. Together with James Westerman, an experienced ironworker and manager of several ironworks in New Castle, the Brown brothers bought the "Old Mill" in March 1855, for $25,000. They operated under the name of Brown,

Bonnell, Westerman & Co.[142] The new proprietors sought to remove the mill's machinery and transfer it to New Castle; however, Youngstown officials demanded that the mill stay in Youngstown, lest the sale fall through. Bonnell and the Browns agreed to keep the mill in the Mahoning Valley and, shortly after the purchase, obtained additional capital, which allowed for several additions to the plant. When the men took over the mill in 1855, it contained only four puddling furnaces, two heating furnaces, one annealing furnace, eight nail-cutting machines, one muck train, and a twelve-inch bar mill train. A recommissioned riverboat steam engine powered the rolls.[143]

The company's daily output remained at roughly seven tons because work was restricted to daylight operations. The mill also had to operate in the summer months exclusively because the canal, which furnished the only bulk means of transportation of fuel, products, and other necessities, froze during the winter. Shortly after repairs began on the mill, all of the partners, along with forty displaced ironworkers and their families, traveled by stagecoach and canal to work at the mill.[144] Each member of the firm played an integral part in operations: Joseph H. Brown managed the sales, William Bonnell the office affairs, and Richard Brown the mill itself. Additional capital ensured enlargement of the facility, and by 1856, it contained nine puddling furnaces, three heating furnaces, three trains of rolls, and sixteen nail machines, all driven by steam. By 1857, Brown, Bonnell, Westerman & Co. manufactured 125 tons of wrought iron spikes and nails per month and 240 tons of various sizes of finished iron. These products were principally shipped to Cleveland, Buffalo, Toledo, Chicago, and other lake ports.[145] The mill sat only 300 yards above the Phoenix furnace and 500 yards from the Falcon, thus guaranteeing a convenient supply of pig iron. The Eagle furnace was an additional source. Brown, Bonnell, Westerman & Co. revived Youngstown's dormant iron industry, provided an economic boost to the village, and lessened the reliance on Pittsburgh rolling mills for wrought iron products, despite the dangerous working conditions common at most similar mills across the country (figure 2.15).

Meanwhile, James Ward and his Niles ironworks expanded almost contemporaneously with Brown, Bonnell, Westerman & Co.'s iron mill. By 1855, Ward's rolling mill grew in size and contained ten puddling furnaces, with four others added in the following year. One of the major additions to the mill was the nail factory, erected in 1854 and 1855, which produced on average 15,000 kegs of nails per year (figure 2.16).[146] Cut nails proved one of the most important products in the mid-nineteenth century. Until new methods of production came in the way of wire-nail manufacturing around 1875, the process of making cut nails remained dangerous and labor intensive and required incredible skill. Puddlers and rollers manufactured nail plate—a flat, slender piece of wrought

FIGURE 2.15. Men and boys of all ages worked in the often dangerous iron mills of Brown, Bonnell & Co. This c. late 1860s image depicts the company's nail mill crew posing with tools and three kegs of wrought iron cut nails. Accidents at the mill were not uncommon. In 1856, a red-hot piece of iron burned through a twelve-year-old boy's thigh, catching his clothes on fire and severely burning his entire body, while in 1867, a boiler explosion at the nail mill killed one boy and severely scalded another. ©*The Vindicator.*

iron—and sent it to the adjoining nail mill. There, men heated the plate and fed it carefully into the cut-nail machines. Most laborers in the mill earned less than two dollars a day, while puddlers averaged between three and five dollars, depending on the selling price of iron.[147]

As was the case with most rolling mill companies in the antebellum period, James Ward & Co. did not yet build their own blast furnace for pig iron consumption. The company used Charles Howard's Falcon furnace from 1851 to 1856, but additions to the mill forced the company to lease the small Mineral Ridge furnace located on a stream in Mercer County, Pennsylvania, three miles from Shakelyville. An old stack, it produced 500 to 600 tons a year but drove transportation costs upward due to its isolation in the hills of western Pennsylvania.[148] Yet, despite its addition, the company still lacked a stable supply of pig iron, as the old Maria furnace, which provided Ward's mill with some iron, was abandoned and put up for sale in January 1856.[149]

Around the same time that Ward's Falcon Iron Works expanded in 1854, the Valley experienced one of the most important geological discoveries since block coal's use as a blast furnace fuel nine years earlier.

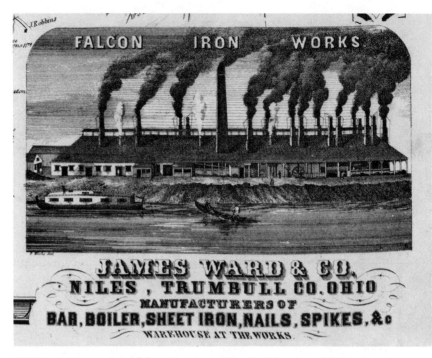

FIGURE 2.16. An 1856 ad depicting James Ward's Falcon Iron Works along the Mahoning River in Niles. From P. J. Browne, *Map of Trumbull County, Ohio, 1856.*

Welshman John Lewis, superintendent of Ward's Mineral Ridge coal mines, approached the office of James Ward with what was thought to be a large piece of slate. "Mr. Ward, I believe what we have been thinking is worthless slate under our coal is black band ore. This looks just like the raw black band ore we mined in Wales where I used to work," stated Lewis.[150] "Let me mine a good sized pile and set fire to it," Lewis continued. "There is enough coal attached to roast it after the fire is started." Ward gave Lewis the go-ahead, and a few days later Lewis returned with a large chunk of genuine calcined black-band iron ore.

Black-band iron ore was an unusual find. In Scotland, Robert Mushet found black-band ore near Glasgow in 1801, and throughout the nineteenth century, ironworks in Scotland worked the ore, which became the basis of the country's industry.[151] Little was found in the United States outside of Ohio, while only two localities in the state—the Mahoning Valley and a region that included northern Tuscarawas County and the southern portion of Stark County—held enough of the ore to merit economic importance.[152] Ohio state geologist J. S. Newberry described black-band ore's unusual characteristics in an 1873 report:

> Blackband ore is a bituminous shale, such as occurs in abundance in
> the Coal Measures, locally impregnated with sufficient iron to make it

valuable as an iron ore. Usually it has almost nothing of the appearance of iron ore, and it is therefore not surprising that its true character should have been for a long time overlooked, and that it is so difficult of recognition even by those who are well acquainted with other varieties of iron ore. Blackband ore contains with its iron so much carbonaceous matter that it is rendered black in color and light in gravity by it, hence its ferruginous character is completely masked. It is not a rich ore in its natural state, as it contains but twenty to forty per cent. of metallic iron, but it forms continuous strata, often many feet in thickness, which are repositories of great quantities of iron, and hence it is practically the most important of our Coal Measure ores.[153]

By 1856, black-band proved a "very favorite ore" among Valley furnace operators, who paid between $2.50 and $3.00 per ton for it in a raw state.[154] In fact, the ore was so valuable that "not only was it mined in connection with the coal, but it was taken up in the old workings where it had been passed over as worthless."[155] Mahoning Valley iron manufacturers mixed black-band with Lake ores, making pig iron that garnered "a high reputation" among its consumers in Pittsburgh and other markets.[156]

By the mid- to late 1850s, iron production in the Mahoning Valley, especially in Youngstown, had grown considerably. Nonetheless, Frederick Overman, in his 1854 treatise *The Manufacture of Iron*, stated that the use of raw coal in Valley furnaces "succeeded exceedingly well; but the demand for pig metal is very limited in the western markets."[157] Indeed, iron manufacturers east of the Appalachian Mountains still largely dominated the country's market because of their location near the industrialized portions of the country. Ironmasters smelting with anthracite coal in the Lehigh Valley in eastern Pennsylvania made 31.6% of the country's pig iron in 1854, while the rolling mills in the same districts, which consumed the pig iron made there, produced about 122,500 tons of finished iron.[158] In contrast, the Mahoning Valley had little in the way of local pig iron consumers. The rebirth and expansion of the Valley's two rolling mills pushed the region's finished iron production to about 4,000 tons a year by 1856 and 1857—still an insignificant number, though, compared to the output of mills in eastern Pennsylvania and Pittsburgh's twenty rolling mills.[159] Nonetheless, the maturing of Youngstown's pig iron manufacture left Pittsburgh reliant on its product. By 1858, the Mahoning Valley contained six blast furnaces using raw coal that produced about 11% (12,000 tons) of Ohio's total pig iron, while the more extensive Hanging Rock Iron Region, with forty-one furnaces over five counties, produced 75% (79,500 tons) of the state's total. Nearly all of them used charcoal.[160] The primary market for the Hanging Rock district at this time was the local mills and foundries in Cincinnati, but after the 1860s, larger firms grew dominant, iron prices fell, and charcoal fuel became increasingly scarce.[161]

It was an era defined by the ironmaster, by individual craftsmen, and by entrepreneurs looking to hone their trade and establish their presence in the Mahoning Valley's young industry. It was not, however, an era of bankers and big business. The latter emerged during the Civil War, when reforms in banking and finance enabled industrial growth to supply the war effort.[162] Nevertheless, future prospects for the Mahoning Valley were high. Cleveland's newspaper, the *Cleveland Leader,* whose home city still struggled to develop a presence in a growing iron trade, proclaimed, "The Mahoning Valley is now the most favored place in the Union for the manufacture of iron."[163] In part, this was true. The Valley had all of the raw materials needed to make iron—black-band ore, Brier Hill coal, limestone, access to Lake ores—and a sufficient transportation network by way of the canal to carry goods and products in and out of the region. In the coming years, there would be a drastic increase in pig iron production in Youngstown, as well as in the nearby village of Mineral Ridge and the growing iron town of Niles. Ironmasters would open more coal and iron ore mines, build blast furnaces, and integrate their rolling mills with pig iron manufacture, thus ushering in a period of great expansion during the war years.

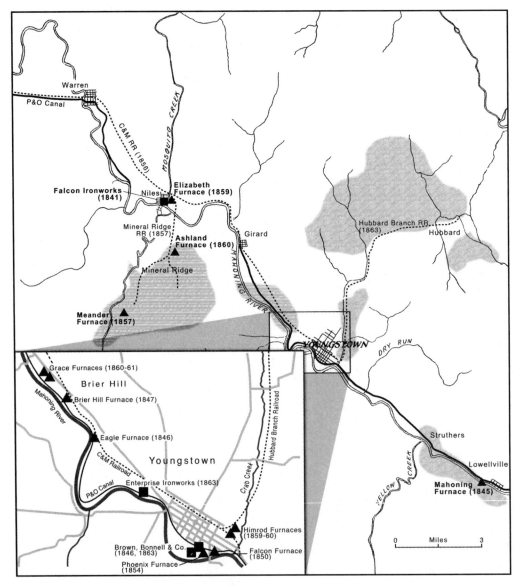

FIGURE 3.1. Map illustrating the Mahoning Valley's industrial growth between 1856 and 1863. The Cleveland and Mahoning Railroad and its Hubbard Branch (dashed lines) intensified the growth of ironworks in Youngstown, leading to a large volume of furnaces (triangles) and rolling mills (squares) built from downtown to the west side in Brier Hill. Coal mining (shaded areas) blossomed as efficient transportation finally arrived to the region.

the National Bank. In 1844, the Democratic Party chose Tod as its candidate for state governor, though Whig candidate Mordecai Bartley defeated him by a slim 1,000 votes.[3] Another attempt for governor in 1846 ended in defeat against another Whig opponent, William Bebb. Despite these setbacks, Tod's work earned him the nickname "giant of Democracy," and President James K. Polk appointed him minister to Brazil in 1847.[4] Tod spent the next five years in Rio de Janeiro. Upon his return to Youngstown in 1851, Tod focused on his business affairs at the Akron Manufacturing Company and coal mines near his homestead, which, due to his long absence, were not left in a "prosperous condition."[5] He was also involved with the promotion of a railroad through the Valley, which would further solidify Tod's coal interests, expand his stakes in the iron business, and benefit the entire Mahoning Valley and the manufacturing centers of northeastern Ohio and western Pennsylvania, including Cleveland and, eventually, Pittsburgh.

COMING OF THE RAILROAD

Ohio's railroads were minimal before 1840, having only forty miles of rails across the state, most of which were short lines that did not cross state boundaries. In 1836, the Erie and Kalamazoo Railroad began as the state's first operating line that connected Toledo to Adrian, Michigan, using cars drawn by horses on wooden rails.[6] The company replaced the primitive horse-drawn cars with a steam locomotive the following year, though the line only went a distance of thirty-three miles. Other small railroads, such as the Little Miami, chartered in 1836, connected Cincinnati and Springfield, Ohio, while the Cincinnati, Hamilton, and Dayton Railroad, built in 1851 for passenger purposes, helped establish and grow communities around the "Queen City." As a result, Cincinnati grew to a population of 161,044 people by the eve of the Civil War.[7] By the early 1850s, Pittsburgh already had two railroads running in and out of the city: the Pennsylvania, which connected Pittsburgh and Philadelphia, and the Fort Wayne Railroad, which was the first to run locomotives out of Pittsburgh in July 1851.[8] Connections were made with the Cleveland and Cincinnati Railroad shortly after, linking rail transportation between Cleveland and Pittsburgh for the first time. Cleveland's first intercity railroad officially began operations when a locomotive of the Cleveland, Columbus & Cincinnati Railroad pulled its first string of wooden flat cars on November 3, 1849. In addition, the Cleveland and Pittsburgh Railroad in 1852 extended a line through the Washingtonville coalfields in southern Columbiana County.[9] The city of Cleveland had amassed a population of over 46,000 by 1860 because of canal, railroad, and Lake harbor operations that attracted small manufactories and immigrants to the city.[10]

Although most nearby regions greatly benefitted from the coming of the railroad, Youngstown and the Mahoning Valley were largely left out of the mix. The first attempt since the late 1820s to charter a railroad in the valley occurred in February 1848 by the Cleveland and Mahoning Railroad Company. The group consisted of eight representatives from Cuyahoga County; eight from Trumbull, which included P&O Canal promoter Leicester King; and six from Mahoning County, including Henry Wick, Henry Manning, and David Tod, who represented the area while in Brazil.[11] The company received a charter and offered the sale of company stocks to investors interested in the project, but it could not sell enough to justify immediate construction. It would be another four years before talk of the project revitalized. In that time, the Valley's iron and coal industry had considerably expanded and matured, and David Tod, one of the major promoters of the project, had returned from his duties in Brazil. In June 1852, the board of directors met to discuss the organization of a corps of engineers to take on the project and survey the road, which was to run from Cleveland to Youngstown. Among the most influential directors of the Cleveland & Mahoning Railroad (C&M Railroad) were David Tod, and the company's president, Jacob Perkins of Warren, a highly devoted member of the company and its financial manager. Other directors included Frederick Kinsman and Charles Smith of Warren and Cleveland representatives Dudley Baldwin, Charles L. Rhodes, and Henry Wick, who had involved himself in financing in Cleveland since the original charter was issued. Incorporated in 1853, the C&M Railroad hired as principal engineer Edward Warner, who had previously worked as the resident engineer on the Eastern Division of the Ohio and Pennsylvania Railroad.[12]

Unsurprisingly, the company's directors believed that "the most important single item in the freight business of our road upon its first opening, will probably be coal."[13] The new railroad would take the brunt of the canal's coal shipments out of Youngstown and Warren. In 1849, shipments amounted to 751,837 bushels of coal; in 1850, 1,389,901 bushels; in 1851, 1,946,225; and in 1852, 2,306,182 bushels.[14] The canal, which proved too slow for the heavy coal and iron trade, transported Brier Hill coal a total of 100 miles from Youngstown to Cleveland, while the proposed railroad cut the distance nearly in half, to fifty-eight miles, thereby greatly reducing costs and transportation times.[15] The second most important item for freight transportation was iron produced in the Mahoning Valley and New Castle. Industry influenced the placement of the line through the Youngstown area and, as the company's directors hoped, further east to the Pennsylvania state border: "Upon its extension beyond the State line, an amount of business in iron . . . will be thrown upon the road from New Castle, the chief business of which town and its 4,000 inhabitants is the production and manufacture of iron."[16]

The company also anticipated the widespread use of Lake Superior ore throughout the Mahoning Valley, as well as other iron-producing centers:

> If the confident anticipations of the iron masters of the country are real-
> ized as to the superior quality of Lake Superior iron ore, a large business
> must hereafter be done in that article, which may be looked forward to as
> likely to furnish back loading to our coal cars, to be reduced and worked
> with Mahoning coal.[17]

The weight of shipping iron and coal on the railroad was a concern for its directors because of steep grades outside of the Cuyahoga Valley, though Edward Warner assured that such an obstacle was not a problem as long as the engineer maintained a safe, practical speed.[18]

Thus, after some difficulty securing an exact route, the general course of the railroad from Cleveland to Youngstown took shape. However, rails were only laid from its lake terminus on the east side of the Cuyahoga River in Cleveland to a point well before entering Warren before funds for the project were exhausted. Terrified at the prospect of the railroad not reaching the Mahoning Valley, considered the most profitable portion of the line, stockholders pledged their own private funds to complete the project, resulting in severe debt and outsiders being unwilling to invest.[19] The company purchased locomotives and cars well before the line was finished; "two of the directors spent two weeks in Philadelphia trying to borrow $20,000, with which to purchase two locomotives, at last succeeding only through personal friendship, and on personal credit."[20] Jacob Perkins was entirely devoted to the success of the railroad and was confident in its potential profitability. He assumed all of the risk and assured the other directors that he would pay the first $100,000 of losses if the venture failed.[21] Perkins regrettably did not live long enough to see the railroad turn a profit for its stockholders and, while on his deathbed in 1859, he told a close friend, "If I die, you may inscribe on my tombstone, Died of the Mahoning Railroad."[22]

With private and personal funding, the railroad finally reached the Mahoning Valley on July 3, 1856, when trains began to run from Cleveland to Warren.[23] The incident was met with enthusiasm by those along the railroad's route:

> At 2 P. M. a train of beautiful new cars from the manufactory of Was-
> son, left the Depot near the Columbus street Bridge, and made the run to
> Warren. . . . The first passenger train excited considerable interest along
> the route, and at Garrettsville . . . the steam horse was welcomed with a
> salute and the cheers of the people.[24]

FIGURE 3.2. This c. 1869 view looking west into
Youngstown shows the Lawrence Railroad and an extension to Brown, Bonnell & Co.'s mill. The Lawrence Railroad, completed in 1867, connected with the Cleveland
and Mahoning on Youngstown's east side. The Mahoning
Valley Historical Society, Youngstown, Ohio.

The tracks crossed the Mahoning River in Warren and turned south
towards Youngstown, following alongside the north bank of the canal.[25]
By August 21, the tracks extended past Niles, where they came within a
few hundred yards of James Ward's Falcon Mill, and reached Girard on
September 18. Officials expected the line to finally reach Youngstown by
October 1, but unexpected weather delays and a lack of iron postponed
completion of the line for another month. Gradually, laborers laid track
through Youngstown, closely and intentionally passing by David Tod's
coal mines and the stock houses of the Brier Hill and Eagle furnaces,
until the tracks hit their eastern terminus in downtown Youngstown at
the railroad depot built just east of Holmes Road (later named Fifth Avenue).[26] All of the Mahoning Valley's operating blast furnaces and rolling
mills stood within several hundred yards of the railroad, with the exception of the Mahoning furnace in Lowellville, which relied on the canal
for another decade. Once completed, the total cost of the railroad was
just over $2.2 million (figure 3.2).[27]

The coming of the railroad, like the canal sixteen years earlier, was
a major event for Mahoning Valley residents. Farmers took the day off
of work to see the first train come into town, and anybody who had the

opportunity to see a train firsthand made it a point to do so.[28] The railroad had plans to extend its tracks to New Castle in the spring of 1857, but building was delayed and the Panic of 1857 halted any further railroad construction for a number of years.[29] As a result, the C&M Railroad had no connection with other railroads at its eastern terminus in Youngstown, and in Cleveland, the track terminated about three quarters of a mile from the harbor landing. There, railroad cars had access to incoming Lake freight and iron ore shipped from the upper Great Lakes, but the railroad was isolated and mainly relied upon local resources for revenue. The furnaces and mills of the Mahoning Valley were the railroad's primary means of making money, a business that served to "swell its earnings."[30] From January to May 1857, the railroad shipped $54,689.76 worth of coal, while passengers contributed only $19,607.15 in the same period.[31] Furthermore, the railroad affected the canal's business drastically. In 1857, coal mines in the Mahoning Valley shipped 110,000 tons of coal by rail, while the canal only carried 12,000 tons' worth, compared to 75,000 tons in 1856.[32]

RAILROAD DRIVES COAL MINING

Amidst the construction of the railroad, David Tod reinforced his industrial holdings by purchasing the Brier Hill furnace from James Wood in April 1856.[33] This was a logical move on Tod's part, as it kept the supply of pig iron in the Mahoning Valley rather than being shipped to Pittsburgh for consumption, giving Tod's railroad more business and largely keeping pig iron shipments off the canal. Simultaneously, Tod began opening up more coal mines in Girard and Weathersfield Township, just northwest of Brier Hill in Trumbull County. Coal had been mined to some degree in Weathersfield Township since the early 1830s, primarily just south of Niles in the towns of Mineral Ridge and Austintown. The first coal mined in the township was on the farm of Michael Ohl, a Pennsylvania native who came to the Valley in 1815. Roger Hill, a coal miner from Beaver County, Pennsylvania, showed Ohl the outcroppings of coal exposed in Coal Run on the south side of Mineral Ridge, convincing him to drill on his property. In 1835, Hill commenced operation of a drift mine on Ohl's farm and found a seam of coal four-feet thick; however, after some tests, the coal refused to burn and proved to be low quality, or bastard coal, and much of it was left unworked, "forming the floor of the excavations."[34]

Some of the coal veins at Mineral Ridge differed slightly from the block coal veins throughout the Mahoning and Shenango Valleys. The coal seam itself was confined to the southernmost border of Weathersfield Township, and crossed over the Mahoning County line into

Austintown. Throughout most of the Ridge, the seam contained five different layers of coal, slate, and black-band iron ore. According to the Geological Survey of Ohio, the first strata of the coal seam contained a grade known as "Mineral Ridge" coal, which belonged to "the general horizon of the Sharon or Block coal," the coal itself being "a different grade from the Block coal, being much softer and more impure, and of a slightly cementing character."[35] The coal was similar to the bituminous coal found in southwestern Pennsylvania and required coking before use in blast furnaces, but it proved suitable for producing steam in rolling mills, in blacksmithing, or for domestic use. Mineral Ridge coal usually had a thickness of zero to four feet within the seam. Underlying that vein was black-band iron ore, which reached a maximum thickness of up to a foot. Another coking coal sat below the black-band ore, two to three feet thick, and below that was a layer of black slate up to two feet thick. Lastly, the deepest strata contained block coal anywhere from six inches to two and a half feet.[36]

Although James Ward & Co. had already established small coal mines in Mineral Ridge in the early 1850s, other small firms began operations around the same time, if not earlier. Some companies and partners who had already opened coal mines in Youngstown established mines in Mineral Ridge, including Morris & Price's Cambria mine, opened in 1850. One of Mineral Ridge's earliest and largest coal developers was Rice, French & Co. of Cleveland, who first opened the Peacock mine in 1853. For years, the primary means of transporting coal from the mines to the canal was by horse and wagon, but in December 1855, Rice, French & Co., along with James Ward & Co. and other coal mining concerns, petitioned the commissioners of Trumbull County to obtain railroad services to their coal mines in Mineral Ridge.[37] The petition called for the "establishment of a steam railway, from the depot grounds of the Cleveland and Mahoning Rail Road" in Niles to the coal mines of Rice, French & Co., James Ward & Co., Betts, Lynds & Co., and Perry Cross & Co.[38] The primary promoters of this small coal railroad were the partners in Rice, French & Co., including Henry F. Rice, Edwin C. French, and Joseph Chamberlain, whose offices stood on the west dock of the Cuyahoga River in Cleveland.[39] Their new enterprise, called the Mineral Ridge Railroad, received a charter in February 1856, and the company completed its two-mile railroad from the Weathersfield coal bank to the Mahoning River in July, when Rice, French & Co. commenced shipments of coal to Cleveland by the canal.[40] Completion of the railroad across the Mahoning River was delayed because flooding of the Ohio River hindered shipments of iron rail coming from the Cambria Iron Works in Johnstown, Pennsylvania.[41] Nonetheless, the bridge across the Mahoning River and the railroad's connection with the Cleveland and Mahoning Railroad was finally finished by April 1857, thus opening the Mineral

Ridge coal mines to Cleveland by rail and greatly expanding coal production in the Mahoning Valley.

After the first few years of operations, the Cleveland and Mahoning Railroad, by virtue of its connection with the Mineral Ridge coalfields, delivered an average of 500 tons of coal per day, 3,000 tons a week, and 150,000 tons per year, while the canal transported half the amount of coal as the railroad.[42] By July 1857, thirty to forty coal mines were opened between Warren and New Castle, Pennsylvania; however, none existed in Warren, as the block coal veins ended just southeast of the town in Niles.[43] Of these mines, only three were connected with the Cleveland and Mahoning Railroad: Rice, French & Co.'s Mineral Ridge Railroad allowed cars to run directly under their coal chutes, a horse-drawn railroad connected David Tod's Girard mines just east of Mineral Ridge, and the C&M Railroad directly connected Tod's Brier Hill mine. Together, these three firms alone mined about 600 to 700 tons of coal per day.[44]

The railroad brought tremendous growth in a short amount of time. Between 1856 and 1858, the small village of Mineral Ridge had essentially become a boomtown. The town expanded considerably, springing "into existence with a rapidity which equals any of the tales told of western city building."[45] Amassing a population of 1,000, Mineral Ridge encompassed 170 buildings, including two churches, schoolhouses, four stores, and saloons "without number."[46] By April 1858, several more coal mines were connected with branches of the Mineral Ridge Railroad, and many companies invested in coal lands yet to be worked. These included a $40,000 investment in 160 acres of land by the firm of Tod, Wells & Co., of which Thomas H. Wells, later to be a Civil War mayor of Youngstown, was a partner.[47] Over eighty acres of coal land belonged to Morris, Price & Co., who held interests in the Eagle furnace in Youngstown in the early 1850s. The company's mine was entered by a slope laid with light rail manufactured at Stone, Chisholm & Jones's Newburgh Rail Mill, built six miles southeast of Cleveland in 1857.[48] The coal cars were drawn up the slope by a chain to about fifteen feet above the ground, where the coal ran over a screen and fell into railroad cars. The capacity of this mine amounted to nearly 200 tons of coal shipped per day to the Cleveland market. Some companies who held coal lands east of the main roadway through Mineral Ridge and Austintown, such as James Ward & Co., still transported coal in wagons.[49] According to an estimate by William G. Darley, supervising engineer of the Mineral Ridge Railroad, one acre of land in Mineral Ridge yielded about 45,000 tons of coal.[50] The extent of the coalfields prompted several more companies to invest in coal lands in Mineral Ridge and nearby Austintown, and by the summer of 1858, speculators expended a total of over $250,000 dollars at the Ridge, 40% of which went into real estate for the sale of lots to miners.[51]

The life of a coal miner, however, was a difficult and grueling one. The mines were a dark, gloomy, lonely, and often dangerous place to work, and workers were subjected to dismal and hazardous conditions such as cave-ins, accidental dynamite explosions, poisonous gasses, and "black lung," a condition that slowly strangled miners after years of inhaling coal particles. The Mahoning Valley saw its share of mining accidents throughout the nineteenth century. In July 1856, for example, three miners were injured and one was killed when they were sent to reopen an entrance to Tod's Brier Hill mine. The roof of the mine caved in, and a basin eight to ten feet deep and partially filled with water formed around the miners: "One of the men went down into the hole to ascertain the depth of the water, but he had scarcely got below the surface of the ground before he fell to the bottom of the hole . . . lifeless," reported the *Western Reserve Chronicle*.[52] The other three men followed, each poisoned by the "foul air there inhaled." In 1870, John Matthews, manager of the Powers Coal Co.'s mine near Hubbard, Ohio, was killed after a dynamite fuse failed to detonate at the expected time. When Matthews moved in to examine the fuse, it suddenly exploded, hurling a large chunk of coal at his head.[53] Accidents such as these were universal among mining towns throughout the United States. In 1858, a correspondent for the *Pottsville Miners' Journal* reported that, of the 12,000 miners in the mining region of Schuylkill County, Pennsylvania, one-third, or 721, were permanently disabled, and another third had been killed, leaving 700 widows and 1,400 orphans without monetary support.[54]

Despite the inherent dangers of coal mining, the dark, subterranean world of the miner was critical to the development of the Mahoning Valley's industry and economy. It also captivated those who worked and lived on the surface of the ground. In December 1859, a party of two women and three men from Warren visited the newly opened coal mine of Tod, Wells & Co. in Mineral Ridge. The company placed mine superintendent Isaac Holford in charge of the group; he gave them caps and hats belonging to the miners, along with small lamps to light their way underground. The group stepped onto the hoisting platform and was lowered into a shaft that measured sixteen by eight feet and 134 feet deep. The *Western Reserve Chronicle* reported the group's experience in the mine:

> Arriving at the bottom of the shaft, we see the lights (which the miners fasten in front of their caps) twinkling in various directions, and the rumbling of a car and trampling of hoofs warns us to step aside from the wooden railroad which stretches back into the darkness. The car is drawn by a mule, which seems quite as much at home in his subterranean dwelling, as his long-eared brethren above ground, and we thought his eyes had a surprised expression as he gravely winked them when we

held our light near his face, though one of our party persists in the assertion that he was stone.[55]

The visitors regarded the miners as "very intelligent, and ready to answer all the questions our curiosity dictated."[56] As in Youngstown, most of the miners in Mineral Ridge were of Welsh origin, many being second-generation immigrants to the United States.

IRON BOOM: NILES AND MINERAL RIDGE

It was not long until investors and ironmasters alike realized the great value in the coal and black-band iron ore mines of Mineral Ridge. In 1857, editors of Warren's *Western Reserve Chronicle* emphatically asked, "With 30,000 acres of coal land, containing 150,000,000 tons of coal, equivalent in manufacturing power to the life long labor of 50,000,000 of men, what is to be the future of the Mahoning Valley?"[57] Though these numbers are most likely embellished to a certain degree, the editors' question would be answered by the ongoing expansion of pig iron production in not only Youngstown but also in the coal and iron ore–rich townships of Weathersfield and Austintown. On May 1, 1857, William Porter, Daniel Smith, and Joseph Barclay began construction on the Mahoning Valley's first blast furnace since the completion of the Cleveland and Mahoning Railroad and the widespread opening of Mineral Ridge's coal mines. In December 1855, Porter, an Irish immigrant who settled in Austintown after coming to the Mahoning Valley in 1837, began securing coal and iron ore mining rights in northern Austintown Township with Barclay and Smith.[58] The new mining concern, known as Smith, Porter & Co., continued to obtain mining rights on adjacent lands throughout the next year and a half, altogether accumulating about 600 acres.[59] With enough iron ore and coal lands to produce pig iron without reliance on outside mining companies, Smith, Porter & Co. constructed the $37,000 Meander furnace (also known as Porter furnace) east of Meander Creek and about a quarter of a mile south of the Trumbull County line in Austintown.[60] The furnace's iron ore and coal supply sat directly underneath the furnace itself. Miners employed by Porter's company worked a seam of coal four feet, eight inches thick, and directly underneath it was a black-band iron ore seam that ranged from twelve to eighteen inches; miners extracted about one ton of black-band iron ore to every three tons of coal. Porter constructed twelve to fifteen company houses for a portion of his 100-man workforce, most of whom worked in the mines and quarries rather than immediately around the furnace.

Despite initial success by Smith, Porter & Co., the construction and commencement of their furnace operations occurred almost simultaneously with the economic panic that swept the nation in the fall of 1857. Though little is known of Porter's partners, Barclay and Smith, Porter himself did not have a background in iron making. Before securing coal and iron ore rights and constructing the Meander furnace, Porter worked as a merchant for twenty years and, though he possessed business sensibility, he likely did not have the ability to manage a highly capitalized iron company during economic downturns.[61] Furthermore, a proposed railroad called the Ashtabula & New Lisbon Railroad—chartered in 1853 for a line from Ashtabula to New Lisbon, Ohio, which would have passed by Porter's furnace—was delayed, leaving the Meander stack without direct rail or water transportation.[62] The company completed twelve miles of the line by 1860 but failed to construct any portion of its proposed line from Ashtabula harbor to Niles, or Niles to New Lisbon, due to the inability to acquire enough money.[63] After about a year and a half of operations, Porter filed for bankruptcy in September 1859, just as the nationwide economic depression began to level off.[64] After three years of legal cases between Porter and his former business partners, as well as several failed attempts to sell the furnace and its mining properties, Mahoning County commissioners finally sold the furnace to George C. Reis, a partner in James Ward & Co. and a director of the Ashtabula & New Lisbon Railroad, for $11,700, a number that was "thousands beyond any figure offered during previous attempts at sale."[65] In 1857, Reis and other directors of the railroad noted that "Messrs. Porter & Co. now have an extensive and very commodious coal opening" and were building a "large furnace . . . upon the immediate line of the road."[66] By either selling the furnace or operating it, Reis likely saw prospective revenue in the complex following the completion of the New Lisbon to Niles portion of the railroad.

An active industrialist and businessman, George C. Reis, born in Pittsburgh in 1823, played a brief but significant hand in developing and expanding Niles's iron industry before the start of the Civil War. Reis began in the grocery business in Pittsburgh with his business partner Andrew B. Berger. Both men sought new opportunities and, in the early 1850s, Reis came to Niles, while Berger remained at his residence in Pittsburgh.[67] Not long after becoming a director of the Ashtabula & New Lisbon Railroad, Reis became a dynamic partner in Niles's James Ward & Co., and by 1855, Berger too had become a silent partner in the company, contributing capital but not taking part in active management. The rolling mill greatly expanded production after the inclusion of Reis and Berger into the firm, increasing the mill's machinery and its production to about 200 tons of finished iron per week (figure 3.3).[68] Such

FIGURE 3.3.
Early stereoview of James Ward & Co.'s Falcon Iron Works along the Mahoning River in Niles, probably sometime in the 1860s. Courtesy of Niles Historical Society.

additions warranted increased consumption of pig iron, but the company continued renting other furnaces in Youngstown and in 1856 entered into a contract with David Tod to lease the Brier Hill furnace for two years.[69] The death of senior partner Thomas Russell in July 1858 left a major void in the company in terms of experience and business management; however, as the 1857 economic depression faded, the company found themselves in a good position financially and with enough working capital to merit the construction of the firm's own blast furnace, thus relinquishing reliance on other units in the Valley.[70]

By April 1859, James Ward & Co. began construction on their new furnace, named the Falcon (not to be confused with Charles Howard's Falcon furnace in Youngstown), which sat on the east side of Mosquito Creek opposite the abandoned Maria furnace, and only a few hundred yards above the company's rolling mill. An advantage for James Ward & Co. was the furnace's immediate connection with the Cleveland and Mahoning Railroad by a switch track about 1,300 feet from the mainline. The company had easy access to incoming shipments of Lake Superior ore from Cleveland docks, as well as other ore from Lake Champlain and Canada. The Lake ore sat side by side with black-band iron ore mined from nearby Mineral Ridge and, like other ironmasters in the Valley, Ward had the furnace's managers mix the ores before charging them into the stack. While visiting the furnace upon its completion, editors of the *Western Reserve Chronicle* witnessed and described the characteristic process of calcining black-band ore before use in the furnace:

> Huge piles of the black band ore were being melted by the simple process of setting one end of the pile of ore on fire (it having sufficient combustible matter combined with the iron to burn slowly,) and before the fire had reached the other end, the first had burned out and cooled, so that workmen were busy carrying the once melted ore into the furnace.[71]

After top fillers dumped the raw materials "into the gaping mouth of the fiery furnace," iron flowed out from the bottom of the stack into the cast house twice every twenty-four hours, producing about twenty tons per day.[72]

James Ward & Co.'s Falcon rolling mill consumed all of the pig iron produced at the furnace, while all of the fuel for the company's mills came from the firm's own mines within two miles of the mill. Such interoperations marked the first instance of integrated iron making in the Mahoning Valley, primarily because of the great amount of capital involved in collectively running these facets of the industry. Rolling mill firms who owned and operated blast furnaces were "among the largest of their day, whether measured by number of employees" or by the "amount of fixed capital."[73] One of the major changes that prompted integration of ironworks and blast furnaces was the transition from small-scale forges to larger-scale rolling mills. The isolated locations of early charcoal furnaces often separated a company's furnace from their forge, if a firm had enough working capital to operate both branches of the industry. Locating primary smelting operations near secondary forges or rolling mills usually proved impossible or, at the very least, uneconomical, as the finished wrought iron product, much like the charcoal furnace, remained isolated from the principal markets.[74] As a shift in furnace fuels took place in the 1840s, many ironmasters began to construct furnaces in more economical locations near transportation outlets, as was the case in Youngstown at Crawford's Phoenix furnace and James Ward & Co.'s furnace. As the need for iron grew in the first half of the nineteenth century, many new iron companies built higher-capacity rolling mills, and the need for more pig iron from blast furnaces became an issue for many firms. Those particularly affected were the rail mills, which grew larger in overall capacity than other rolling mills, thus necessitating integration of pig iron, wrought iron, and finishing facilities.[75]

Ultimately, the growth and expansion of railroads in antebellum America required a greater amount of iron, despite much of the country's iron rail being imported from England. Companies in the United States began building rolling mills for producing rails in the mid-1840s and 1850s, most of which were integrated operations. An example of the earliest such integration was the Brady's Bend Iron Company in Brady's Bend, Pennsylvania, which combined operations of two blast furnaces and a rolling mill to roll iron rails in the early 1840s.[76] The Mount Savage Iron Company, composed of British and American capitalists, built two blast furnaces nine miles northwest of Cumberland, Maryland, in 1840. In 1843, the company added a rolling mill, and in the following year, the mill, described as "an immense establishment in the wilderness," rolled the country's first heavy iron rails.[77] The Cambria Iron Company, established in Johnstown, Pennsylvania, in 1852, exhibited integration when the company acquired and operated several old charcoal furnaces in the Johnstown area.[78] In 1853, the company began construction on four coke furnaces and a rolling mill for the production of rails, leading to the company's designation as one of the most important rail mills of the nineteenth century.[79]

Before the Civil War, no other iron firm in the Mahoning Valley other than James Ward & Co. combined pig iron production with rolling mill operations, and only two firms in the Shenango Valley operated rolling mills and owned their own blast furnaces, both of which were located in New Castle. These included the Orizaba Rolling Mill, whose owner constructed the Sophia furnace in 1853, and the Cosalo Iron Works, operated by Alexander Crawford and his brother James.[80] Along with their rolling mill, the Crawford brothers also owned the Mahoning furnace in Lowellville, and the Tremont furnace in New Wilmington, Pennsylvania, though these works were separated by up to twenty miles and were sold off or abandoned during the 1860s.[81] The Cosalo Works was also the only mill in the Mahoning or Shenango Valleys to manufacture rails of any kind. In 1853, the mill produced 4,000 tons of "Winslow split rail," a type of rail invented by Mount Savage Iron Co. president John F. Winslow.[82] Three years later, the Crawford brothers entered into a contract with the Cleveland & Columbus Railroad to produce 10,000 tons of iron rails at their mill.[83]

The ability of Mahoning Valley ironmasters to own and operate localized sources of raw materials and blast furnaces, as well as finished iron production, would drastically change the region's industrial structure in the coming decade, and James Ward & Co. acted as a pioneer in this looming era of bigger business. However, it would not be until the height of the Civil War that other Valley iron manufacturers began combining blast furnace and rolling mill operations, primarily through the means of horizontal integration. Nonetheless, the continued

FIGURE 3.4. Portrait of Jonathan Warner (1808–1895), as he appeared around the 1880s or early 1890s. The Mahoning Valley Historical Society, Youngstown, Ohio.

development of Mineral Ridge's coal and iron ore mines presented more opportunities for growth, particularly in the realm of pig iron production, just as William Porter and his Meander furnace attempted and failed to do in Austintown.

Veteran iron manufacturer Jonathan Warner, who at the age of fifty had earned himself the title of "Furnace Builder," established a commercial presence in Mineral Ridge about the same time other speculators began developing the town's coalfields (figure 3.4).[84] In 1856, he invested $50,000 in 130 acres of coal land on the east side of Mineral Ridge, only two years after acquiring an interest in the coal mines of Davidson, Green & Co. in the borough of Wampum, Pennsylvania, with Henry Manning of Youngstown.[85] Before establishing coal mining interests in the Mahoning Valley and western Pennsylvania, Warner had already accrued a long history in the iron manufacturing and mercantile businesses. Born in 1808 near Oaks Corners, Ontario County, New York, Warner lost his father at the hands of a British fleet that entered Sodus Point, New York, during the War of 1812. Following his father's tragic death, Warner became an orphan, as his mother died three years before the start of the war. Warner's father was a poor man, and young Warner himself had little to his name: "I had no shoes and very little of anything to wear. My uncle was also poor, but he had a kind heart, and did what he could do for me."[86] Warner made enough money buying and selling grain that he eventually opened a store in the village of Sodus and another at Sodus Point in the 1820s.[87] Myron Israel Arms, a Sodus Point

FIGURE 3.5. James Wood (1789–1867), a pioneer in the Pittsburgh and Youngstown iron trades, played a brief but significant role in developing Mineral Ridge's pig iron industry with Jonathan Warner. From *Century Cyclopedia of History and Biography of Pennsylvania*, vol. II (Chicago: The Century Publishing and Engraving Company, 1904), 96.

native, became a partner with Warner in the mercantile business there until 1843, when Warner, his wife Eliza, and their five children briefly moved to Pittsburgh before coming to Youngstown, where Warner opened a dry goods store on the southeast corner of East Federal Street and Central Square. Arms, who married Warner's daughter Emeline in 1848, followed Warner to Youngstown a few years later and joined him in his mercantile business, formally named J. Warner & Co.[88] Shortly after, Warner aided in the construction of Youngstown's second blast furnace with William Philpot. After the death of Philpot in 1851, Warner left the furnace and commenced the construction and management of several other iron enterprises in western Pennsylvania.[89] After selling his interests in these ironworks, Warner moved back to Youngstown, where he developed his coal mining interests in Mineral Ridge.

After successfully operating coal mines at the Ridge for three years, Warner's interests shifted once again toward making iron. James Wood (figure 3.5), Warner's new partner in the iron and coal business, no longer had any industrial interests in the Youngstown area. After selling his Brier Hill furnace in 1856, Wood, who owned and operated the Eagle Iron Works in Pittsburgh, constructed the Homewood furnace in northwestern Beaver County, Pennsylvania, in 1857, and put it into blast in 1858.[90] Warner purchased an interest in Wood's Homewood stack, and the two men developed a mutual business relationship.[91] The need for additional pig iron for Wood's Pittsburgh rolling mills probably prompted him to go into business with Warner in Mineral Ridge, where his coal and iron ore mines, along with the rail connection to Cleveland for Lake ores, promised profitability.

Together, the two experienced ironmasters formed the company of Wood, Warner & Co., taking in Wood's son as a junior partner. The company began construction on the $60,000 furnace in the winter of 1859, naming it Ashland in honor of the estate of Henry Clay.[92] The company placed Warner in immediate supervision of the stack, marking the fourth furnace built under his management. The forty-five-foot tall Ashland furnace, the second raw coal stack built in Trumbull County, sat along the line of the Mineral Ridge Railroad on the east side of the town. When blown in on June 18, 1860, the stack produced eight and a half tons of iron in its first cast and averaged about eighteen to twenty tons of iron per day.[93] The entire furnace complex was sizeable compared to others in Youngstown, as most of the furnace's source of raw materials, like Porter's Meander furnace, sat on and around the site itself. About 200 feet behind the furnace complex was a boarding house and six two-story double houses occupied by the company's furnace workers and miners. Each house had a "whitewashed" fence in between, with the houses being painted, "giving to them that neat and tidy appearance which characterizes the whole works."[94] Wood and Warner created an intrinsic, self-contained community around the furnace. Altogether, the Ashland furnace complex presented a similar appearance and structure to the traditional charcoal furnace plantations common in the Hanging Rock region of southern Ohio.

A unique aspect that set the Ashland furnace apart from other stacks in the Valley was its partial use of coke in the furnace burden. In general, widespread use of coke in Mahoning Valley furnaces did not occur until 1869, when ironmasters gradually began to mix Connellsville coke with Brier Hill coal.[95] However, marginal use of coke in the Valley likely began shortly after Wood and Warner blew in their furnace. The company constructed a row of six beehive coke ovens behind the furnace, each four and a half feet tall and eleven feet in diameter.[96] The ovens were the only ones of their kind built in the Mahoning Valley, and were used to coke the less pure Mineral Ridge coal found in the topmost strata of Ridge's coal veins, constituting the first use of coke in the Valley, nine years before it was introduced in other Youngstown area furnaces. An elevated trestle 360 feet in length allowed miners to deliver Brier Hill coal and black-band iron ore straight from the mines and into the stock house, "saving a large amount of labor" and the "wasting of coal" from repeated handling.[97] The Cleveland and Mahoning Railroad delivered Lake Superior ores by way of its connection to the Mineral Ridge Railroad, which ran directly past the furnace's stock trestles, giving the company "unsurpassed facilities for making a good article of metal."[98]

The establishment of Wood, Warner & Co. and the Ashland furnace created a model iron and mining town amidst the Valley's growth into one of Ohio's premiere industrial regions. Less than a year after the company started operations at their furnace, Wood, Warner & Co. secured an additional coal and iron ore supply by purchasing the old Peacock mines abandoned by Rice, French & Co. in 1858.[99] The first battles of the Civil War gradually brought an increase in pig iron prices, and soon demand grew enough to merit the construction of a second furnace in Mineral Ridge. In the winter of 1861, James Wood, while continuing his separate partnership with Jonathan Warner, formed with his two sons, James T. and Charles A., the company of James Wood & Sons. The company built the Wheatland furnace, named after the estate of President James Buchanan, whom Wood admired for his stance on nurturing American industry.[100] Wood began construction on the furnace in the summer of 1861, while at the same time he also purchased interests in nearby coal mines to secure additional fuel and ore sources. The new stack resembled the Ashland furnace in every respect, and by December 1861, work on the trestles, stock house, and furnace proper were completed; however, Wood delayed operations until the spring in order to secure reasonable rates on iron ore shipments from the Cleveland & Mahoning Railroad.[101]

By the spring of 1862, Wood had still not blown in his furnace owing to Cleveland & Mahoning Railroad superintendent Charles Rhodes's refusal to give the same rates to James Wood & Sons as he did to other companies in Niles and Youngstown.[102] Wood rejected the rates proposed by Rhodes, who responded by saying, "You have built your furnace and can't help yourself."[103] Wood replied to Rhodes's obstinacy by stating that he "wouldn't live in a country where he couldn't have half the say in making a bargain," and in June 1862, Wood dismantled the Wheatland furnace, removing it to Mercer County, Pennsylvania.[104] In 1863, Wood, along with his sons and prominent Shenango Valley iron manufacturer John J. Spearman, rebuilt the furnace at a location two miles south of Sharon along the Erie Extension Canal and the newly completed Pittsburgh & Erie Railroad, which secured reasonable and reliable freight rates. In past business trips between Pittsburgh and the Mahoning and Shenango Valleys, Wood had often passed between New Castle and Sharon along the canal and always admired the "beautiful site" where he reconstructed his Wheatland furnace.[105] Between 1863 and 1865, Wood built an additional three stacks and developed an extensive blast furnace plant in what became Wheatland, Pennsylvania, incorporated in 1872 on the foundation of James Wood & Sons and their extensive ironworks.[106]

WAR BRINGS GROWTH

Mineral Ridge's residents considered the removal of the Wheatland furnace from their town a significant commercial setback. A correspondence from the town to the editors of the *Western Reserve Chronicle* stated, "It is

TABLE 3.1. Pig Iron Prices and Corresponding Annual
Pig Iron Production in the United States, 1860–1867

Year	Price (per ton)	Total Annual Production (net tons)
1860	22.75	919,770
1861	20.25	731,544
1862	23.88	787,662
1863	35.25	947,604
1864	59.25	1,135,996
1865	46.12	931,582
1866	46.88	1,350,343
1867	44.12	1,461,626

Source: William T. Hogan, *Economic History of the Iron and
Steel Industry in the United States*, vol. 1 (New York: Lexing-
ton Books, 1971), 14; Peter Temin, *Iron and Steel in Nineteenth
Century America: An Economic Inquiry* (Cambridge: The MIT.
Press, 1964), 283.

a great loss to Mineral Ridge, that has been hoping some day to rival
any other portion of the country in manufacturing."[107] Another con-
sequence was the removal of James Wood from the Mahoning Valley,
as well as from active management at the Ashland furnace, prompting
Jonathan Warner to purchase Wood's interest in the company in June
1862.[108] After simultaneously buying out the interests of Wood's son,
Warner became the sole proprietor of the Ashland furnace, losing the
experience provided by a seasoned ironmaster as well as a large amount
of working capital. However, wartime conditions caused a rise in pig
iron prices, from $23.88 in 1862 to $35.25 in 1863, the highest since 1854,
driving Warner to expand his interests even further (table 3.1).[109]

Warner looked to boost pig iron production with the addition of a
second blast furnace next to his already extensive Ashland plant. Building a blast furnace from the ground up was an expensive process, and
Warner sought to avoid such a situation. Just two miles southwest of
Warner's furnace sat the long-idle Meander furnace, owned but never
operated by George C. Reis. In November of 1862, Reis and his longtime
business partner Andrew Berger sold their interests in James Ward &
Co. to James and William Ward.[110] Reis and Berger left the Mahoning
Valley for New Castle, where they purchased and rehabilitated the
old Orizaba rolling mill.[111] Though sources are not entirely clear, Reis
sold the Meander furnace to Jonathan Warner sometime in 1863 or
early 1864.[112] Shortly after the sale, Warner dismantled the furnace and

FIGURE 3.6. An 1873 painting of Jonathan Warner's Ashland and Porter furnace complex in Mineral Ridge. ©*The Vindicator*.

removed it by way of the Ashtabula & New Lisbon Railroad to Mineral Ridge, rebuilding much of the equipment and the furnace to similar specifications as the Ashland furnace (figure 3.6). In a letter written by Warner's daughter Emeline to her husband Myron I. Arms in May of 1864, she notes that her father's "new furnace" was working well and producing an "exceptionally fine grade" of pig iron.[113] Ultimately, Warner's success in Mineral Ridge forged one of the largest furnace complexes in the Mahoning Valley in terms of size, labor force, and output, consistently manufacturing "metal copious in quantity and superior in quality" throughout the war years.[114]

In Youngstown, the growth of Weathersfield Township's coal mines and the need for iron for the war had a significant effect on the continued development of the Mahoning Valley's largest town. In 1860, Youngstown had a population of 2,759, while the township itself contained 5,377 residents.[115] Though Youngstown village had seen significant industrial expansion in the decade before the Civil War, visitors to the Valley's manufacturing center by way of the canal or railroad witnessed a fairly clean, growing town that had not yet been completely enveloped in soot and smoke, unlike Pittsburgh and other major urban areas. Federal Street, for example, described by one citizen as an unpaved "mudhole," contained several sizeable hotels and hostels for visitors (figure 3.7).[116] On the westernmost end of Federal Street near Spring Common sat the Mansion House, opened in 1842. The American House also sat on West Federal Street, while several others, including the Central House and the extravagant Tod House, the latter built

FIGURE 3.7. West Federal Street in Youngstown, as it appeared around 1870. Author's collection.

in 1869, flanked the Public Square.[117] However, by the start of the Civil War, blast furnaces surrounded the central business district on its south, west, and east sides. By 1861, Youngstown produced pig iron from eight blast furnaces, four of which were constructed just as the country began to recover from economic depression and progressed toward war.

The success of the Cleveland and Mahoning Railroad and the extensive use of Lake Superior iron ore among area blast furnaces meant potential profit for iron ore and pig iron speculators, particularly those in Cleveland with stakes in Lake Superior ore lands. In June 1859, the Lake Superior Iron Co., a New York firm established for the mining of iron ore around the Lake Superior region, purchased ten acres of land on the west side of Crab Creek along Youngstown's east side. The company's purpose was to reserve the land so that its stakeholders and partners could form a separate company for the construction of a blast furnace with the purpose of only smelting the former company's Lake ores, while also using the Mahoning Valley's coal supply.[118] The new firm was the Himrod Furnace Company, incorporated under the laws of New York and named after its primary promoter and renowned Shenango Valley iron manufacturer David Himrod. After Himrod sold his Clay furnace in Mercer County, Pennsylvania, to the Sharon Iron Co. in 1851, he traveled to upper Michigan to inspect the region's vast iron ore mines, and subsequently became an agent for the Jackson Iron Company, located in Marquette, Michigan.[119] Due to his financial involvement, Himrod became a major advocate for the use of Lake Superior ores in iron making and, following the trial and error of the use of Lake ores in blast furnaces in the mid-1850s, believed Youngstown's fuel sources and railroad connections the perfect amalgamation for producing pig iron. After moving to Youngstown in the late 1850s, Himrod became the firm's resident manager, while bookkeeping responsibilities belonged to Augustus B. Cornell, also of Youngstown.

Himrod and Cornell were the only two local partners in the firm; the rest of the stockholders consisted of men from Cleveland and New York. Samuel H. Kimball of Cleveland served as the Himrod Furnace Company's treasurer and general manager. Previously, Kimball was president of the Sharon Iron Company in the mid-1850s, when the firm successfully smelted Lake ores at the Clay furnace. He was also the Cleveland agent for the Lake Superior Iron Co., which, in 1857, constructed a new type of ore dock in Marquette, Michigan, to reduce bottlenecking when loading iron ore, thereby decreasing costs and increasing shipments to Cleveland docks.[120] George Greer of New York was the Himrod Furnace Co.'s vice president. He too invested largely in Lake Superior iron ore in the 1850s and was one of the many stockholders who personally traveled to Sharon to congratulate the Sharon Iron Co. after the company succeeded in smelting Lake ores in the Clay furnace. Sharon Iron Co.'s success guaranteed Lake Superior Iron Co.'s stockholders that "the large amount they had invested in it [Lake ore] would not be lost."[121] Lake Superior Iron's president and largest stockholder was William Kelly, a former New York state senator who, in 1860, was nominated on the Stephen Douglas Democratic ticket as a candidate for New York state governor. Kelly was known throughout New York as one of the foremost ironmasters of his day. Besides the Himrod Furnace Co., he owned interests in the Iron Cliff Co. and the Kemble Coal and Iron Co. in Pennsylvania.[122] In the 1850s, Kelly purchased interests in the Lake Superior iron country and, like Himrod, strongly promoted the idea of using Lake ores in blast furnaces in northeast Ohio and western Pennsylvania.[123]

In August 1859, the company commenced construction of its first furnace, which was blown in on February 15, 1860. The Cleveland and Mahoning Railroad had ceased construction on the opposite side of town in the fall of 1856 and now deliberately extended its tracks to the stock house of the furnace to deliver Lake ore, thus terminating just after crossing Crab Creek. Himrod Furnace Co. received its coal from a bank two miles away owned by Andrews & Hitchcock, a coal mining partnership started in 1859 by Valley industrialists Chauncey H. Andrews and William J. Hitchcock.[124] To deliver coal from the mine to the furnace, four mules or horses hauled a train of four to five cars, each holding two and a half to three tons of coal at a time. The furnace itself employed thirty men daily and, upon the hour of casting, a steam whistle alerted a team of fifteen workmen, who cast twice a day, at 9 A.M. and 9 P.M. Every twelve hours, a second turn of fifteen men relieved the men on the first turn. After a cast, the iron was cooled and loaded on a buggie inside the cast house, where mules transferred it to Andrews & Hitchcock's main coal track, from which it was then transported to the canal wharf for shipment to Pittsburgh foundries and rolling mills.[125] After six months of profitable operations, the company commenced the construction of a second furnace in August of 1860. Together, the two furnaces manufactured about forty-seven tons of pig iron per day.[126]

With their new furnaces, Himrod and Kelly succeeded in profiting from Lake Superior ore. This was evident as in 1864 the Cleveland and Mahoning Railroad alone shipped 90,775 tons of Lake ore into the Mahoning Valley, compared to 3,000 tons brought by the canal in 1853.[127] The company also benefitted from the extension of the Cleveland and Mahoning Railroad northeast to the extensive coalfields of Hubbard Township in Trumbull County. Completed in 1863, the Hubbard Branch Railroad turned north from the Himrod furnaces, following Crab Creek until turning east toward Hubbard and, after a few years, extending to Sharon, Pennsylvania (see figure 3.1).[128] The line opened up another large source of block coal for Youngstown's rolling mills and blast furnaces, and for the first time linked the Mahoning Valley with the Shenango Valley via rail. Andrews & Hitchcock was one of the first partnerships in the Valley to largely exploit Hubbard's vast coal mines and was the first to ship coal over the Hubbard Branch Railroad in 1863 from its Burnett mine.[129] The Himrod Furnace Co. soon took advantage of its position directly on the Hubbard Branch's line and, together with Youngstown's Brown, Bonnell & Co., established large coal mining properties in Hubbard Township with the formation of the Mahoning Coal Co. at the height of the Civil War. Among its properties, the Mahoning Coal Co. worked four large mines along the Hubbard Branch Railroad, all located within two miles of Hubbard, with an annual capacity of over 125,000 tons.[130]

The large investment from Cleveland capitalists in Youngstown pig iron manufacturing substantiated Cleveland's inability to establish a strong foundation in the industry before the Civil War. Between 1859 and 1862, iron-mining companies near Marquette, Michigan, shipped 343,828 tons of ore to Detroit, Cleveland, Erie, and Huron, with Cleveland taking in the vast majority of the raw material.[131] By the end of the war, the *Cleveland Leader* reported that over half of the iron ore produced from Lake Superior mines came through the docks in Cleveland.[132] Still, Cleveland businessmen hesitated to invest in building blast furnaces in the city. Although iron ore was plentiful coming from the Lake Superior region, coal or coke for the smelting process was not, nor was it economical to continuously ship block coal from the Mahoning Valley. However, one Clevelander believed that "the advantages one place has in coal, the other has in ore," and questioned the city's "Rip Van Winkle policy" of promoting heavy industry: "Would it not be well for those interested in the welfare and advancement of Cleveland, to take a trip to the Mahoning Valley . . . to see whether towns of about a thousand inhabitants have sprung up in a single year under the influences of the mining and iron business?"[133] Nonetheless, Youngstown's coal trade still proved one of its most attractive assets to outside industrialists.

FIGURE 3.8. The Brier Hill Iron Company's Grace furnace as it appeared in 1889, rebuilt and modernized since its original construction. In the center of the image, notice the old-style iron-pipe hot-blast stoves sitting side by side with the new tall firebrick stoves that would replace the former. These new, more efficient stoves became the standard among Valley blast furnaces by the late 1890s, well over a decade after other major iron- and steel-producing centers adopted the technology. From *Youngstown Illustrated* (Chicago: H. R. Page & Co., 1889).

As the east side of Youngstown slowly developed during the war, the town's west side, which had been a source of industry since 1846, persistently grew. Like the Himrod furnaces, other industrialists from outside the Mahoning Valley sought an active role in pig iron production in the Youngstown area. The Eagle furnace, well known as one of the Valley's largest and highest producing stacks, became a target for investors because of owner Lemuel Crawford's success in using only Lake ores in the furnace by 1859. A syndicate of Cleveland businessmen took over operations of the Eagle furnace by creating the Eagle Furnace Company in 1862.[134] The Eagle Furnace Co. was highly profitable, and in 1869, their furnace had reportedly been in continuous operation "longer than any other in the valley."[135] Such external interests stimulated Valley industry, but David Tod would begin to solidify his control over much of Youngstown's iron trade, as his investments in Girard and Mineral Ridge coal mines, as well as their direct connection with rail transportation, opened more opportunities for pig iron production near his old Brier Hill homestead. In 1859, Tod took over the presidency of the Cleveland and Mahoning Railroad after the death of its first president, Jacob Perkins.

Consequently, Tod's control over the railroad and several large coal mines elicited the removal of his Akron Manufacturing Company from Akron to the Mahoning Valley. Tod not only moved the company's

FIGURE 3.9. Portrait of John Stambaugh (1827–1888), who began his career working under David Tod as bookkeeper for his coal mines and manager of the Grace furnaces. From *The Iron Trade Review* 44 (January 28, 1909): 225.

offices to Brier Hill but also dismantled the firm's old Etna furnace in Akron, removing most of its usable machinery to Youngstown. The company began construction on a new $75,000 furnace in the summer of 1859, located in Brier Hill between the railroad and canal and only a few hundred yards northwest of Tod's Brier Hill furnace. During the furnace's construction, the company filed for a change in its corporate title. In December 1859, the state legislature approved the company's new name, Brier Hill Iron Company.[136] The new forty-six-foot furnace, named Grace in honor of Tod's daughter, was one of the largest furnaces yet constructed in the Mahoning Valley (figure 3.8). The Grace furnace first began production in January 1860 at about twenty to twenty-two tons per day and utilized modern construction techniques, similar to those of Jonathan Warner's Ashland furnace, such as a sheet-iron exterior and a steam-powered hoisting device.[137]

The Grace furnace was under management of thirty-two-year-old John Stambaugh, who had aided Tod in managing his coal business throughout the 1850s (figure 3.9). Before entering the iron business, Stambaugh began his career as a bookkeeper for the coal mining firm of Tod & Ford, and became associated with others in the mining business in Brier Hill and Mineral Ridge before opening a dry goods store in Youngstown.[138] Stambaugh's business background proved valuable. Under his management, Brier Hill Iron Co.'s new furnace produced 10,517 tons of iron in its first fourteen months of operation, and in one week alone manufactured 207 tons, one of the highest marks yet by a Youngstown furnace.[139] Pleased with Stambaugh's work, David Tod commissioned the construction of a second, twin furnace directly adjacent to the first, called Grace No. 2, which began operations in 1861. With the management of his furnaces in the reliable hands of Stambaugh, Tod

FIGURE 3.10. David Tod, shown here at the 1860 Democratic National Convention in Baltimore (standing on the first veranda, second from the left), took a brief hiatus from his iron and coal works in the early 1860s to focus on politics. After his term as Ohio governor ended in 1864, he returned to Brier Hill to manage his industrial interests, never again serving in a public office. ©The Vindicator.

began to refocus his efforts politically after taking a decade-long hiatus. In 1860, he served as chairman of the Northern Democratic National Convention held in Baltimore, where he actively campaigned for Illinoisan Stephen Douglas's presidential nomination on the Democratic ticket (figure 3.10).[140] After Douglas's defeat at the hands of Republican candidate Abraham Lincoln in the 1860 presidential election, Tod joined the Union Party, which consisted of pro-war Democrats and Republicans, and became a supporter of Lincoln's administration. Tod helped in the organization of a company of troops in Youngstown as part of the Seventh Ohio Voluntary Infantry, and in 1861, the Union Party nominated him as candidate for Ohio state governor, which he easily won over Democrat Hugh J. Jewett.[141] Tod took office as Ohio's twenty-fifth governor in 1862, leaving much of his industrial interests in Youngstown under Stambaugh and other trusted business associates.

During his two-year term as governor, Tod was active in helping troops in the Union military by supplying sufficient transportation, rations, health care, and most importantly, pay. However, Tod was not an open supporter of the Emancipation Proclamation or of the Union war effort, and many Union Party members believed he was still too close with the old Democrats. Consequently, the Union Party did not choose

Tod for renomination in 1863, but instead turned to John Brough, a more outspoken pro-Union War Democrat from Marietta, Ohio. In 1864, Tod rejected an offer from President Lincoln to serve as secretary of the treasury, and instead devoted his waning years to managing his iron, coal, and railroad interests.

Before the start of the war, Tod leased his old Brier Hill furnace to Brown, Bonnell & Co. for a period of seven years from 1858 to 1864.[142] Shortly after the lease expired, Tod created a joint stock company called the Tod Iron Company, which engaged in the manufacture of pig iron using the Brier Hill furnace and coal from Tod's mines throughout Brier Hill and Weathersfield Township. Incorporated in June 1865, the new company had an extensive capital of $150,000, well above the average for a blast furnace company at the time, and consisted of Tod, John Stambaugh, and Nelson Crandall, who had worked under Tod as a bookkeeper and weigh master at his coal mines.[143] The remaining members of the firm were David Tod's four sons, Henry, John, George, and William Tod. The Brier Hill Iron Company, of which any reorganization was still restricted by the original thirty-year charter of the Akron Manufacturing Company, remained a separate interest from the Tod Iron Co. Not long after the war, however, Tod would combine all of his coal and pig iron manufacturing interests, creating one of the Mahoning Valley's largest and most significant pig iron firms of the nineteenth century and, at the same time, allowing Tod and his business partners to take control of Youngstown's heavy industry.

INDUSTRIAL INTEGRATION AND EXPANSION IN YOUNGSTOWN

The great expansion of the Mahoning Valley's pig iron production in the early 1860s catapulted the region to the second largest pig iron manufacturer in the state, behind southern Ohio's Hanging Rock Iron Region. In 1864, the Mahoning Valley contained twelve blast furnaces that produced an average of 275 to 300 tons per day.[144] However, southern Ohio continued to lead the state in pig iron manufacturing by a vast margin, with only one of the top six pig iron–producing counties, Mahoning County, located in the northern part of state.[145] The United States government preferred the high-quality charcoal iron produced in the Hanging Rock region for the manufacture of guns, cannon, and armor plates for gunboats during the war, and railroads used the iron for the production of car wheels and axles. One of the Hanging Rock region's most famous ironworks was the Hecla furnace, an old charcoal stack originally built in 1833: "During the Civil War the Hecla furnished armor for the gunboats that stormed Forts Henry and Donelson, and all the metal that could be spared was engaged by the Government in the manufacture of ordnance at Pittsburgh."[146] Some cannon used in the siege of Charleston, South Carolina, in 1863, was manufactured from Hecla furnace pig iron, one reportedly being the "Swamp Angel," a cannon that allegedly had the ability to shoot a 100-pound shell five and a half miles.[147] Iron produced at the Hecla furnace regularly sold for $80 to $90 a ton.[148] As in other prominent iron-producing regions, several men in Ironton, Ohio, promoted the idea of a National Armory in the Hanging Rock area during the first year of the Civil War, albeit unsuccessfully.[149]

Similarly, Youngstown's growth into one of Ohio's major industrial centers gave the region a reputation as a significant purveyor of iron used in the production of war armaments for the North and, especially, for railroad construction. In 1860, Ohio had more miles of railroad lines than all other states, and they linked passengers and freight to nearly every region, including the East Coast, the Mississippi Valley, and the Great Lakes.[150] Consequently, Ohio's expansive transportation network was a potential target for Confederate forces. From June 11 to July 26, 1863, Confederate general John Hunt Morgan led a cavalry on a raid across southern Indiana and through Ohio. As the Confederates prepared to enter Ohio on the night of July 12, Governor Tod issued a proclamation to the Ohio militia to protect the state's southern counties, and called upon citizens to "take their axes and obstruct the roads over which Morgan's troops would be compelled to pass."[151] However, most militiamen did not receive Tod's proclamation in time, and Morgan crossed into Ohio on July 13, destroying bridges and railroads, terrorizing civilians, raiding government stores, and taking supplies from warehouses. Morgan continued through portions of the Hanging Rock region and eventually led the cavalry north and east, nearing the Mahoning Valley. Joseph G. Butler Jr., at the time only a young man living in Niles, recalled these events in July of 1863:

> One Sunday in mid-summer, when the weather was extremely warm, a horseman rode into Niles with the news that Morgan had crossed into Columbiana County and was headed north, of course directly for Niles. It was generally believed that he meant to raid the Mahoning Valley, destroy the iron mills, and capture the money in the banks. The money was not such a great amount, perhaps, but the iron mills were of immense value to the government, as from them and from the blast furnaces came a great deal of material needed to win the war.[152]

It is likely that much of Youngstown's pig iron was sent to the Fort Pitt Foundry in Pittsburgh during the war.[153] However, some iron went into making rails. One of the Mahoning Valley's most notable furnaces used to make pig iron for rail production was Alexander Crawford's

Mahoning furnace in Lowellville. Morgan's raid supposedly targeted the Mahoning furnace, which would have cut off Crawford's raw pig iron source used to manufacture rails at his rolling mill in New Castle.[154] But, Union forces captured Morgan's remaining cavalry at Salineville in southern Columbiana County, only thirty-three miles south of Lowellville, leaving the Mahoning furnace, as well as the Valley's industrial infrastructure, intact.

Exactly one month after Morgan's capture, Youngstown's Republican newspaper, the *Mahoning County Register*, reported, "Despite the war, great activity in business enterprise has been apparent in Youngstown this season."[155] A number of new stores and private homes were completed across the town, and additions and enlargements were made to existing businesses, including Homer Hamilton & Co., one of Youngstown's largest foundry and machine shops, founded in 1861 by Homer Hamilton, William Tod, and John Stambaugh.[156] One of the new and more radical industrial enterprises to enter Youngstown was a firm started by Youngstown native Lorenzo Lane and Christian Shunk of Canton, Ohio, who built the Brier Hill furnace for James Wood in 1847.[157] In August of 1863, the firm of Shunk & Lane began construction of an "establishment for the manufacture of ingot or bulk steel . . . by a new process, known as the Bessimer [*sic*] patent," on the town's east side near the Himrod furnaces on the banks of Crab Creek.[158] Though the newspaper reported that the new firm would use the Bessemer process, the company's intentions were to use Shunk's own patented method. In 1856, he developed a process for "improvement in refining iron by means of blasts of air."[159] His process was as follows: "Blowing atmospheric air into and through a mass of molten crude iron from the ore, or from the remelted pig iron . . . for the purpose of decarbonizing and converting the same into refined iron or steel, and malleable semi-steel, without the use of fuel to keep up combustion."[160]

Shunk was not the first to patent such a process for pneumatic steelmaking, nor was he the first in the United States. Around 1847, William Kelly of Pittsburgh began to experiment with pneumatic refining of pig iron as a means to save fuel and speed production of wrought iron at his forge and blast furnace near Eddyville, Kentucky.[161] He continued perfecting this device in the early 1850s, when, at the same time, Englishman Henry Bessemer began similar experiments, but with the intention of producing steel rather than wrought iron. Bessemer, who had little background in iron making and achieved only some success as a mechanical inventor, took out a patent for his process in 1855, while Kelly still looked to commercialize his converter to ironmasters in the United States.[162] Kelly patented his "pneumatic converter" in 1857 for "Improvement in the Manufacture of Iron," and began promoting his process to ironmasters, one of whom was James Ward in Niles.[163]

Around 1857, Kelly travelled to Niles in an attempt to sell Ward his converter, which Kelly promised would decrease the cost of puddling at his Falcon Iron Works. Ward, considered an "authority on the iron question," had several dinners with Kelly at his home in Niles to discuss the converter. Ward failed to see the value in the pneumatic process and called Kelly "crazy" after he left the dinner table.[164] Ultimately, Bessemer's process won out, and with the help of engineer Alexander L. Holley, several iron and steel companies in the United States acquired the rights to Bessemer's patent after the Civil War. Bessemer's converter had the ability to mass-produce up to ten tons of steel in twenty minutes, depending on the size of the converter. By contrast, a puddling furnace produced just under a ton of wrought iron in an hour. Ultimately, the Bessemer process became the most important technique for steelmaking in the nineteenth century, and surpassed all other similar processes in importance.[165] In Youngstown, however, Shunk & Lane's steelworks proved a failure. All of the buildings built by the company were made of wood instead of brick and, according to Lane, "the plant would have burnt up if we had tried to operate it, as the fire flew so when the cold blast was introduced into the converter that a wooden structure could not have withstood it."[166] Lane lost $2,500 and Shunk sold the plant's machinery to Brown, Bonnell & Co. before fleeing east.[167]

Though an early attempt at steel manufacture in the Mahoning Valley floundered, finished iron production, which had sat moderately stagnant since the end of the 1850s, began to dramatically increase. Prices for bar iron skyrocketed to $91 a ton, while iron rails sold for nearly $77 per ton, leading several Youngstown businessmen and iron-workers to take advantage of the late-war iron boom by establishing Youngstown's second rolling mill and third overall in the Mahoning Valley.[168] In 1862, James Cartwright and William Clark, puddlers and rollers at Brown, Bonnell & Co., noticed the sharp increase in demand for iron from the war and were inspired by the success of the Brown-Bonnell interests.[169] Cartwright was an England native who, before coming to Youngstown, worked as an experienced puddler for the Cambria Iron Company.[170] Clark, much like Cartwright, was a man of working-class social origin. He was born in England in 1831 and immigrated to Pittsburgh with his parents in 1841, where he became a skilled ironworker, like his father, at the young age of sixteen. Clark worked at various iron mills throughout the country before coming to Youngstown, where he cofounded the Enterprise Iron Works with Cartwright.[171] The company used the old Dabney farm just west of town as the location for their mill, and intended to produce only smaller classes of merchant bar iron.[172] Cartwright and Clark built their mill just east of the Eagle furnace, in between the canal and railroad, securing pig iron from Youngstown's merchant furnaces, particularly the Eagle.[173]

Youngstown residents and ironworkers alike referred to the Enterprise Iron Works as the "Little Mill," while Brown, Bonnell & Co.'s substantial works was given the differentiating title of the "Big Mill."[174] Despite the differences in size, the Enterprise mill's product contrasted with that of Brown, Bonnell & Co., which manufactured sheet iron, spikes, and other equipment for the railroad.[175] The Enterprise mill exclusively operated in the production of light bars, hoop, and band iron used for oil barrels, beer kegs, whiskey barrels, cotton ties, and packaging pails, allowing the firm diversity in the region's iron market. In June 1864, the Enterprise Iron Works had only one heating furnace and three puddling furnaces, along with two trains of rolls for producing five tons of bar, band, and hoop iron per day, all encompassed in a small frame building sixty by ninety-six feet.[176] Despite the mill's size, the company's hoop iron quickly established a reputation as a "very superior article" that stood "as high as any in market."[177]

As was the case with the Enterprise Iron Works, the entrepreneurial, small business–like approach to iron manufacture before and during the war not only helped Youngstown's iron industry grow in terms of output during the 1860s but also aided in the transformation of the region's business structure. As the war dragged on, Congress in 1863 passed the National Banking Act to create and distribute a national, uniform currency, which served to help finance the war. In June 1863, the Mahoning County Bank, established in 1850, acquired a national charter and became the First National Bank, with a capital of $156,000, while other Youngstown banking concerns, such as the Wick Brothers & Co., continued to thrive.[178] The refinancing of the country's banks helped back manufacturers, ultimately aiding in the growth of the Valley's existing industrial concerns. It was during this period that Youngstown's Brown, Bonnell & Co. solidified its position as one of the Mahoning Valley's most important and dynamic iron firms.

Throughout the early 1860s, Brown, Bonnell & Co. gradually expanded their existing operations. In 1859, the company added an addition to their nail mill to accommodate the installation of fourteen more nail machines built by the company at the mill.[179] By 1860, the company employed over 200 men, both young and old, and produced from 55,000 to 60,000 kegs of nails and 3,000 tons of finished iron per year.[180] The lease of David Tod's Brier Hill furnace by the company allowed some control over their pig iron supply, which supplied enough for the firm's ten puddling furnaces, a modest amount compared to the nation's largest iron producers. In 1862, James Westerman, an original member of Brown, Bonnell & Co. who developed practical machinery within the mill, such as new types of shears for cutting the ends of bar iron, disposed of his interest in the company to Chicago iron merchants John Ayer and Sam Hale, who established the Hale & Ayer Iron Company

FIGURE 3.11. Brown, Bonnell & Co.'s No. 2 rolling mill, as seen looking southwest from the north bank of the canal, after a boiler explosion in 1872. The company built several bridges to transport finished material and pig iron across the canal from each mill, which included this drawbridge. ©*The Vindicator*.

in 1859.[181] Ayer quickly became known for his "shrewd and successful management" of the company, which had nearly unlimited credit following the Chicagoans' inclusion into the firm.[182] High wartime demand prompted Hale and Ayer, as well as the remaining original founders Richard and Joseph H. Brown and William Bonnell, to add an extensive addition to their existing rolling mill. The company broke ground on a new mill in March of 1863, and by September of the same year, the new rolling mill, known as the No. 2 mill, began operations on the south side of the canal opposite the original No. 1 mill (figure 3.11).[183] The mill rolled sheet iron from two inches thick to the width of paper, as well as boilerplate iron and heavy armament plate for Union gunboats.[184]

In preparation of their need for more pig iron, Brown, Bonnell & Co. purchased Lemuel Crawford's Phoenix furnace in March of 1863, giving the company control over their supply, while also reducing freight charges from constantly sending their pig iron down the canal from the Brier Hill furnace to the rolling mills.[185] In the following year, the company acquired the Falcon furnace, which for several years had been a direct supplier of pig iron to Jones & Laughlin's American Iron Works in Pittsburgh, after previous owner Charles Howard fell into bankruptcy in the midst of the 1857 Panic."[186] Jones & Laughlin's need for a direct pig iron source for their rolling mill prompted use of the Falcon furnace for a number of years, until the company in 1860 formed a subsidiary called Laughlin & Co. to build their own blast furnaces across the Monongahela River opposite the American Iron Works.[187] Completed in 1861, the Eliza furnaces were the first in Pittsburgh built expressly to use Connellsville coke, and were also the second and third blast furnaces constructed in the Pittsburgh region overall. The first was the Clinton furnace, built in 1859 by the rolling mill firm of Graff, Bennett & Co. at the present site of Station Square, which pioneered the use of coke as fuel in the Pittsburgh region. Graff, Bennett & Co. initially used Connellsville coke in the Clinton furnace but afterward attempted to economize by using local coke, though they found the local product to be inferior and quickly reverted back to the former.[188] The Clinton and Eliza furnaces ushered in the era of pig iron production in Pittsburgh after decades of relying on the Mahoning Valley's product. However, it would be well over a decade before Pittsburgh overtook the Mahoning Valley in pig iron production, though Pittsburgh iron and steel men continued purchasing Youngstown pig iron in some degree until after the turn of the twentieth century.

Laughlin & Co.'s Eliza furnaces rendered the Falcon furnace redundant, leading to its sale to Brown, Bonnell & Co. following a brief, two-year ownership by another Pittsburgh firm, Canfield & Alford. After acquiring the Phoenix and Falcon, Brown, Bonnell & Co.'s appropriately named Mahoning Iron Works became the largest integrated iron producer in the Mahoning Valley. The firm's prominent position and seemingly unlimited credit resulted in "immense fortunes" for its owners, leading to expensive additions to the mill immediately following the war. In the summer of 1865, Brown, Bonnell & Co. imported a "ponderous steam hammer" with a force of twenty tons from Glasgow, Scotland, which merited the construction of a separate building "for its reception."[189] Perhaps one of the most important additions to the company was not a machine, but the knowledge and capability of Job Froggett, whom Brown, Bonnell & Co. hired in October 1865 to manage the Phoenix and Falcon furnaces after longtime manager William Pollock accepted a "lucrative position" at David Tod's Tod Iron Co. (figure 3.12).[190] Froggett, born in Southfield, New York, in 1829, came to Youngstown in 1848 when he was only nineteen years old. Although little is known of Froggett's early years, he likely obtained experience in blast furnace practice under his father, who worked at a charcoal furnace near his hometown of Southfield.[191] Upon coming to Brown, Bonnell & Co., Froggett had twelve years of experience working at various blast furnaces. In 1868, he patented several improvements in auxiliary equipment used at the Phoenix and Falcon, including a hot-blast apparatus, whereby the heating cylinders were placed diagonally across the

FIGURE 3.12. Portrait of Job Froggett (1829–1909), who successfully managed Brown, Bonnell & Co.'s Phoenix and Falcon furnaces for over twenty years. From *The Iron Trade Review* 44 (January 28, 1909): 226.

furnace. The air was heated by passing through those cylinders, which connected together on the outside of the heating chamber.[192] Froggett also widened the bottom of the blast furnace so a larger charge and more blast air could be used, therefore producing more iron.[193] Froggett became one of the most successful blast furnace operators in the Mahoning Valley and later contributed tremendous technological innovation to the region's pig iron industry. He and Brown, Bonnell & Co. worked together to pioneer technological efficiency among blast furnace operators in the Valley after the Civil War.

Indeed, by the end of the Civil War, Youngstown and the Mahoning Valley had grown significantly. Youngstown's population steadily increased and eventually exceeded 5,000 residents, resulting in its incorporation as a city in 1867. Although finished iron production continued to mature slowly in the region, the Mahoning Valley still led the rapidly developing cities of Cleveland and Pittsburgh in pig iron production. At the close of 1865, Pittsburgh contained seven blast furnaces with an annual capacity of about 48,000 tons. Cleveland, on the other hand, only had a single furnace built by the Cleveland Rolling Mill Co. in 1864. Unlike Pittsburgh furnaces, which used coke from the start, the managers of Cleveland's Newburgh furnace used Brier Hill coal shipped from the Valley to smelt Lake ores, doing so until the mid-1870s.[194] However, the aggregate pig iron production of the Mahoning Valley's two rival industrial centers was marginal compared to the Youngstown region's production of over 100,000 tons. Nonetheless, the official end of the Civil

War in April of 1865 prompted many ironmasters in the Mahoning Valley to blow out their furnaces for lack of demand. Only one furnace, the Falcon in Youngstown, remained in operation. In addition, coal mine operators began reducing wages in April, and for months afterward continued to drive miners' wages down until local strikes multiplied and stopped nearly all coal production in the Valley, leaving furnace operators without a source of fuel; the Valley's three rolling mills still operated, but they were "not working to near their full capacity."[195] Furnaces in the Shenango Valley fared just as badly, as six of the eight furnaces in Sharpsville, Wheatland, West Middlesex, and Sharon were out of blast. Editors of Youngstown's *Mahoning County Register* attributed the abrupt decline in demand to the "sudden termination of the war," but the lull in the market would not last long.[196] Over the next decade, a new market for Youngstown's pig iron appeared by way of Bessemer steel making, and the explosive growth of Pittsburgh's rolling mills merited a "tenfold increase" in output for Mahoning Valley blast furnace manufacturers.[197] Moreover, mass railroad construction, compounded by a 26% increase of the United States' population, from 31.5 million in 1860 to nearly 39 million in 1870, elicited iron production beyond the boundaries of Youngstown and into the city's neighboring towns and villages.

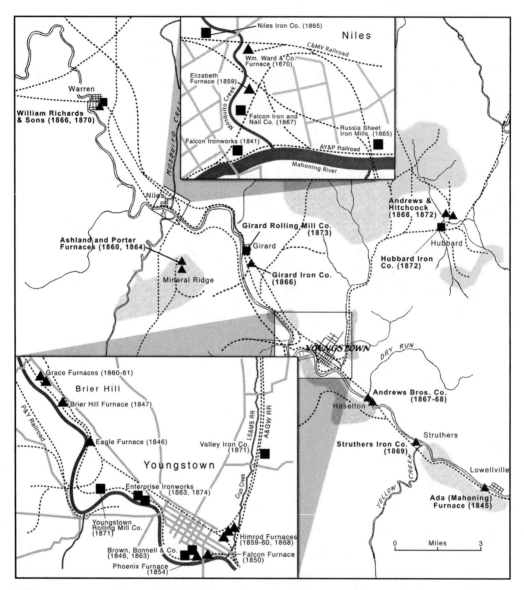

FIGURE 4.1. As in other industrial centers in the country, significant growth occurred in the Mahoning Valley between 1865 and 1874. Major iron-producing centers like Youngstown and Niles experienced heavy construction of blast furnaces (triangles) and rolling mills (squares). Industrialists also branched outside of these towns, taking advantage of major coal mining areas (shaded regions) and new railroad construction (dashed lines) that eliminated the once-important canal.

EXPANSION AND DEPRESSION, 1865–1879

> The westward bound trains wing their way through these valleys between a fiery gauntlet of blast furnaces and mills. Barely are we beyond the flicker of the last furnace in our rear till the flames of the next fiery torch looms up in the midnight darkness to light us and beckon us onward.
> —*American Working People*, April 1872.

AS THE CIVIL WAR ended, the United States entered the Reconstruction Era. Much of the American South sat in ruins because of the war, including the larger cities of Atlanta, Richmond, and Charleston. The war took an enormous toll on the South's economic infrastructure and its industrial foundations, while the region's railroads, most of which were destroyed by Union forces, became a major priority for Reconstruction state governments. In the North, however, Reconstruction efforts were not entirely focused upon material reclamation. In the Mahoning Valley, Reconstruction took on a different meaning. During the war years, Youngstown and most other Valley towns grew in their industrial capacity, but failed to incite much municipal growth. In 1866, Youngstown's village council convened to discuss improvements to the town, thus bringing it out of the backwater, pioneer era that consisted mostly of a collection of houses and old hostelries lining Federal Street. The council, which included wealthy coal operator Chauncey. H. Andrews, Richard Brown, William Wirt, Homer Hamilton, and George Baldwin, authorized $80,000 for improvements, much to the dismay of Youngstown's infuriated taxpayers.[1] The council's intention was to, as one citizen put it, "transform Youngstown . . . to a decent place in which to live."[2] The initial improvements to the town consisted of the grading of Federal Street and the construction of flagstone sidewalks.

Perhaps one of the most significant contributors to Youngstown's transformation into a modern city was architect Plympton Ross Berry, a free African American born in Mount Pleasant, Pennsylvania, in 1834.

FIGURE 4.2. The extravagant Tod House, built in Youngstown's Public Square in 1869 as part of the city's modernization campaign. Hidden behind the hotel and other buildings are the sprawling iron mills of Brown, Bonnell & Co. Northwestern winds often blew the company's black smoke over downtown, covering the Tod House and other buildings with soot. Author's collection.

At the age of sixteen, Berry became a master brick- and stonemason, and by 1851, he moved with his family from New Castle, Pennsylvania, to Youngstown. He became a major contributor to the town's foremost building projects until the early 1880s.[3] Among his first buildings following his arrival to Youngstown were the Rayen School and the foundry of Homer, Hamilton & Co. In 1863, Berry designed the First Baptist Church and St. Columbia Cathedral and, in 1869, the Tod House hotel in Youngstown's Central Square (figure 4.2). In 1867, David Tod commissioned Berry to build a $50,000 mansion along Fifth Avenue that consisted of nine rooms on the first floor and eleven on the second.[4] Unfortunately, the mansion caught fire only a few months after its completion, but Tod insisted "no one other than Berry rebuild the mansion."[5] On a visit to Youngstown in 1872, editors of the Pittsburgh newspaper *American Working People* described the city as "beautiful" and "no longer a country village visited every Saturday by a few hard fisted farmers." It had "wide, well paved streets . . . long costly residences," and an "air of comfort and thrift."[6]

The contributions of Berry to Youngstown's rise into a modern city after the Civil War were great, but the subsequent expansion of iron manufacturing throughout the entire Mahoning Valley brought the region to Ohio's industrial forefront. Although the Cleveland Rolling Mill Company constructed Ohio's first Bessemer steel works in 1867 and 1868 (the fourth such company to produce Bessemer steel in the United States), Mahoning Valley blast furnace and rolling mill operators continued to grow with iron.[7] The labor-intensive method of making wrought iron in the puddling furnace had changed little, if at all, since its introduction into the Mahoning Valley by James Ward in the early 1840s. However, the Bessemer process was still a novelty in the United States, where the traditional puddler and his furnace vastly dominated over the new steel-making technology. Rolled iron production in the United States, nearly all of which was produced by the puddling furnace, totaled 765,000 tons

FIGURE 4.3. Joseph G. Butler Jr. (1840–1927), shown here in the early 1890s, was well known among his contemporaries as one of the finest minds in the iron and steel industry. Under James Ward, he learned both the business and the operating end of iron making, leading to Butler's ascension to one of the Mahoning Valley's top blast furnacemen, industrialists, and philanthropists. From *The Iron Trade Review* 29 (January 2, 1896): 12.

in 1865, while the country's lone Bessemer steelworks did not produce enough of the metal to even merit recording tonnages by the American Iron and Steel Association.[8]

TRUMBULL COUNTY'S "SMOKY CITY"

Outside of Youngstown, ironmasters began building more ironworks or adding to their already extensive mills, none of which, however, were more important than James Ward & Co. in Niles in the decade after the Civil War. Editors of local newspapers began giving Niles the title of Trumbull County's little "smoky city," a moniker shared by Pittsburgh and an endearing term that alluded to prosperity and affluence more so than dirt and smoke.[9] It was under James Ward that Valley industrialist Joseph G. Butler Jr. began his distinguished and lengthy career in iron and steel (figure 4.3). A man of Scotch-Irish lineage, Butler was born on December 21, 1840, "within twenty feet of a blast furnace" in Mercer County, Pennsylvania.[10] His father, Joseph Green Butler Sr., built this blast furnace in 1837 or 1838, naming it "Temperance" after his wife, Temperance Orwig.[11] After a fire ravaged his furnace in the early 1840s, Joseph Green Butler Sr. sold what remained of the property and moved his family to Niles, but soon afterward he went back to western Pennsylvania, where he managed Alexander Crawford's Tremont furnace for a short period.[12] There, young Butler followed his father around the company's charcoal pits, familiarizing himself with the industry and the processes needed to make pig iron in the early charcoal days.

After moving back to Niles in 1854, Butler's father received a job at James Ward & Co.'s company store. In 1856, a distressed James Ward entered the store and stated to Butler's father, "Our shipping clerk is on a drunk and I want one of the boys to come and help me out. There is a boat due soon which we must load for Cleveland."[13] "Mr. Ward, there are my three boys. Take your choice," said Butler's father. Ward asserted at once, "I will take Joe." Anxious to start his new job, a sixteen-year-old Joseph Butler Jr. rushed down to the company's warehouse, recruited other men to help, and loaded the canal boat with iron destined for the George Worthington Co. and William Bingham Co. in Cleveland. The speed and determination of Butler's work prompted Ward to discharge the company's former shipping clerk and hire Butler in his place.[14]

Accordingly, Butler's appointment into the company proved beneficial to its overall operations. After solving production and efficiency problems that had plagued the mill for a number of years, Butler was promoted to rolling mill manager by 1860.[15] During his employment at James Ward & Co., he established a close paternal relationship with James Ward, whom he affectionately called "Uncle James." When his father was appointed as Trumbull County sheriff in 1858, Butler remained in Niles rather than moving with his family to Warren. Ward took young Butler into his home, caring for him like a member of the family, and Butler eventually became the financial manager of James Ward & Co.[16] However, at the height of the Civil War, Butler accepted a position as the Youngstown district representative for Chicago iron merchants Hale & Ayer after they purchased a large interest in Brown, Bonnell & Co. Butler left his position at the Niles firm, much to James Ward's disappointment. Nevertheless, Ward's influence upon Butler was great. "There is no doubt in my mind that 'Uncle James' . . . changed the whole course of my life and that I owe to him the opportunity to succeed, as well as much of whatever is worth while in my later life," wrote Butler.[17]

Despite the loss of Butler's managerial and financial skills, wartime demand elevated James Ward & Co.'s ironworks in production and value, although much of the mill was over twenty years old. Writing in his later years, Butler described Ward's ironworks during the Civil War as a "small plant" without a fence: "Anyone could go in and talk to the workmen at any time."[18] The mill did not have a time clock, and the puddlers working in front of the furnaces constantly had "sweat running out of the tops of their shoes."[19] Butler noted that most men working at the mill drank a good deal while on the job—the very reason he had received his original position as shipping clerk from Ward:

> [The workers] were in the habit of sending out to a nearby saloon for a bucket of beer whenever they wanted it. Sometimes they drank a mixture of barley flour and water, but more generally beer. Nearly always these men stopped at a saloon on their way home and drank a glass of whiskey, with a glass of beer. This was known as a "Puddler and a Helper."[20]

The harsh conditions in the mills and the resultant imbibing among its workers were common throughout the Mahoning Valley and ironworks everywhere. "One effect of the severe heat and exertion is the creation of a craving for stimulants, such as beer, which at once cool and support the workmen," wrote Dr. David Bremner in his 1869 book, *The Industries of Scotland.*[21]

By the mid-1860s, however, James Ward's Falcon Iron Works had an estimated worth of $500,000 and employed 400 to 500 workers, while 1,200 of Niles's 1,500 residents depended on the company for financial support.[22] Yet, nine months before the end of the war, tragedy struck James Ward & Co. and the town of Niles: Niles resident Frank O. Robbins violently gunned down the company's prolific founder and president, James Ward, near the company's blast furnace on July 24, 1864. Described as a large man weighing 240 pounds, Robbins engaged in sexual relations with a woman named Lydia Stevenson, who lived in Ward's tenant houses near the ironworks after her husband and former James Ward & Co. employee was killed fighting for the Union Army in Keller's Bridge, Kentucky. A religious man, Ward was unhappy with the situation and ordered Stevenson to vacate the house. Angry and intoxicated, Robbins pulled out a revolver and shot Ward in the forehead in front of the company's workers as well as his only son, James Ward Jr.[23] Ward died a few hours after the incident.

The murder of James Ward would have a far-reaching and damaging effect on the industrial structure of Niles, as well as most of the Mahoning Valley and several companies outside of the region. Initially, however, this effect seemed highly positive. William Ward managed James Ward & Co. for a year and a half before his late brother's estate was legally settled. Yet, less than a year after James Ward's murder, twenty-three-year-old James Ward Jr. began to take over his father's position as one of the Mahoning Valley's leading iron manufacturers. Young Ward worked at his father's mill for a number of years, learning the ins and outs of iron making and the methods of running an extensive business, while also becoming good friends with Joseph Butler (figure 4.4). In 1863, Ward married Elizabeth L. Brown, known to many as "Lizzie," daughter of the wealthy Pittsburgh coal merchant William Hughey Brown, at the Monongahela House in Pittsburgh. Butler, who served as Ward's groomsman, described the wedding as a "brilliant affair for that time."[24] The marriage served to link two prominent and wealthy industrial families, as Lizzie's father had made numerous connections to supply coal to Pittsburgh-area ironworks before the Civil War.[25] The personal and financial connection Ward now held with the Brown family gave him the ability to drastically expand his iron interests in the Mahoning Valley. Accordingly, in January 1866, Ward and Brown purchased the Falcon Iron Works in Niles, while the company retained the corporate title of James Ward & Co.[26]

In the summer of 1865, however, Ward and Brown had already begun a business partnership. Together, this dynamic and eccentric duo started

FIGURE 4.4. This 1861 daguerreotype taken by a traveling canal boat artist in Niles shows childhood friends James Ward Jr. and a twenty-year-old Joseph G. Butler Jr. In the front row first on the left are J. M. Brush, William Ward Jr., Butler, and John Dithridge. Standing from left to right are James Ward Jr.; George L. Reis; Irvin Butler, the younger brother of Joseph Butler, who died serving in the Civil War; and William B. Berger. Courtesy of the National McKinley Birthplace Memorial.

construction on Niles's second rolling mill a quarter of a mile east of the Falcon Iron Works and just below the north abutment of the Mineral Ridge Railroad. It was exclusively built for the production of Russian sheet iron (figure 4.5). In the early half of the nineteenth century, several Russian forges and rolling mills manufactured iron sheets with a fine, blued finish favored in Europe and the United States for covering boilers and cylinders of steam engines.[27] Even famed writer Mark Twain claimed Russian sheet iron to be "the best," as it was "slicker and more showy than the common kind."[28] Many ironmasters believed that artisans in Siberia made Russian iron by a "highly profitable, secret process" owned by the Russian government. Rumors asserted that "when a workman enters the service he bids farewell to family and friends . . . and is never heard from afterwards." Many desperate attempts to steal the secret were said to have ended in death.[29]

James Ward Jr. and William H. Brown were not the first to attempt to emulate the Russians' process. In the early 1840s, Alan Wood experimented with making a similar imitation brand of Russian sheet iron at

FIGURE 4.5. Russia Sheet Iron Mills, Niles, c. 1890. Piles of pig iron for refining in the puddling furnaces (stacks with dampers) are seen to the left. Courtesy of the Youngstown Historical Center of Industry and Labor (SMC 5093).

his Delaware Iron Works, and within a few years, customers began purchasing his product in place of the imported type.[30] In the late 1840s, an unknown American ironmaster claimed to know the process of manufacturing the authentic product, which, according to Philadelphia's *Niles' National Register*, "would contribute more to our national wealth than hundreds of ordinary inventions made at home."[31] Despite this, American ironmasters' attempts at manufacturing the real Russian product before the Civil War failed, and, like Wood, many produced only an American glazed iron.

In the years after the war, however, many ironmasters and metallurgists rekindled their efforts to search for and experiment with the Russians' secret process. In 1871, eminent Russian scientist N. de Khanikof wrote a letter in German to British metallurgist John Percy describing the process of producing authentic Russian sheet iron.[32] In the process, the iron sheets, derived from pig iron smelted with charcoal, were converted into malleable iron in a charcoal finery. Ironworkers then rolled the malleable iron into plates fifty-six inches long and twenty-eight inches wide. Some ironworks attempted to use puddled iron, but the sheets did not possess the same structure and composition as malleable iron made with charcoal. The chief characteristic of the Russian method consisted of producing a mirror-like glaze of a brown, bluish, or smoke-gray color on the surface of the sheets. After rolling, the sheets were sheared and arranged in packets of anywhere from 50 to 100, with the surface of each sheet dampened and dusted over with charcoal powder before packing. Each packet was enclosed with waste sheets and heated in an annealing furnace for five to six hours and immediately taken to the trip hammer before cooling. All of the charcoal dust was instantly wiped away and each packet hammered uniformly over its entire surface for ten to fifteen minutes and, after cooling, was annealed. Ironworkers repeated this process four to five more times, but during the last annealing, they hammered the packet for twenty-five to thirty minutes with a glazing hammer that had a much larger striking face.[33]

Although John Percy reported several variations of this method in his 1871 book *The Manufacture of Russian Sheet-Iron,* the overall processes were relatively similar. In Niles, Ward and Brown hired a Pittsburgh sheet mill worker named John Noble to manage and superintend the mill. They believed Noble held the secret to manufacturing Russian iron; however, it is unclear as to how Noble obtained the secret. It is likely that Ward himself sent Noble to Russia to report on the possibility of manufacturing Russian iron in Niles.[34] Other reports indicated that Noble came to Niles claiming to know the secret to manufacturing rust-proof Russian iron and enticed Ward to construct a potentially profitable new plant that turned out an extremely sought-after product.[35] Whatever the reason, the new mill, like its counterparts in Russia, was kept a secret. Editors of the *Western Reserve Chronicle* reported that "employees in these works will all be hired for a certain term of years, and all sworn to secrecy as to the manufacturing of the same, and no admittance be allowed to outsiders."[36]

The mill itself was extravagant and expensive. It was 206 by 60 feet and consisted of a foundation of lavish stone walls ten or twelve feet deep with heavy timbers embedded into the ground, a design that one Niles correspondent assessed "must have cost a small fortune."[37] Yet, despite the secrecy and extravagance of the Russia mill, Noble's process for making authentic Russian sheet iron was ultimately unsuccessful. Still, the failure to produce the genuine Russian product did not affect the mill financially, as Ward and Brown maintained operation by producing regular black sheet iron.[38] At the same time, other companies in Ohio and the Mahoning Valley claimed to produce the genuine Russian product. In Cleveland, C. C. Hinsdale in August 1866 asserted that he "succeeded in producing sheet iron of a quality that equals . . . the best Russia sheets," and in 1868 established the American Sheet and Boiler Plate Company for producing "Hinsdale iron."[39] Like Noble's process in Niles, Hinsdale's was a failure. Several newspapers from cities such as Philadelphia, Washington DC, and Columbia, South Carolina, reported that in February 1867, Brown, Bonnell & Co. successfully made Russian sheet iron at their mill in Youngstown under George C. Kungonchieff's process.[40] Many of these reports claimed that it was the first time Russian iron was made in the United States. Nevertheless, this cannot be verified any further.

The failure of Noble's process in Niles, however, did not faze the mill's owners, and Ward and Brown's expansion of the town's already extensive iron industry did not cease with the Russia Sheet Iron Mill. Shortly after purchasing his father's Falcon Iron Works, Ward rebuilt the company's Elizabeth furnace, an expensive process that required tens of thousands of dollars. Not long after, local journalists noted the furnace as the "largest in the Mahoning Valley" at a towering sixty-five feet in height, consuming 10,000 tons of Lake and black-band iron ore annually.[41] Many larger anthracite furnaces in the eastern part of the country reached a height of only forty-five to sixty feet, including the Thomas Iron Company's furnaces in Hokendauqua, Pennsylvania, along the Lehigh River, considered by some observers at the time to be "the largest and most productive Furnaces in the United States."[42] Some of the biggest furnaces in the world built in the prominent iron-producing district of Cleveland, England, reached sixty to seventy-five feet.[43] Therefore, the need for more raw pig iron for manufacturing iron sheets at the Russia mill required the increased capacity of the Elizabeth furnace, which produced 150 tons a week.[44] Yet, not long after the completion of the highly secretive Russia mill, Ward, Brown, and a new partner, Michael Greenebaum, an experienced tinner who owned a small hardware store and tin shop in Chicago, began construction on yet another rolling mill in late 1866.[45] The mill, named the Falcon Iron and Nail Co., sat on the east side of Mosquito Creek between the Elizabeth furnace and the original Falcon rolling mill. Completed in June of 1867, local journalists heralded the mill as "the finest works of the kind west of Pittsburgh," producing 6,000 kegs of nails per month.[46]

By 1868, James Ward Jr., with the aid of the exceedingly wealthy William H. Brown, had expanded his father's iron mills over twofold. The growth was extremely swift and elevated Niles to among the top iron-producing towns in not only the Mahoning Valley but also the state of Ohio. As the Ward family name maintained its dominance over Niles's iron trade, William H. Brown also extended his industrial reach in the region by purchasing Jonathan Warner's Ashland and Porter furnaces in Mineral Ridge. Warner, the sole operator of the extensive iron manufacturing and coal mining plant at the Ridge, began to feel the financial pressure of managing such an extensive ironworks, particularly during the postwar lull in the iron market. Warner overshot his capital and was on the verge of financial collapse—in his own words, "afloat on the ocean without a compass or rudder."[47] To counter potential bankruptcy,

Warner enlisted the aid of Cleveland banker Lemuel Wick; Judge Milton Sutliff, a Republican and staunch abolitionist from Warren who served as an Ohio Supreme Court Judge from 1858 to 1863; and Joseph H. Brown of Youngstown's Brown, Bonnell & Co. In 1865, Sutliff and Wick, among others, helped organize the Leetonia Iron & Coal Company in Columbiana County. Warner and Joseph Brown had a personal connection through the marriage of their children, Edward J. Warner and Mary Jane Brown.[48]

In 1866, Warner, Joseph Brown, Sutliff, and Wick organized the Mineral Ridge Iron and Coal Company with an extensive capital of $200,000.[49] Warner remained resident manager of the two furnaces, which produced about 12,000 tons of iron per year.[50] The capital invested by the other partners in the firm helped Warner fight off bankruptcy, and the company and its furnaces prospered as the iron market gradually, but not exceptionally, improved. Mineral Ridge Iron and Coal Co.'s success and value led its partners to sell out after only two years to William H. Brown, who paid $200,000 along with several thousand acres of undeveloped mineral lands in the Lake Superior region. Brown organized the Brown Iron Company in August of 1868, with a capital of $250,000 (figure 4.6).[51] Its partners included, among others, James Ward Jr. and another of Niles's leading socialites, Thomas Carter, who started the Globe Foundry and Machine Works in Niles in 1858.[52] Furthermore, in 1870, Carter gave James Ward Jr. his Globe Foundry in exchange for an interest in the entire firm of James Ward & Co.[53]

By the early 1870s, Ward and Brown held a commanding presence over Niles and Mineral Ridge's iron industry. The two companies served as one of the more extensive manufactories of its kind in the state, with a combined three blast furnaces, three rolling mills, coal mines, and a large foundry. Accordingly, editors of the *Western Reserve Chronicle* wrote: "We believe there is no firm in the United States embracing more extensive iron business than this [James Ward & Co.], and we believe it is more extensive than the entire iron business of Youngstown."[54] However, at the age of fifty-five, William H. Brown's health began to decline, and in 1870, he withdrew from James Ward & Co., leaving as remaining partners only James Ward Jr., Thomas Carter, and Jacob Greenebaum Jr. of Chicago, who purchased an interest in the firm in 1870.[55] Likewise, in February of 1871, Brown dissolved the Brown Iron Co. in Mineral Ridge and presented the furnaces to his daughter and son-in-law. The wealthy coal and iron mogul continued to manage his Pittsburgh enterprises until his death in October 1875. The loss of such prominent financial backing from Brown, combined with overzealous postwar expansion, left James Ward & Co. financially vulnerable in the event of economic downturns, which occurred almost cyclically in the nineteenth century.

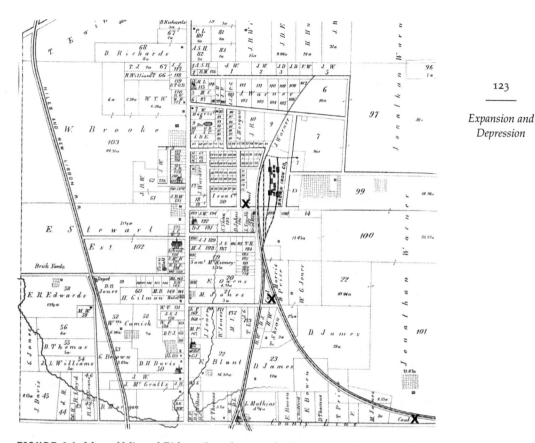

FIGURE 4.6. Map of Mineral Ridge, 1874, showing the Brown Iron Co. and coal line railroads that connected the furnace stock houses to the nearby Niles and New Lisbon Railroad after crossing into Mahoning County. Coal mines (**X**) in the immediate vicinity of the furnace complex provided a convenient and relatively cheap source of fuel. From *Atlas Map of Trumbull County* (Chicago: L. H. Everts, 1874), 108.

YOUNGSTOWN IRON MANUFACTURERS

The postwar industrial growth of Youngstown-based iron manufacturers contrasted with that of Niles. Rather than the extreme measures of expansion taken by James Ward Jr., many Youngstown iron producers took a more conservative approach, with a focus on technological innovation and efficiency in the postwar period. Economic conditions in the five-year period following the Civil War grew at a dawdling pace, and Youngstown iron companies such as Brown, Bonnell & Co., Brier Hill Iron Co., and the Enterprise Iron Works largely mirrored this pecuniary development. Production of pig iron in the United States gradually increased, from 1.3 million tons in 1866 to 1.8 million tons in 1870. Pig iron prices fluctuated between thirty-five and forty-five dollars a ton, but they never reached the high mark of nearly sixty dollars per ton seen during the war years.[56] However, the demand for pig iron from

Pittsburgh rolling mills grew rapidly during the postwar period, and by 1870, the city's mills consumed upwards of 380,000 tons of pig iron per year.[57] A statement from an 1870 report titled *Pittsburgh: Its Industry & Commerce* exhibited the city's extensive industrial progress:

> In 1850, Pittsburgh numbered thirteen rolling mills, with a capital of about $5,000,000, employing three thousand five hundred hands. In 1854, there were nineteen rolling mills, having one hundred and seventy-six puddling furnaces and two hundred and fifty-three nail machines. To-day there are thirty-two iron, nine steel, and two copper mills. A survey of the chronological table of these forty-three establishments shows that the increase, since 1830, has been at the rate of one per year. Out of forty-six mills erected since 1813, forty-three are still in operation, the greater portion of them having increased their dimensions and capital from four to eight fold. Six mills were erected in one year.[58]

Among Pittsburgh's most prominent mills in the postwar period was Jones & Laughlin's American Iron Works, which produced upwards of 50,000 tons of wrought iron products per year.[59] Others included Alexander M. Byers's comparatively new wrought iron pipe mill, built at the height of the Civil War, and the Clinton and Millvale rolling mills of Graff, Bennett & Co., which became Pittsburgh's first rolling mill firm to engage in pig iron manufacture within the city following the construction of the Clinton furnace. Yet, of the thirty-two iron rolling mill companies referenced in the 1870 report, only three were connected with their own blast furnaces. In the Mahoning Valley, the opposite occurred, albeit on a much smaller scale. By 1868, two of the Valley's five rolling mill firms owned their own blast furnaces, while the remaining eight iron companies only produced pig iron for outside consumption.

As a result of Pittsburgh industrialists' expansion of finished wrought iron production, Youngstown ironmasters continued to build blast furnaces after the Civil War in order to feed the over 600 puddling furnaces and nearly forty foundries in Pennsylvania's Smoky City.[60] Aside from James Ward Jr.'s construction of rolling mills in Niles, the Mahoning Valley would see widespread growth in pig iron production rather than finished iron in the late 1860s, a phenomenon largely unlike that of Pittsburgh. Most of Youngstown's old-time ironmasters remained steadfast in their investments in coal mining and blast furnace construction, particularly those who already had extensive mineral and iron developments within Youngstown and the western portion of the Mahoning Valley.

Youngstown's Brown, Bonnell & Co. was still the Valley's single most important iron manufacturer in terms of size—its mills covering a total of twenty-five acres—and technological innovation. "Far to the

FIGURE 4.7. This 1883 photo, once owned by Brown, Bonnell & Co. puddler William Davis (who appears in the top row with his hand on top of a long iron rod), shows the day-turn workers at the company's No. 1 mill. Davis claimed that he reworked "kegs of nails melted solid" and "stacks of sheets welded together a foot thick" that were shipped to Youngstown from Chicago after the Great Chicago Fire of 1871. The Mahoning Valley Historical Society, Youngstown, Ohio.

east and west of us it stretched, while below us the iron castles filled by their American soldiery were puffing and roaring away by night and day," wrote one visitor to the ironworks in 1872.[61] Brown, Bonnell & Co. employed 800 men in its mills alone, while countless others toiled in the coal mines six miles away (figure 4.7). Footbridges spanned the length of the canal to connect the No. 1 and No. 2 rolling mills, while tramways and rail lines extended in every direction around and through the mills. Indeed, not only was Brown, Bonnell & Co. Youngstown's largest employer, but the company's owners were also some of the Valley's most progressive industrialists. In summer 1868, Brown-Bonnell began innovations for further utilization of blast furnace gas, an advancement of great importance to iron manufacturers around the country. Since the late 1840s and 1850s, ironmasters utilized blast furnace waste gases to power boilers and to heat the hot blast almost exclusively.[62] In 1856, Frederick Levick and John James of the Cwn Celyn and Blaina Iron Co. in Monmouth, Wales, began experimenting with utilizing blast furnace gas to generate steam for puddling and rolling mills, but any such innovations failed to materialize on a large scale in the United States.[63] Brown-Bonnell's experiment, however, involved conveying the gases from the top of the Phoenix furnace down through a wrought iron pipe three feet in diameter, where a brick culvert then conveyed the gases underground to bars below a special battery of boilers designed by

Joseph Brown. These generated steam to power the engines driving the rolls throughout the company's two rolling mills.[64] The gas traveled in iron pipes throughout the plant and underneath the canal to any area that needed power. Brown's new system of using the gases to power the entire mill was well in advance of many ironworks in the country, as it was both efficient and economical, saving thirty tons of coal per day.[65]

One of the more significant technological improvements introduced to the Mahoning Valley by Brown, Bonnell & Co. was the closed-top, or "bell and hopper," method on blast furnaces. In the mid-1860s, closed tops were not "enthusiastically received" by many ironmasters, as they believed it affected the quality of the iron by only producing "white" iron, which had an unusually high carbon and sulphur content.[66] However, in 1860, the Ormesby Iron Works in Middlesborough, England, experimented with using a closed top and reported saving about 150 tons' worth of coal per week without affecting iron quality.[67] The trials performed at ironworks in Great Britain throughout the 1860s assured American ironmasters that sealing the furnace had its cost advantages. According to Edwin and Joseph Froggett, sons of Brown, Bonnell & Co. blast furnace manager Job Froggett, their father and Joseph Brown began experimenting with a bell and hopper on the Phoenix furnace in 1870 (figure 4.8). Brown received a handbook in English blast furnace practice that illustrated the different advantages and engineering of the bell and hopper used at several English and Welsh ironworks. He gave the manual to Froggett, who, with machinist and foundryman Homer Hamilton, forged a cast-iron bell for the furnace's top in Hamilton's small foundry off Phelps Street.[68]

Froggett was still uncertain of the bell and hopper's practicality. He insisted on seeing a similar installation that was simultaneously taking place at one of the Eliza furnaces owned by Jones & Laughlin in Pittsburgh. After taking a train to Pennsylvania's Smoky City and witnessing the success of the Jones & Laughlin experiment, Froggett immediately completed the installation of the bell and hopper on the Phoenix furnace the next day. However, several of the furnace's top fillers refused to work after the installation of the bell, while bottom fillers and other workers deserted the stock and casting houses for fear that the stack would explode when sealed. Ultimately, the new technology did not cause the stack to explode and proved highly efficient. The bell not only saved fuel but also helped top fillers distribute the charge more efficiently because iron ore and fuel did not accumulate in the center of the furnace.[69] Brown, Bonnell & Co. set a precedent for all other Mahoning Valley blast furnace operators. Not long after the success of the bell and hopper on the Phoenix furnace, many Valley iron companies converted from opened to closed tops, although some continued to use the former until the early 1880s, when competition in the industry forced

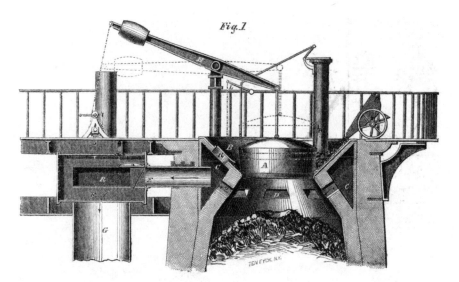

FIGURE 4.8. Blast furnace cross-section showing the bell and hopper top installed in the mouth of the Ormesby Iron Works' furnace in England. Brown, Bonnell & Co. installed a similar top on the Phoenix furnace. From *Scientific American* 4, no. 4 (January 26, 1861): 49.

modernization. By 1882, however, all blast furnaces in the Mahoning Valley had closed tops.

Another one of Youngstown's old industrialists who instituted efforts in innovated business tactics and industrial efficiency was David Tod. After leaving politics, he began to aggressively invest in more pig iron manufacturing after the Civil War, despite a decline in health after years of holding public office and managing his already extensive industrial interests. Rather than rapid expansion and construction, Tod reorganized his existing companies while also investing in new ones. In November 1867, Tod consolidated his Brier Hill Iron Co. and the Tod Iron Co. to organize the Brier Hill Iron and Coal Company with a capital stock of $432,000, the largest ever for a Youngstown area iron firm.[70] The large-scale merger combined Tod's three blast furnaces along with the extensive coal mines of the Tod Iron Co. under one management. The incorporators of the Brier Hill Iron and Coal Co. were all Youngstown men: along with Tod, who took the position of the company's first president, they included William Pollock, Nelson Crandall, John Stambaugh, and David Tod's son, Henry. David Tod's new enterprise swiftly became the largest pig iron producer in the Mahoning Valley, manufacturing 600 tons of pig iron per week, while employing 120 men at the furnaces and 150 at the mines.[71]

The wealth and political standing associated with the Tod family name aided in the Brier Hill Iron and Coal Co.'s well-known reputation and strong financial success for decades after the firm's formation.

FIGURE 4.9. Slope mines such as the Sodom (or McCurdy) mine, seen here in the 1870s, were among the Valley's most productive. This mine alone produced 25,767 tons of coal in 1872. Miners often dumped fine coal particles and shale into nearby streams, causing serious damage to animal life and vegetation. The long-term effects on former coal mining areas like Mineral Ridge are well documented and include reports of numerous sinkholes opening in residential areas. Courtesy of Girard Historical Society.

David Tod's iron and coal interests, however, did not stop at Brier Hill. In 1866, he extended his reach in the industry just northwest of his Brier Hill homestead and into the town of Girard, just across the Trumbull County border, which had reached a population of about 1,000 by 1860.[72] Although Tod had opened coal mines in Girard in the late 1850s, no other ferrous industries had taken advantage of the small town's mineral or transportation resources, which included both the canal and railroad. Tod himself aided in the laying of Girard's town plat in the late 1830s, but only small settlements began after the completion of the canal. Its pioneer industry began after Jesse Baldwin and Abner Osborn built a flouring mill around 1840, which flourished along the Mahoning River for nearly 100 years.[73] Frederick Krehl, a German immigrant who came to the United States in 1853, established Girard's first tannery in 1860, but until after the Civil War, heavy industry remained mostly confined to nearby Youngstown, Niles, and Mineral Ridge.

During the Civil War, Liberty Township (in which Girard is located) and Hubbard Township directly to its east experienced a significant coal mining boom. In 1864, David Tod and his son John, among others, established the Church Hill Coal Company in Church Hill (now

TABLE 4.1. Coal Mines in the Mahoning Valley, 1872

Location	Number of Mines	Tons Mined in 1872
Mineral Ridge	7	232,847
Austintown	3	62,144
Youngstown (south side of Mahoning River)	6	154,884
Hubbard Township	16	557,291
Liberty Township	7	294,501
Total	39	1,301,667

Source: *Western Reserve Chronicle*, April 30, 1873.

Note: It is likely that a number of active mines were not documented.

Liberty), a few miles east of Girard.[74] The company constructed a coal mining railroad from their banks to the Cleveland and Mahoning Railroad in Girard and produced 97,459 tons of coal in 1872.[75] In the same year, Tod, John Stambaugh, and Myron Arms opened the Veach Coal Works in Hubbard Township, a mine that employed forty men and produced about 120 tons per day, with most of the coal shipped to Brier Hill Iron and Coal Co.'s Grace furnaces.[76] The density of mining in this region just one mile northwest of the town of Hubbard, which included sixteen different mines, created the small village of Coalburg, a "booming, rip-roaring mining community of 1,500 people."[77] Similar to Mineral Ridge in the late 1850s, this once small agricultural village of no more than 100 people transformed into a community with a saloon-to-church ratio of eight to three. Workers in Coalburg's mines were paid fairly well because of the purity and high demand of the block coal found in Hubbard Township.[78] Other mines in southern Trumbull County, such as the Sodom (or McCurdy) mine, operated by John Stambaugh and located on the Vienna Branch of the Cleveland and Mahoning Railroad in northeastern Liberty Township, became some of the highest coal producers in the county (figure 4.9). Together, the twenty-three coal mines in Liberty and Hubbard Townships produced 851,792 tons of coal in 1872 (table 4.1).[79]

This great growth in coal mining after the Civil War promised the ability to profitably manufacture pig iron in southern Trumbull County. As the Tod and Stambaugh families became the region's primary coal developers, David Tod envisioned the construction of a blast furnace in Girard, which he believed could be one of the most efficient and state-of-the-art stacks of its day.[80] To help develop his vision, Tod recruited some of the most knowledgeable minds available in the region's iron business. As a veteran himself in the iron and coal industry, Tod enlisted another longtime veteran of the business in William Ward. Since James

FIGURE 4.10. William Richards (1819–1876), a Welsh immigrant, was one of the Mahoning Valley's most distinguished ironmasters before and after the Civil War. He helped organize Girard's lone blast furnace and developed an extensive ironworks in Warren. Courtesy of Shelley Richards.

Ward Jr.'s takeover of his father's iron mills in Niles in early 1866, sixty-year-old William Ward was not active in his nephew's iron business, but he provided Tod's new enterprise with over thirty years of experience in both rolling mill practice and pig iron manufacturing. In addition, Tod convinced the young Joseph G. Butler Jr. to join him and Ward in the new enterprise. Excited for the opportunity to work with some of the finest business minds in the Mahoning Valley, Butler immediately left his position as district representative for Hale & Ayer in the fall of 1865 and joined Tod's new venture.

One of the more noteworthy members of the firm, however, was iron industry veteran William Richards (figure 4.10). Since the early 1850s, Richards, known by his contemporaries as an "iron monger" who "stood at the head of his class in the Mahoning Valley," played an important role in managing several of Youngstown's blast furnaces.[81] Richards was no stranger to the smoke, fire, and steam of the iron industry. Born near Quaker's Yard, Wales, in 1819, Richards grew up only ten miles from Cyfarthfa, which, along with the town of Dowlais, contained one of South Wales's most prolific and largest nineteenth-century iron companies.[82] The hillsides around Richards's childhood home in Wales were often "scorched" and had a "blackened look." However, Richards himself did not work for the iron companies in Wales. In 1840, he and his mother and sister, both of whom were dependent upon Richards for financial support, came to the United States.[83] He worked as a blacksmith in numerous Ohio towns before coming to the Mahoning Valley in 1845, where he worked as a blacksmith in several coal mines and finally landed a position in James Ward & Co.'s rolling mill in the late 1840s.

Richards first familiarized himself with blast furnace practice at William Philpot's furnace in Brier Hill when, in 1850, William Philpot named Richards manager of the stack. Philpot's death one year later left the furnace idle, resulting in Richards establishing his own blacksmith shop in Girard where he furnished boilerplates, thereby earning enough money to purchase a farm near his property. In 1853, Richards returned as manager of Philpot's furnace and later became superintendent of the nearby Brier Hill furnace. After accumulating a small fortune in a rolling mill venture in New Castle, Pennsylvania, Richards took his family on holiday to his home country of Wales in 1865, which served as more of a business trip than leisure time.

Before David Tod's new pig iron enterprise was officially organized and incorporated in Girard, the members of the proposed company asked Richards to study blast furnace design of newly erected furnaces while on his trip to Wales.[84] Of particular interest to Tod, Ward, and Butler were the new furnaces constructed in Dowlais, where Richards found a stack "that seemed to meet the specifications required" by the other partners.[85] Welsh immigrants often held important positions at many American ironworks, reflecting the dominance and influence of Welsh iron companies in the United States.[86] In addition, many correspondences between American ironmasters referenced ironworks in Wales, such as Dowlais and Cyfarthfa, as was the case with William Richards and David Tod. The influence that Welsh-born ironmasters and coal miners and their industrious mentality had on the Mahoning Valley's industrial growth is clear, especially since Wales largely dominated as one of the world's largest iron producers in the 1860s.[87]

Upon Richards's return to the Mahoning Valley in early 1866, he presented his plans for the new blast furnace to Tod, Ward, and Butler. Pleased with the design, the group of men then formed their new enterprise, called the Girard Iron Company, incorporated in February 1866 with a capital of $100,000.[88] The company began construction on the Mahoning Valley's first blast furnace in the postwar period on May 1, 1866. The owners chose a portion of William Richards's farmstead just west of the Cleveland and Mahoning Railroad and east of the canal to construct the stack, which was intended to be a model of blast furnace construction and engineering. In February 1867, editors of Youngstown's *Mahoning Register* were taken on a tour of the new furnace by the owners and reported that the works were "constructed on a principle different from the other furnaces in the Mahoning Valley, and, as we understand, in this country entirely."[89] Completed in approximately eight months and blown in on January 16, 1867, the fifty-six-foot tall Girard furnace contained all of the normal accoutrements required for furnace operations (figure 4.11). However, the major difference in design implemented by William Richards was the interior of the furnace: "Rising from the

FIGURE 4.11. The original Girard furnace, c. 1867. Composed entirely of brick, a construction principle brought from Wales by William Richards, the stack was the first attempt by Youngstown iron manufacturers to expand the Valley's pig iron production after the Civil War. The Mahoning Valley Historical Society, Youngstown, Ohio.

hearth, the old furnace [traditional construction] expands its diameter rapidly and is widest at about 8 feet from the bottom and then contracts gradually to the tunnel head. The new furnace [Girard] expands gradually to near the top and then contracts gradually."[90]

Unknown to local journalists and the furnace's own proprietors, both of whom praised the "radical" new design, the engineering of the Girard furnace was a major setback and a costly oversight. Although the furnace manufactured about twenty-five tons of iron per day, its production was still on par with several of the older furnaces in the Mahoning Valley, far from being the efficient stack that Tod envisioned.[91] Only after William Richards finished construction on the furnace did he realize the failure of his design of the stack, which Joseph G. Butler Jr. recounted nearly sixty years later:

> It seemed that in crossing the ocean, which was a rough trip in those days, Mr. Richards must have got the drawings mixed, for the furnace, planned to be the most efficient of its day, was a failure. We found that it had been practically built upside down, or at least the bosh was inverted. Our capital was not sufficient and when we had to rebuild the stack, it was exhausted.[92]

This drastic loss in capital left the Girard Iron Company in dire financial circumstances, and it was forced to secure additional capital on notes endorsed by David Tod, marking a withdrawal of two of the company's original partners.[93] In November 1868, Tod suffered a fatal stroke, leaving the Girard Iron Company open to additional investors as well as a void in the presidency of the Brier Hill Iron and Coal Co., the latter being filled by Tod's longtime friend and business partner John Stambaugh. Butler remained manager of the Girard Iron Co., despite his feeling that "no concern could prosper under such conditions" because the company was usually short on funds and paid as high as 10% interest on Tod's notes. However, it still managed a $305,000 profit in 1869.[94] William Ward left the Girard Iron Co. to establish his own pig iron firm in Niles in 1870. Around the same year, Pittsburgh iron baron Alexander M. Byers—worth an estimated $7.5 million at the time of his death in 1900—and his brother-in-law Joseph Fleming purchased a large interest in the Girard Iron Company, which stabilized the firm financially for the next fifty years.[95] Increase in capital allowed for technological improvements that pushed the Girard furnace's daily production to forty tons by the mid-1870s.[96]

William Richards set out to invest in his own iron company in Warren after selling his interest in the Girard Iron Co. in September 1867. The seat of Trumbull County long neglected industrial growth and, as Butler notes, "showed no desire for material progress" in the latter half of the nineteenth century.[97] Contradicting Butler's words, however, post–Civil War journalists in Warren expressed a desire to rival the manufacturing town of Niles and the city of Youngstown. In 1866, editors of the *Western Reserve Chronicle* asserted that Warren was "destined to take a position among the iron manufacturing towns of the Mahoning Valley."[98] For years, Warren's citizens hoped for outside capitalists to see the town's potential to manufacture iron, but very few men in Warren had the technical or business expertise to produce the metal. Because Warren Township and city lacked block coal, industrialists were hesitant to build mills there, and some residents felt that the smoke and grime of industry would tarnish the town's clean, aesthetic appeal:

> Warren . . . had become one of the most beautiful cities of Ohio. There was a New England-like air to the community. The unusually large public square had become the beautiful City Park and furnished a setting to the stately Trumbull County courthouse. The residence streets were lined with magnificent shade trees. There was more cleanliness than was common in Mahoning Valley municipalities. Warren had become a seat of culture and its people not only had deep appreciation of the better things of life but a dignified respect for things of the past. There was a marked conservatism in business and in other respects.[99]

In the years after the Civil War, however, Youngstown, with its financial, municipal, and especially its industrial progress, became the envy of almost every other Mahoning Valley town. Comparatively, Warren's growth lagged. The town had a population of 3,457 in 1870, while the city of Youngstown grew to 8,075 citizens.[100] Nonetheless, the postwar reconstruction mentality motivated Warren administrators to develop "a spirit of internal improvement."[101] The town first began to make minor industrial progress in the spring of 1865, when R. H. Barnum and Warren Packard established a small steam forge called the Warren Iron Company at the corner of South and Pine Streets. The forge, built to make rail-car axles, gradually grew in production, and in 1866, a number of industrialists from Youngstown, Warren, and Greenville, Pennsylvania, formed the Packard & Barnum Iron Company. The firm added a rolling mill onto the existing forge and opened the mill to the public in June 1867, when it was "thronged all day . . . by visitors, curious to observe the process of making iron."[102]

The question of building Warren's first blast furnace soon arose. Just months before the opening of Packard and Barnum's rolling mill, the *Western Reserve Chronicle* printed an emphatic headline: "Shall Warren have a Blast Furnace?"[103] The newspaper's editors stated that the question in hand was one that "the citizens of Warren have in their power to decide affirmatively."[104] Warren administrators cautiously looked at the possibility of attracting outside industrialists and other capitalists to construct a furnace, similar to "casting bread upon the waters." Other manufacturing cities used this method to bring in industry, including Youngstown, which gave $40,000 to secure railroad machine shops and a roundhouse on its east side; the city of Akron, Ohio, raised $20,000 to $30,000 to entice manufacturers, and Sharon, Pennsylvania, raised $100,000.[105]

Warren's town-wide program of subscribing a loan for the construction of a blast furnace was a success. By October 1869, Warren citizens and town administrators secured $75,000 for the furnace loan, while Warren Packard sold his rolling mill property, which included the "rolling mill . . . its machinery, tools, fixtures and attachments," to William Richards for $35,000 in January 1870.[106] Richards's blast furnace was blown in on September 16, 1870, with its first cast of iron made six days later. Richards, along with his two sons, Samuel Allen and William Jr., incorporated the new firm as William Richards & Sons; it immediately became the most extensive manufactory in the seat of Trumbull County. The fifty-five-foot-tall furnace, which produced about thirty to thirty-five tons of iron per day, employed twenty-six men, while another 175 toiled at the adjacent rolling mill.[107] Warren's largest employer, Richards's ironworks provided financial support to one-eighth of the total population of the town.[108]

Visiting the ironworks in late August 1870, Reverend N. P. Bailey, pastor of Warren's First Presbyterian Church, described the furnace there as a "splendid structure, having all the newest and most improved apparatus . . . a thing of which Warrenites feel proud, and from which we expect great things." Bailey also alluded to it being the seed for industrial growth, saying, "It is the acorn; the big oak will grow from it; and in a few years its spreading branches will cover all the place."[109] Likewise, several of Richards's contemporaries believed that the success of his ironworks would bring to Warren "other iron mills, blast furnaces, and various manufacturing establishments [that] will cover the flats adjoining the southeast part of the town."[110] Despite its promise and early success, Richards's new enterprise was the only integrated iron or steel manufacturer in Warren until after the First World War. The town would not come close to the industrial progress made by Youngstown or Niles in the nineteenth century.

THE NEW IRON BOOM

Although the opening of William Richards's integrated ironworks failed to attract other major companies to Warren, it reflected the growth of iron manufactories elsewhere in the Mahoning Valley. New ironmasters, industrialists, and financiers began to take the place of the Valley's pioneer coal and iron operators, and many expanded from just blast furnace construction to building rolling mills. In 1869, Valley manufacturers of all kinds made $4.9 million in profit, while each iron company earned between $118,000 and $1.3 million that year.[111] This progress was spurred by a growing economy and the industrial development of the west, causing a dramatic increase of railroads in the country, which expanded from 52,922 miles of track in 1870 to 93,267 miles in 1880.[112] Like the majority of Ohio, the Mahoning Valley saw significant railroad growth after the Civil War, including the construction of the Lawrence Railroad, which finally connected Youngstown with New Castle, Pennsylvania.

The opening of this new rail line in 1867 was of great importance to both Pittsburgh and the entire Mahoning Valley. It finally increased the railroad facilities of the region and improved the coal and iron trade in the Mahoning and Shenango Valleys. In New Castle, the line connected with the Pittsburgh, New Castle & Cleveland Railroad, which then linked with the Pittsburgh, Fort Wayne & Chicago Railroad in Darlington, Pennsylvania, thus allowing passenger and freight trains to run on a regular schedule between Pittsburgh and Youngstown. Additionally, the Lawrence Railroad gave rail transportation to the eastern portion of the Mahoning Valley, a region industrialists and ironmasters neglected

FIGURE 4.12. An artist's depiction of the Mahoning furnace in Lowellville, 1867, the same year that the old plant finally received rail connections. The furnace itself is the broad, stone-masonry feature with a handrail along it in the upper center of the sketch. Drawing in vertical file under "Railroads" at the Public Library of Youngstown and Mahoning County, Youngstown, Ohio.

to develop for many years. This included Lowellville and its Mahoning furnace, which was the last blast furnace plant in the region to obtain railroad connections after relying on canal shipments for over twenty years (figure 4.12).[113]

By 1870, the need for new rail systems in the Mahoning Valley was apparent, as the Valley's industrial output had increased dramatically. "If we make any more iron, we'll have to buy a place to put it," said one official of Youngstown's Himrod Furnace Company.[114] In addition, the need for another blast furnace fuel was becoming a necessity. Iron and mining companies had continually unearthed the region's block coal reserves for over twenty-five years, and although many antebellum surveyors believed the coal supply to be infinite, by 1874, the Ohio inspector of mines reported, "The coals of . . . the Mahoning Valley . . . are yearly approaching exhaustion. Ten years hence, and the magnificent block coal of the Mahoning Valley will be as rare as it is valuable."[115] In 1869, Thomas H. Wells, one of Youngstown's longtime coal operators, warned that coal mines in the Mahoning Valley should be consumed there rather than shipped to other markets. He argued that if Valley coal companies continued to ship their product to outside consumers in large quantities, a day would come "when this exhaustive process will be regretted by Youngstown as a manufacturing community."[116] Still, companies opened several small, unworked coal reserves east and south of Youngstown in the early 1870s, and in 1872, total coal production for both Trumbull and Mahoning Counties was just over 1.3 million tons.[117]

Production gradually decreased to 1.2 million tons by 1877, and to 1 million tons in 1880.[118] In fact, in 1872, Mahoning County was not even ranked among the top sixteen coal-producing counties in Ohio, whereas Trumbull County came in at sixth; Stark and Columbiana Counties took the top two spots in the state.[119] Consequently, Wells's bold prediction had come to fruition. Coal mined for iron production in the Mahoning Valley ceased to be economically practical, and Valley ironmasters had to consider a new fuel: coke.

The Connellsville coalfields in southwestern Pennsylvania were the center of the coke industry in the United States. The region had only seventy coke ovens at the start of the Civil War, but by 1873, it had more than 3,600 due to the tremendous increase in pig iron production in the United States.[120] One observer noted that the coke produced in the Connellsville region was "very . . . pure, and is, in fact, equal to any in the world. Its excellence is attested by the fact that it is largely shipped to supply furnaces even as far west as St. Louis."[121] Ironmasters in Pittsburgh began using Connellsville coke almost immediately after the city's first furnaces were constructed, primarily because of the fuel's low cost and the furnaces' close proximity to the coke ovens. Mahoning Valley ironmasters, however, neglected the use of Connellsville coke until 1869 (Jonathan Warner coked Mineral Ridge coal at his Ashland furnace a decade earlier, as mentioned in chapter 3). Not only did they have to use a large amount of block coal per ton of iron, but larger furnaces could not support the immense weight of the coal in the furnace burden. In addition, the density of raw coal in the furnace, whether bituminous or anthracite, prevented the use of "hard-driving," a distinguishing feature of American furnace practice that enabled higher blast pressures in larger coke furnaces, causing "spectacular increases" in output:

> Anthracite was a dense fuel, and although the lack of bituminous gas is common to anthracite and coke, the often-heard characterization as a natural coke is not exact. For the lack of gas makes coke a very porous one. When bituminous coal is coked, its volume does not fall. Rather than collapse upon itself, the remaining material in a good coking coal forms a fine honeycombed structure with air spaces in between. This means that coke has a higher surface area in relation to its volume than anthracite, and that it can burn faster. If a greater volume of air is blown into a blast furnace, coke will burn faster, while anthracite will speedily reach a limit set by the maximum rate of combustion on its restricted surface area.[122]

This analysis can be directly applied to Mahoning Valley ironmasters and their use of block coal. Continued use of the coal ultimately resulted in smaller stacks and lower blast pressures into the early 1880s, which inhibited greater output in Valley furnaces. The standard pressure for a slow-driven furnace was about four pounds and for a fast-driven one from seven and a half to seventeen and a half pounds per square inch.[123] Many furnaces in the Mahoning Valley had blast pressures ranging from three to six pounds per square inch.[124]

By the early 1870s, the merits of using coke and hard-driving for achieving greater output were well understood in Pittsburgh. In 1872, the Carnegie interests, specifically the firm of Carnegie, Kloman & Co., constructed the Lucy furnace on 51st Street to supply the city's vast puddling mills and their own Union Iron Mills, which furnished wrought iron shapes for Piper & Shiffler, the forerunner to Carnegie's Keystone Bridge Company.[125] The Lucy furnace was immense, towering over the Allegheny River at a height of seventy-five feet, larger than any other furnace previously constructed in the United States.[126] Using Connellsville coke as fuel, managers of the Lucy furnace achieved tremendous outputs and were among the first to implement the use of hard-driving, attaining blast pressures upwards of nine pounds per square inch. By the end of 1872, the furnace averaged seventy-two tons of iron per day—eclipsing any single daily production record set in the Youngstown region.[127] In addition, the Isabella Furnace Company, rivals of Carnegie's Lucy furnace, put into blast an equally large furnace on the opposite side of the Allegheny River in the same year. Similarly, by the end of 1872, the Isabella furnace achieved an output of seventy-one tons of iron per day.[128]

However, the inspiration for Carnegie and his Lucy furnace and the Isabella Furnace Co. to build such massive stacks came from the Mahoning Valley. In 1865, Thomas Struthers, whose father John helped establish the Montgomery furnace along Yellow Creek in Struthers in 1806, purchased his late father's farmstead on the south side of the Mahoning River, just west of its junction with Yellow Creek. A lawyer who practiced in Warren, Pennsylvania, Struthers dabbled in the railroad business before coming to the Mahoning Valley. After the Lawrence Railroad completed its line west from the Pennsylvania border, Struthers in 1869 constructed the Anna furnace, named for his only living daughter (figure 4.13). Thomas W. Kennedy, a native of Mahoning County and father of prominent Pittsburgh blast furnace engineer Julian Kennedy, designed and engineered much of the Anna furnace, which stood fifty-six feet tall with a nine-foot hearth diameter—the hearth size being equal to that of the Lucy furnace built three years later. When put into production, the Anna furnace produced 1,600 tons of iron per month, making it "much talked about in Pittsburgh," particularly because such a large output was achieved using raw coal.[129] In October 1871, *The Mahoning Register* reported that the Anna furnace manufactured "the largest yield of iron ever produced by a single stack furnace."[130] In a three-day span, the stack yielded 210 tons, or seventy tons per day, and in six days produced

FIGURE 4.13. The Anna furnace in Struthers and its high production rates proved the envy of Pittsburgh's iron barons when it first went into production in 1869. This photograph of the Anna furnace dates from 1905. Author's collection.

382 tons, an average of sixty-three tons per day.[131] However, the use of block coal leveled off production. In Pittsburgh, smelting with coke and continued experimentation with hard-driving at the Lucy and Isabella furnaces allowed steady increases in output over the next twenty years, and reached 114 tons per day by 1878. The same year, the Anna furnace, still using mostly block coal, averaged fifty-five tons per day.[132]

Nevertheless, decreasing block coal reserves and their simultaneous increase in cost per ton prompted several Mahoning Valley blast furnace operators to begin mixing Connellsville coke in the burden. In the late 1860s, Joseph G. Butler Jr. made the region's first coke contract with a young and struggling Henry Clay Frick, who would later become one of the United States' foremost steel and coke magnates with his business partner, Andrew Carnegie. In reference to the price Butler paid for the coke, he wrote, "I would be ashamed to tell you the price, as I think he [Frick] would also."[133] Butler used the coke in the Girard furnace in 1869, mixing it with block coal, a combination that he thought was a "very satisfactory and economical fuel, the coal adding to the surplus gas production" of the furnace.[134] In the following years, Butler purchased "thousands of tons of good beehive coke" at eighty-five cents per ton; by 1880, the average price per ton in the Connellsville coke region was $1.79.[135] In 1871, block coal from the Valley sold at $3.00 per ton.[136] Brown, Bonnell & Co. was the first Youngstown iron firm to establish coke ovens in the Connellsville region. In 1878, they started the Mahoning Coke Company just south of Dunbar with a capital of $40,000 and produced

FIGURE 4.14. Chauncey H. Andrews (1823–1893) helped shape the Mahoning Valley's industrial progress in the late 1860s and early 1870s by developing extensive coal, iron, and railroad interests. From Butler, *History of Youngstown and the Mahoning Valley,* vol. 3 (Chicago: American Historical Society, 1921), 529.

about 137 tons of coke daily.[137] Other Youngstown blast furnace managers, including John Stambaugh, Augustus B. Cornell, and Henry Bonnell, son of Brown, Bonnell & Co. founder William Bonnell, established the Youngstown Coke Company in 1879.[138] Many Valley furnace companies began mixing coke with raw coal in the mid-1870s, such as the managers of the Anna furnace, who used one-sixth coke and five-sixths block coal; furnace managers in Cleveland, who used the Valley's block coal almost exclusively, followed suit and began mixing coke by the late 1870s.

To transfer cheap coke from southwestern Pennsylvania to the Mahoning Valley and Lake regions required efficient railroad services. To ensure this, Chauncey H. Andrews, one of Youngstown's most prolific postwar industrialists and the city's first millionaire, began promoting the construction of additional railroads in the Valley as early as the late 1860s (figure 4.14). Andrews, described as "gruff in manner" and "known by many, but understood by few," was a wealthy coal operator from Vienna, Ohio.[139] His father was in the farming and mercantile business and opened the Mansion House on West Federal Street in Youngstown in 1842. Educated in Youngstown schools, Andrews worked at his father's hotel and later opened the mercantile business of Brenneman & Andrews, which fell into bankruptcy in 1853. Looking to counter his losses, Andrews returned as manager of the Mansion House until deciding to invest in the Valley's coal mines. He took a gamble and pushed his investments, opening the Thorn Hill coal bank northeast of Youngstown near Hubbard Township, a highly successful mine that produced half a million tons of coal in nine years.[140] After establishing the coal mining firm of Andrews & Hitchcock with business partner William J. Hitchcock, a former bookkeeper for Mackintosh, Hemphill & Co. in Pittsburgh, Andrews expanded his investments throughout the Mahoning and Shenango Valleys during the Civil War. Among them was the opening of the Oak Hill and Coal Run mines in Mercer County, Pennsylvania, with his brother Wallace. He also purchased a large interest in the Westerman Iron Company in Sharon, which included blast furnaces, coal mines, a rolling mill, and an interest in the extensive Brookfield Coal Company.

The railroad coming to the eastern portion of the Mahoning Valley prompted immediate investment in mining and industry from Andrews. After the completion of the Lawrence Railroad in 1867, he and his two brothers, Wallace and Lawrence G., who was educated as a physician, purchased several coal mines on the east side of Youngstown near the new railroad's connection with the C&M Railroad. Together, they formed the firm of Andrews Bros. Co. with a capital of $150,000. In 1867 and 1868, the company constructed two blast furnaces, collectively named the Haselton furnaces. In addition to this plant, Chauncey Andrews in 1868 constructed a large furnace just north of Hubbard with W. J. Hitchcock. Andrews & Hitchcock also opened another extensive coal works known as the Stewart Coal Company in Hubbard Township, employing 200 miners and producing 200 tons of coal daily that supplied the Hubbard furnace and industries in Cleveland.[141] With the establishment of two separate coal mining and pig iron manufacturing plants within two years, Andrews quickly elevated himself to being one of the Mahoning Valley's wealthiest and most well respected businessmen.

Andrews, however, also sought to control the shipment of his coal and pig iron. By the early 1870s, small coal railroads snaked their way across the Mahoning Valley and became the primary method of supplying coal to blast furnaces and rolling mills (figure 4.15). In 1869, Andrews and his brother Wallace provided capital to finally complete the Niles and New Lisbon Railroad. After the financial failure of the New Lisbon Railroad Company, which leased the Niles to New Lisbon portion of the Ashtabula & New Lisbon Railroad in 1864, the track south of Niles remained incomplete for another six years.[142] After Andrews finished construction on the narrow-gauge railroad from Niles to New Lisbon, Ohio, in 1870, the coalfields in southern Columbiana County were opened up. Hence, Andrews began investing in additional coal operations south of the Mahoning Valley, giving him cheap transportation of coal to his mills in Youngstown, as well as to customers along the Great Lakes.[143] Although the bituminous coal in Columbiana County was not of the same grade as Youngstown's block coal, it provided a cheap alternative for providing fuel to rolling mills and other industrial equipment. Andrews and L. E. Cochran purchased the defunct Niles Iron Works in 1871, a small rolling mill built by Harris, Blackford & Co. after the Civil War.[144] With Andrews as the primary shareholder, they incorporated the Niles Iron Company in 1873 with a capital of $500,000.[145] Thus, through progressive forward integration, Andrews created cohesive and

FIGURE 4.15. The Youngstown Railroad, a small coal line connected with the Powers Coal Co. on the south side of the Mahoning River, allowed coal delivery to various blast furnaces and rolling mills, such as the Youngstown Rolling Mill Co., faintly seen in the center background. The Mahoning Valley Historical Society, Youngstown, Ohio.

self-sustaining ironworks spread across the Mahoning Valley, owning his own fuel sources, pig iron, finished wrought iron production, and several transportation sources.

In the early 1870s, Andrews expanded his interests even further in the Mahoning Valley and helped increase the region's pig iron production. After a disastrous and fatal explosion in 1871 that all but destroyed his two Haselton furnaces, Andrews immediately rebuilt the plant, taking every measure possible to render it fireproof and streamlining it to modern standards (figure 4.16). The costly investment ultimately paid off for Andrews. In 1874, distinguished British ironmaster Sir Lowthian Bell toured the major coal and iron regions of the United States. Upon reaching the Mahoning Valley, Bell visited several of the district's coal mines and blast furnaces, noting that the furnaces he visited were "not so large as those at Cleveland City [Ohio], nor is the blast so well heated."[146] However, after Bell's visit to Andrews's Haselton plant, he "spoke in very high terms of these furnaces, ranking them among the best in the country."[147] Such success warranted Andrews to continue expanding his interests. In 1872, he added a second large furnace to the existing Hubbard stack, while at the same time increasing the size of the first furnace (figure 4.17). Both stood sixty feet tall and were among the largest and highest producing units in the Mahoning Valley, together manufacturing about 600 tons of iron per week.[148]

Although Andrews had invested hundreds of thousands of dollars in coal mining operations throughout the Mahoning Valley and Columbiana County, he too believed that coke was to take the place of block coal. By January 1875, Andrews began mixing one-fifth coke

FIGURE 4.16. C. H. Andrews's Haselton furnaces, photographed shortly after reconstruction finished following the devastating explosion at the plant in 1871. The rail line extending out of the image crossed to the south side of the Mahoning River and connected the furnaces with Andrews's coal mines. Hot-blast stoves, seen in the center, burned both coal and blast furnace gas and generally emitted more smoke than the furnaces themselves. The Mahoning Valley Historical Society, Youngstown, Ohio.

with four-fifths block coal in the Haselton furnaces.[149] In the same year, he became an avid promoter of the newly projected Pittsburgh & Lake Erie (P&LE) Railroad and had a significant influence on the line's extension into the Mahoning Valley. Proposed by Pittsburgh businessman William McCreery, the P&LE received a charter in May 1875, but construction was delayed until 1878. After learning that the line was to connect Youngstown to Pittsburgh, railroad mogul William H. Vanderbilt purchased $300,000 worth of stock in the P&LE after he made large investments in the Lake Shore and Michigan Southern Railroad, which, in 1874, leased the Mahoning Coal Railroad, a line built from Youngstown to Andover, Ohio, in 1873.[150] This ensured that the P&LE would be affiliated with Vanderbilt's New York Central system rather than the B&O, as McCreery had originally intended. Once the P&LE was

FIGURE 4.17. Along with the Haselton furnaces, Andrews built these two large stacks in Hubbard with William J. Hitchcock, which were among the most financially stable in the Mahoning Valley. This view shows the Hubbard furnaces before 1880, with hot-blast stoves on the left and stacks of pig iron ready for shipment in front of the boiler buildings on the right. The Mahoning Valley Historical Society, Youngstown, Ohio.

completed in 1879, Chauncey Andrews's gamble on it paid off. The railroad's eastern terminus was at the Pittsburgh and Connellsville Railroad Company's depot in Pittsburgh, and from there it went north and west until reaching its western terminus in Haselton near Andrews's furnaces, which, by 1879, were at the junction of three different railroads. The P&LE's connection with the Pittsburgh and Connellsville Railroad, leased by the B&O, was its most attractive feature. As Mahoning Valley furnace managers began increasing coke ratios in their furnaces, the P&LE provided a more direct and cheaper method for shipping the fuel from the Connellsville region; between January 1, 1879, and June 30, 1880, the line shipped 83,360 tons of coke.[151]

By the early 1870s, Ohio had over 4,100 miles of railroads, and the number in the Mahoning Valley had considerably increased.[152] The old canal was no longer a factor. Its business drastically declined after the Civil War because of the growth of railroad transportation; its last shipment was in 1872, when a barge delivered limestone from Lowellville to the Brier Hill furnace.[153] In 1873, the Board of Health declared the canal's stagnant water a "public nuisance" owing to the "decaying and putrid condition of animal and vegetable matter therein," ordering it to be filled in with dirt.[154] Naturally, the amount of materials produced and shipped in and out of the region via rail experienced a significant boom, particularly as the Valley's industrialists continually developed and revitalized old mills and built new ones. In 1866, the Mahoning Valley only produced 70,000 tons of pig iron and 20,000 tons of bar iron. In contrast, in 1872, the region manufactured 1.3 million tons of coal, 212,000 tons of pig iron, and 68,100 tons of bar iron, and 371,000 tons of iron ore was shipped from Cleveland. Local railroads handled a total of 1,952,794 tons of coal, iron ore, pig iron, and finished wrought iron products that year.[155] The Valley saw rolling mills built between 1871 and 1873 by entrepreneurs who sought to take advantage of the new iron market. Unlike C. H. Andrews and his brothers, most had little skill in managing such an extensive business, but they found that the plethora of blast furnaces in the Mahoning Valley could reliably supply new rolling mills without the great expense of transporting pig iron long distances via railroad.

The westward movement of the United States gave rise to the need for new wrought iron agricultural implements such as mowers, reapers, threshers, and plows to tame the Wild West. In 1870, Ohio statistician E. D. Mansfield wrote a letter to the *New York Times* regarding the manufacturing presence in the west:

> The great interior is fast emancipating itself from the dependence upon Eastern or European fabrics. Within a few years the new deposits of the finest iron, and of the best coal to manufacture it, have been found, opened and worked in the Mahoning or Shenango Valley. . . . In the same period immense exports of iron, amounting to 700,000 tons a year, are got out on Lake Superior. . . . The result of all this is the erection of a great number of new furnaces, and the employment of great numbers of people and a vast capital.[156]

Whole towns began to develop around railroads, and the manufacture of wrought iron for construction and farming purposes was a rapidly growing market. After the last "golden" spike was nailed in the first Transcontinental Railroad in May 1869, it became commonplace for railroads to link the east to the west, while also connecting mills with distant raw materials and opening regional markets, freeing plant owners from relying on purely local markets. As a result, Mahoning Valley iron manufacturers focused on making agricultural implements, bar iron, hoop and band iron, sheets, nails, and products used for railroad construction, such as spikes.

Along with Brown, Bonnell & Co. and James Ward & Co., the Enterprise Iron Works in Youngstown quickly became one of the Mahoning Valley's most important finished iron manufacturers. After the company's humble beginnings during the Civil War, it continually grew and expanded from 1868 to 1870 under manager William Clark, who was a recognized expert in rolling mill practices. Clark's technological innovations pushed the mill's production to thirty tons of finished iron daily.

Fittingly, the company took home the first prize for iron hoops at the 1872 Cincinnati Industrial Exhibition.[157] Amidst the firm's expansion in 1869, Clark sold his interest in the mill and went to Pittsburgh, where he established the Solar Iron Works in partnership with his son. After Clark's departure, the Enterprise Mill reorganized under the firm of Cartwright, McCurdy & Co. Gradually, partners in the company purchased shares in the old Eagle furnace just west of the mill and, by 1872, took full possession of the stack.

New rolling mill companies, however, would not integrate their operations with blast furnaces during this period. Just below the Enterprise Iron Works, brothers Paul and Henry Wick established the Youngstown Rolling Mill Company in 1871, which produced hoop and band iron along with "wagon box, horseshoe bar, hame, and lock iron."[158] Other industrialists built rolling mills in Girard and Hubbard that produced muck and bar iron; nut, bolt, and horseshoe iron; and pins and links used in railroad construction. In addition, the Girard mill produced agricultural implements such as guard and finger iron, drag and brace bars, knife-back iron, cylinder bar, and tooth iron for threshers. These new companies completely diverted from sheet iron and nail production, which had been a staple product for older firms such as Brown, Bonnell & Co. and James Ward & Co. New market trends, such as the rise of John D. Rockefeller and the oil industry, required an increase in muck bar used for making pipe and iron bands and hoops for oil barrels.

By the fall of 1873, the Mahoning Valley had reached its nineteenth-century industrial peak in terms of the volume of its mills, not their capacity. Nine Valley iron companies operated rolling mills within Warren, Niles, Girard, Hubbard, and Youngstown, while Mineral Ridge, Struthers, and Lowellville each contained companies operating blast furnaces but lacking rolling mills (table 4.2). The Valley boasted twenty-one blast furnaces and achieved the highest total pig iron production in the state, with an annual capacity of 219,500 tons, ousting the Hanging Rock Iron Region from the top spot. Cleveland industrialists had built only four blast furnaces by this time, but the city featured the only Bessemer steelworks in Ohio. Like iron manufacturers in the Mahoning Valley, those across the state border in the nearby Shenango Valley produced a significant amount of merchant pig iron and experienced substantial blast furnace construction in the postwar period. In 1873, the region boasted thirty furnaces, twelve of which were built between 1870 and 1873, with an annual capacity of 290,000 tons. Pittsburgh, too, had made great progress in pig iron manufacturing during this period. Despite both the Mahoning and Shenango Valley outnumbering Pittsburgh by ten and nineteen blast furnaces, respectively, Pittsburgh's use of coke and hard-driving pushed its annual capacity to 200,000 tons, within only 20,000 tons of the Mahoning Valley. Pittsburgh blast furnaces

TABLE 4.2. Rolling Mills in the Mahoning Valley, 1873

Company Name	Location	Number of Puddling Furnaces	Capacity in Tons (annually)
William Richards & Sons	Warren	20	14,600
James Ward & Co.	Niles	43	32,500
Niles Iron Co.	Niles	22	8,000
Girard Rolling Mill Co.	Girard	8	3,600
Cartwright, McCurdy & Co.	Youngstown	20	6,000
Brown, Bonnell & Co.	Youngstown	54	25,000
Youngstown Rolling Mill Co.	Youngstown	10	5,000
Valley Iron Co.	Youngstown	24	20,000
Hubbard Iron Co.	Hubbard	8	4,000
Total		209	118,700

Source: *The Ironworks of the United States* (Philadelphia: The American Iron and Steel Association, 1874), 83–84; *Western Reserve Chronicle*, April 30, 1873.

outproduced the Mahoning Valley for the first time in 1873, with actual production reaching 158,789 tons, compared to the Valley's 136,972 tons.[159] Pittsburgh pig iron production would not overtake that of the Shenango Valley until 1878.

With a focus on manufacturing pig iron, the Mahoning Valley's new iron companies did not embrace iron rail production on any significant level. Rail, a market long dominated by British iron and steel manufacturers, was the United States' single most important operation of the iron and steel industry after the Civil War, and more than doubled in output, from 318,000 tons in 1865 to 809,000 tons in 1872.[160] Iron and steel rail imports from Britain decreased by 6% from 1871 to 1872; however, there was a 78% increase in the importation of Bessemer steel rails during the same period.[161] In 1872 and 1873, there were 335 rolling mills in the United States; forty-eight made iron rails and seven manufactured both Bessemer steel and wrought iron rails.[162] The largest density of rail mills was in Pennsylvania, which contained seventeen; Ohio was second in the country with nine. The Shenango Valley embraced iron rail production to a small degree, with the Westerman Iron Co. in Sharon and James Wood's Sons & Co.'s Wheatland rolling mills making the product.[163] In Pittsburgh, Jones and Laughlin's American Iron Works began manufacturing eight- to forty-pound iron rails, while the Kensington Rolling Mill and the Wayne Iron and Steel Works produced light rails. The Cleveland Iron Company fashioned light iron rails, while the Cleveland Rolling Mill Co. had been manufacturing Bessemer steel rail since 1868.[164]

Such growth in the manufacture of rails was not a coincidence after the Bessemer converter appeared in the United States. Larger steam locomotives and heavier trainloads, combined with constant use and the fragile composition of wrought iron rails, caused breakage, especially in the winter months, thus forcing railroad companies to constantly replace their track. Not only was Bessemer steel rail declining in price as imports decreased and domestic production increased in the late 1870s and 1880s, but it also proved its worth as the stronger metal. In the late 1860s, the North British Railway Company experimented with Bessemer steel and wrought iron rails, laying nine solid steel rails and nine iron rails alternately along one of the company's main lines. The result was considerably more wear on the iron rails, leading one of the company officials to conclude that "even granting that wrought iron of great excellence could be produced in large quantities, which he much doubted, even the best wrought iron ever made was far inferior to Bessemer metal, at least for rails."[165] As Temin notes, "The change from iron to steel was the grand achievement of the rail mills."[166] Of the first twelve companies to construct Bessemer steel plants in the United States between 1865 and 1876, eight added steel production to an existing iron rail mill. Only the Edgar Thomson Steel Works in Braddock, Pennsylvania, the Joliet Steel Company in Illinois, and the Pennsylvania Steel Company in Steelton built entirely new works for making Bessemer steel rail.[167]

It would not be until 1877 that steel rail production surpassed iron rails in the United States. However, as railroads multiplied in the early 1870s, several Mahoning Valley industrialists and bankers felt that iron rail production was the ticket to great wealth, despite the ever-increasing production of Bessemer steel rails and the overall higher capital costs associated with an efficient rail mill. One of the Mahoning Valley's most vocal promoters of a rail mill in the region was Youngstown native Caleb B. Wick. In 1846, when Wick was only ten years old, his father helped organize Youngstown's first rolling mill and predecessor to Brown, Bonnell & Co. But rather than going into the iron business, Wick went into finance like the majority of his family.[168] He became the director of the Ashtabula, Youngstown and Pittsburgh (AY&P) Railroad after its formation in February 1870 and believed that owning the means of rail production would significantly cut costs for the railroad. In March 1871, Wick devised a plan to procure $75,000 to move the old Portsmouth, Ohio, rolling mill to Youngstown and convert it exclusively to iron rail production.[169] Wick was successful and, in August 1871, incorporated the Valley Iron Company with several other members of his family. One visitor to the mill in 1872 emphasized that the officers and directors of the Valley Iron Company were nearly all *"young* men."[170] The new mill—the largest single rolling mill in the Valley—sat along the

FIGURE 4.18.
Wick's Valley Iron
Co. rail mill, shown
here shortly after
operations began
in 1872, was one of
the most ambitious
projects undertaken
by Mahoning Valley
iron manufacturers
in the early 1870s.
The Mahoning Val-
ley Historical Soci-
ety, Youngstown,
Ohio.

edge of Crab Creek on Youngstown's east side and had a daily capacity of roughly 150 tons of large rail and fifty tons of small rail for coal
mining operations (figure 4.18).[171] Finally, after what Pittsburgh journalists described as a "necessity" for the Mahoning Valley, the rail mill
increased the region's standing, placing it among the ranks of the most
important iron and rail producers in the country.

After going into production in April 1872, the company's success
merited almost constant expansion and improvements to the mill, and
Caleb B. Wick's connection with the AY&P Railroad was one of the major
reasons for the mill's profitability. Almost immediately in the beginning
of 1872, the AY&P Railroad put in an order to the company for 1,000
tons of iron rail.[172] The Atlantic & Great Western Railroad also placed an
order to reroll 800 tons of old wrought iron rails from their lines, while
the Painesville and Youngstown Railroad signed a $95,000 contract for
1,000 tons of thirty-five-pound rail.[173] "Almost every day some new
addition or improvement is suggested and at once acted upon. The mill
is doing a splendid business, of course, else these extensive improvements would not be required and made," reported *The Vindicator* in
August 1873.[174] However, such prosperity was fleeting. The acceptance
of $100,000 worth of bonds from the Northern Pacific Railroad and the
devastating financial crisis that followed in the fall of 1873 left the Valley

Iron Company—one of the Mahoning Valley's most ambitious and seemingly successful new iron firms—in financial ruin. The company was one of 5,183 businesses in the United States that filed for bankruptcy in 1873.[175] The liabilities of the firm approached $300,000, resulting in the failure of many of its smaller shareholders, who had mortgaged their homes to buy stock in the company.[176]

The financial ruin of the Valley Iron Company was not an isolated incident in the Mahoning Valley. The Great Panic of 1873 began as a bank crisis, during which the stock market closed for the first time in its history. Several banking houses failed one after the other, creating a domino effect that many observers believed was only a "temporary depression in the stock market" and was of "no important significance."[177] The outcome was much worse. The culprit was overzealous expansion of undercapitalized and unprofitable railroads in the United States. After New York banker and Northern Pacific Railroad promoter Jay Cooke's investment house failed in September 1873, the financial crisis bankrupted a quarter of the country's 360 railroad companies and 20,000 other businesses. The country's unemployment rate rose from 3.99% in 1873 to 8.25% by 1878, and wages for those who did work were reduced significantly.[178] Several historians believe the 1873 Panic was "second only to that of 1929–32."[179]

The Mahoning Valley was hit hard by the financial crisis. Companies reduced wages 10% to 15%, and many of the banking firms in the Mahoning Valley were unable to keep pace with the rapid growth of the region's industries in the years before the Panic. This resulted in employers being unable to pay their workers in currency, but rather in notes bearing 8% interest. Common laborers at the Struthers Iron Company's furnace saw their daily wages reduced from $1.75 in 1873 to $1.12 by 1876. Top and bottom fillers' wages dropped nearly one dollar, from $2.25 in 1873 to $1.30 in 1877.[180] By November 1873, eleven of the twenty-one blast furnaces in the Mahoning Valley stopped production, throwing 800 men out of work. Coal mining operations suffered immensely. The mines in the Mahoning Valley, on average, employed 4,000 to 5,000 men, and about half, or 2,500 miners, lost work. Of the Mahoning Valley's total population of 60,000, about 3,600 men employed in the mines, furnaces, and rolling mills were left without jobs.[181] Thus, for much of the 1870s, the Valley was filled with "ghostlike smokestacks, idle men," and "relief societies." One writer noted that "machinery was shut down and acquired a coat of rust; money, credit, and even confidence almost disappeared."[182]

Most of the Valley's rolling mills, however, were not immediately fazed by the economic conditions. Caleb B. Wick's Valley Iron Co. was the first rolling mill firm in the region to feel the Panic's effects. The company tried to reorganize in 1874 by bringing in an iron rail specialist from Danville, Pennsylvania.[183] Despite hiring seasoned experts in the industry, the decision to maintain production of iron rails over steel caused Wick to declare bankruptcy once again in 1875. The price of iron rails dropped to $47.75 per ton by 1875, a decrease of almost twenty-five dollars in five years; Bessemer steel rails still commanded upwards of seventy dollars per gross ton. However, demand for rails, especially those made of iron, continually fell. The opening of Andrew Carnegie's 106-acre Edgar Thomson Steel Works did not help the cause for small-scale iron rail manufacturers, either. Carnegie's massive Bessemer steel plant, "a technological and organizational marvel of the modern world," began producing steel rail on a spectacular scale.[184] With an annual capacity of about 45,000 tons, its Bessemer converters had the ability to produce five tons of steel in thirty minutes, whereas hand puddling the same amount of wrought iron in a single furnace would take five days. Iron rail manufacturers could not compete with such massive enterprises and economies of scale, and many did not have the capital to invest in converting to Bessemer steel rail production.

THE GREAT COLLAPSE

The declining significance of wrought iron rails left Wick's rail mill idle for over four years. Although many of the Valley's other iron firms staved off bankruptcy for the time being, the coming effects on the region would be severe. In his 1927 book *The Emergence of Modern America*, historian Allan Nevins wrote that "the gloom over the country thickened, and still firms continued to fall like lines of dominoes, each toppling over some neighbor."[185] This domino effect was especially true for the Mahoning Valley, its focal point being the firm of James Ward & Co. in Niles. After a period of significant postwar expansion, the company had increased the standing of their scrip to a value equaling that of government-issued greenbacks. However, in 1868, under protest of Niles's merchants, the use of Ward's scrip was abolished, and for five years, the company used actual currency to pay their workers. Most Valley iron manufacturers reverted to paying their workforce with bonds and scrip at the onset of the economic depression. In Niles, this was disastrous and indirectly caused the bankruptcy of one of Ohio's largest iron companies, James Ward & Co. Joseph Butler recalled that the company's financial collapse in February 1874 "struck Niles with all the force of a tornado."[186]

For months, James Ward & Co.'s bankruptcy was the major discussion topic among businessmen and citizens alike throughout the Mahoning Valley. Like many in the industry, Ward's actions during the Panic were not unfounded, as the withering of the economy threw most industrialists and businessmen into desperate circumstances.[187] As they did throughout the country, the decline of wholesale prices through constant agricultural and industrial production affected nearly all businesses in the Mahoning Valley, but James Ward & Co. happened to be one of the largest, and thus garnered the most unwanted attention. Still, the company's failure was daunting. Ward's liabilities reached $1 million, and the shuttering of his mills threw hundreds of men out of work. Desperate, the people of Niles held public meetings to furnish relief, not only for the company but also for the thousands who depended on paychecks to feed their families. The employees of Ward's ironworks subscribed a loan of $63,295, while the citizens of Niles contributed an additional loan of $60,080 in an attempt to reopen the mills.[188] "It is the uniform desire of the community that some plan may be devised which will enable the firm to resume operations at the earliest possible day," reported the *Western Reserve Chronicle*. Despite their efforts, the $120,000-plus loan was not able to save the company, and Ward's property went into the hands of his assignees. The entire ironworks, which included three rolling mills, a blast furnace, a nail factory, a foundry, a machine shop, a roll-turning shop, warehouses, railroad track and switches, and eighty-eight acres of land, was valued at $1 million.[189]

The colossal financial failure of James Ward & Co. was not only a calamity to those immediately involved with the company, but it also severely affected every business interest, more or less, in the Valley. In 1859, John Murray Forbes, an investor in the Mount Savage Iron Works in Maryland, wrote a letter to his brother Paul, who himself had considered investing in the iron industry. "The Iron trade requires a combination of skill & capital (& both are very rare) with general business ability to boot," Forbes wrote.[190] His words rang true for many so-called ironmasters and businessmen of the nineteenth century; even Andrew Carnegie took caution, calling iron a "most hazardous enterprise."[191] Ultimately, James Ward lacked the business ability that his father and a few other entrepreneurs and mill owners had (figure 4.19). Those who invested in James Ward & Co. suffered equal, if not more devastating, financial losses. According to *The Vindicator*, Ward's failure had a debilitating effect on the Hubbard rolling mill and its owner Jesse Hall, who, "through the failure of James Ward during the Panic of 1873, was driven to the wall."[192] Both the Hubbard and Girard mills instituted efforts to reduce costs and avoid looming bankruptcy by introducing worker-owned cooperatives, whereby they leased mill operations to workers on the condition that the companies "withhold 20 percent of the year's earnings until the end of the year." If the company showed a profit at year's end, workers received the withheld wages. Nonetheless, this method, described as a "simple wage cut," was unsuccessful, resulting in the owners both being unable to sustain profitability and filing for bankruptcy.[193]

FIGURE 4.19. Without the aid of his late father, James Ward Jr. (1842–1919) expanded his ironworks in Niles too quickly, thus leading to disastrous results in times of economic depression. Large companies with limited credit, like James Ward & Co., were especially susceptible to the ebb and flow of the iron trade. From *Book of Biographies, Lawrence County, Pennsylvania* (Buffalo: Biographical Publishing Co., 1897), 532.

Another new enterprise named the Mahoning Iron Company, which formed in 1871 to operate Lowellville's Mahoning furnace, ceased operations in 1874, and the furnace remained idle for another five years.[194] The most drastic effects of Ward's failure, however, were in the towns of Mineral Ridge and Warren. After making a small fortune in the iron mining district near Marquette, Michigan, Mineral Ridge iron manufacturer Jonathan Warner bought back his furnaces in Mineral Ridge for $100,000.[195] Operations at his furnaces and mines in Mineral Ridge gradually declined until he was forced to cease operations in 1875. Despite Warner being one of the Mahoning Valley's "oldest, ablest . . . and one of the most successful iron masters," he had never experienced such a severe financial situation as that which occurred during 1874.[196] Warner's accumulated losses throughout the year approached a quarter of a million dollars, primarily due to the failures of other manufacturers, including Ward. The furnaces at Mineral Ridge never operated again, as Warner liquidated the entire property to various groups in the late 1870s.

Like Warner, Warren iron manufacturer William Richards held various interests in James Ward & Co. Richards began to feel the pressure of the economic downturn in September 1874, when he asked for an extension for payments on the furnace loan given to him by the city of Warren.[197] Consequently, Richards drove his puddlers' wages down, resulting in a strike in January 1875. The lack of profit and the failure to pay off his mortgage pushed Richards to declare bankruptcy the same month. With liabilities of over $235,000, Richards was unable to dig himself out of such a large financial hole and was incapable of paying back any of the $75,000 furnace loan. Richards died at the age of fifty-six in February 1876, and the ironworks he built would also fade from existence. A fire broke out at the idle mill in 1878, which destroyed the furnace and nearly wiped out the rolling mill, although the latter was eventually rebuilt and operated sporadically through the remainder of the nineteenth century.

As these weaker iron companies in the Mahoning Valley failed, those with strong capital and high credit weathered the economic storm. Most iron manufacturers were hesitant to invest in and enlarge operations during the Panic; however, in Pittsburgh, Andrew Carnegie and Henry Clay Frick believed that economic downturns were the ideal time to expand because prices were low and competition almost nonexistent. Some of Youngstown's successful iron companies shared in Carnegie and Frick's somewhat unorthodox ideology, although on a much smaller scale. Youngstown's Brown, Bonnell & Co. was one of these financially sound enterprises. After the death of the company's founding partner William Bonnell in May 1875, Brown, Bonnell & Co. incorporated their longtime copartnership with a capital of $1.5 million in September the same year. Richard Brown served as the company's first president. The board of directors appointed two of Bonnell's sons, William Scott Bonnell and Henry O. Bonnell, who had worked in their father's mill for a number of years, vice president and secretary, respectively (figure 4.20). Although Brown, Bonnell & Co. held their ground and did not physically expand during the Panic, company veteran Joseph H. Brown founded the Joseph H. Brown Iron and Steel Works in South Chicago with long-time business partners and Brown-Bonnell shareholders Samuel, George W., and Charles B. Hale.[198] In 1881, Brown sold the mill—the first built in the Calumet region—to the Calumet Iron and Steel Company for $2 million, allowing him to retire comfortably in Youngstown.[199]

Other Youngstown iron manufacturers successfully expanded production and invested in their mills during the Panic without terrible financial losses. In 1874, Cartwright, McCurdy & Co. added a second rolling mill to their Enterprise Iron Works. The company's investments continued with the overhaul of the old Eagle furnace, which they tore down in August 1874 and rebuilt as a modern, iron-clad stack, pushing its daily production to thirty-five tons. One of Youngstown's largest pig iron producers, the Himrod Furnace Co., wavered on the brink of financial failure. Under scandalous circumstances, company president R. A. Wight issued $100,000 worth of bonds to keep the company afloat financially, but he distributed them without the approval of two-thirds of the company's stockholders.[200] The court finally charged Wight with fraud in 1887, but he had secured enough funds in the time being to entirely rebuild two of the three Himrod furnaces in 1876 and avoid bankruptcy. In addition, Wight drove his employees' wages down and

FIGURE 4.20. Brown, Bonnell & Co. officials pose for a photograph outside the mill's main office, possibly after the incorporation of the company in 1875. Sitting from left to right are W. Scott Bonnell, Richard Brown, and Henry O. Bonnell. Standing from left to right are William Battelle, Martin Bonnell, and John Brown. Battelle was a master nail maker at the mill. ©*The Vindicator*.

only operated his furnaces an average of eight months out of the year in the thick of the depression.[201] Most merchant pig iron companies in the Valley remained stagnant and decided not to expand during the 1870s, often banking their furnaces or operating on a limited basis. The Brier Hill Iron and Coal Company was "making no money" but had "large capital and unlimited credit," waiting until after economic conditions improved to modernize and increase output.[202] Well-managed blast furnace companies in Hubbard, Haselton, and Girard followed suit by cutting back production.

The Panic of 1873 ultimately served to weed out the vulnerable iron companies and restructure the Mahoning Valley's role in the national context of the growing iron and steel industry. As the Panic subsided in 1878, the United States boasted 698 blast furnaces, 340 rolling mills, eleven Bessemer steelworks, and fourteen open-hearth steelworks, the latter a technology that steel companies would develop further in the coming years. First introduced by German and French engineers in 1868, the open hearth was a rectangular chamber lined with firebrick and charged with cold scrap and pig iron, either in a cold or molten state, to make steel. The open hearth achieved a high enough temperature to melt steel (2,600 to 2,800 degrees Fahrenheit), burning away the impurities in the metal in the form of slag. The process, which used unskilled and semiskilled labor, was slow and took up to twelve hours. It was also much more expensive than the Bessemer converter, but it produced a higher quality of steel.[203] Many iron and steel works were initially hesitant to adopt the open-hearth furnace for large-scale steel production, but with constant improvements by British engineers Sidney Gilchrist Thomas and Percy Gilchrist in the late 1870s and early 1880s, the productivity of the open hearth increased and the overall costs of production decreased. Firms such as the Cleveland Rolling Mill Co. and the Otis Iron and Steel Co. in Cleveland began producing open-hearth steel in the mid- to late 1870s. The Pittsburgh Bessemer Steel Company, predecessor to Andrew Carnegie's Homestead Steel Works, made its first Bessemer steel in 1881 and first open-hearth steel five years later. Youngstown industrialists, however, continued to produce iron. Thus, they had to develop and adapt to changing market trends in the closing decades of the nineteenth century, much like the many independent iron enterprises in Pittsburgh who did not adopt the economic techniques of steel magnates such as Carnegie.[204] To Valley industrialists, iron was still king until steel proved its enduring value over the romance and tradition of iron making.

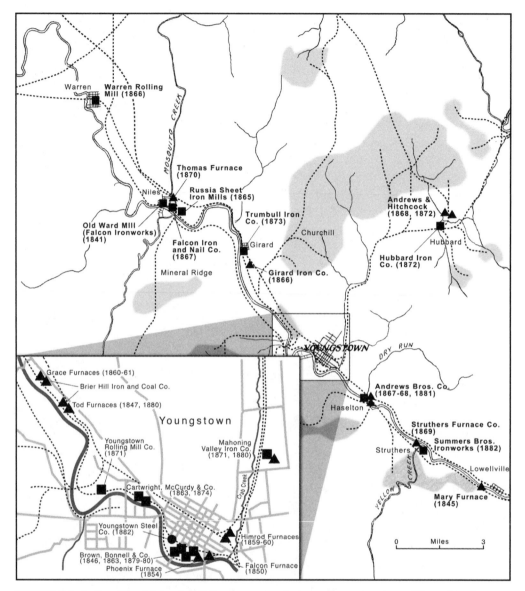

FIGURE 5.1. In the 1880s, Mahoning Valley iron manufacturers began to rebound from the 1873 Panic, as shown in this map illustrating the Valley's industrial progress through 1885. Railroads (dashed lines) still snaked their way through the region's coalfields (shaded areas), despite coal mining being on the decline throughout the decade. Many new companies took over shuttered rolling mills (squares) and blast furnaces (triangles), while the Valley saw the formation of its first steelworks (circles) during this period.

THE PRESSURE OF STEEL, 1879–1894

> Since the center of iron production moved out of Pittsburgh . . . Mahoning iron manufacturers have been hedged up on the one side by the rapid cheapening of steel and its consequent displacement of iron, and on the other hand by the natural slowness of the men who puddled iron to grasp the fact that the preference for steel was not a passing phase but a permanent condition.
> —*The Iron Trade Review* 27 (August 16, 1894): 9.

THE MAHONING VALLEY emerged from the economic depression badly beaten but not entirely broken. Although there was little progress made in the civic, municipal, and industrial development of the region in the 1870s, the population of Youngstown alone had reached 15,435 by 1880, an increase of 90% in ten years, making it the state's tenth most populated city.[1] The growth of Youngstown reflected the steady increase in population of other major industrial cities in the region, including Cleveland, which reached just over 160,000 people in 1880, making it the eleventh largest city in the United States. The Mahoning Valley's other industrial rival, Pittsburgh, ranked just behind Cleveland as twelfth largest in the country, with 156,389 citizens. However, cities with little industrial growth saw their populations stagnate. Warren, the Mahoning Valley's second largest city, developed much slower due to its failure to attract industry, leading to almost static growth from 1870 to 1880. The seat of Trumbull County only managed to draw 971 citizens during this ten-year period, resulting in a population of 4,428.[2] Likewise, Canfield, the seat of Mahoning County, with a population of only 650 in 1880, remained an agricultural and rural country village, with industrial growth nonexistent other than a booming farm trade. Canfield's position at the geographic center of Mahoning County was the primary reason that it won the county seat over Youngstown in 1846. However, the latter's incredible growth over thirty years incited a public meeting headed by John Stambaugh to outline a plan for removal of the county seat to Youngstown. To support their plan, Stambaugh and others stated that Youngstown paid one-half of the county's tax bills and contained one-third of its population, while most litigation and business involving property transfers took place in the city, prompting the state legislature in Columbus to approve the plan in 1874. The citizens of Canfield resisted, however, and took the case to the Supreme Court of Ohio, where, in 1876, the decision to remove the county seat from Canfield to Youngstown was upheld.[3] The Mahoning Valley's leading coal, railroad, and iron operator, Chauncey Andrews, fronted the majority of the money to construct the new courthouse, county jail, and sheriff's residence.

The removal of the county seat to Youngstown made it one of the most important civic and manufacturing centers in northeast Ohio, encouraging many entrepreneurs to consider locating in and doing business within the city. However, other towns and villages in the Mahoning Valley suffered greatly from the economic depression. The industrial progress made in Warren and Mineral Ridge during and after the Civil War was almost entirely lost, resulting in those towns losing their economic foothold in the region, while promise of growth in the remaining decades of the nineteenth century faded. Of the Valley's eighteen iron companies, nine failed financially in the 1870s, while three blast furnaces were torn down. Although many smaller, less-capitalized businesses and iron mills floundered during the Panic of 1873, the period served to develop some of the country's largest and most powerful iron and steel companies, largely based on the hard-nosed business mentality of Andrew Carnegie. Investing while weaker companies cut back allowed many iron and steel firms to expand and become more profitable than ever. The Cambria Iron Company in Johnstown was one of the largest in the United States and set records for rail production in the mid-1870s. Carnegie's Edgar Thomson Steel Company made almost a half a million dollars in profit by 1878, and net earnings rose to $2.1 million in 1882.[4] Despite operating at a loss during much of the 1870s, the Bethlehem Iron Company emerged as one of country's leading producers of iron and steel products, including rail.[5] Ohio's largest single iron and steel producer, the Cleveland Rolling Mill Co., produced 90,000 tons of steel rails and 40,000 tons of open-hearth ingots annually by 1880.

By 1878 and 1879, smaller companies began investing and increasing production. The national economy and iron market improved and production numbers of all facets of the industry greatly increased. National pig iron production jumped from 2.5 million tons in 1878 to 4.2 million just two years later.[6] Demand for Bessemer steel rail more than doubled production from 1878 to 1881, and even the manufacture of rolled iron products continually increased despite competition from steel manufacturers. The demand for iron rails, however, steadily decreased, just as they had for much of the previous decade. In 1881, American Iron and Steel Association secretary James Moore Swank wrote,

Iron rails can not hereafter be made by our manufacturers in competition with steel rails, except under favorable circumstances . . . [this] being the comparative cheapness with which old iron rails may be rerolled at iron rail mills located in sections of the country where there are no steel rail mills.[7]

Consequently, most railroad companies no longer had any use for wrought iron rails. Those that continued to produce iron rails usually did so for a strictly local rather than national market, which included light rail used for mining operations.

In terms of production, Ohio retained its position as second in the United States only to Pennsylvania in manufacturing rolled iron, a number that increased by the mid-1880s. Although the Mahoning Valley did not produce any rail, it led all other iron-making regions in the state—Cleveland included—in rolling iron products such as bar, bolt, rod, skelp, hoop, and shaped iron, but it was a distant second to the Lake counties in sheet and plate iron manufacturing. Similarly, Ohio was second only to Pennsylvania in total pig iron production, and the Mahoning Valley led all of Ohio, with 226,877 tons produced in 1880. However, unlike the Mahoning Valley, industrialists in Cleveland began to invest in their plants on a larger scale, while many Valley iron manufacturers were still trying to rise out from the ashes of the Panic and start anew. The capital costs of building an efficient and fully equipped steel rail mill were in the millions, a price that many Mahoning Valley iron manufacturers could not afford, nor did they feel the need to.[8] Because Bessemer steel largely dominated the rail trade, there was still a strong market for all other kinds of shaped iron. Likewise, Bessemer steel facilities in Pittsburgh, Cleveland, and Chicago often required upwards of 50,000 tons of pig iron a year. Other pig iron consumers, such as foundries, needed the metal to cast heavy equipment used by iron and steel companies in building massive new steelworks and blast furnace plants throughout the 1880s.

RISING FROM THE PANIC

Like the majority of the country, the Mahoning Valley's industrial rejuvenation began in the late 1870s. Despite the steady exhaustion of the region's coal reserves, there were still thirty coal banks operating in the Valley in 1880 that employed 3,157 miners. The twelve rolling mills, which required a large labor force, employed 3,293, while the Valley's nineteen blast furnaces needed 755 laborers to operate at capacity levels.[9] The 1880 census reported that Brown, Bonnell & Co. was the region's largest employer, with a workforce of 900, while Cartwright, McCurdy & Co.'s rolling mills employed 600. In total, over 7,200 men were again working full force in the Mahoning Valley's iron and coal industries. The largest iron companies in the Mahoning Valley in the 1880s were mostly centralized in Youngstown, while some new companies invested in other Valley mills shuttered by the Panic. Most of this revival came by the way of Youngstown financiers and some of the area's leading businessmen who had weathered the economic depression by stopping production, running at less than half capacity, or investing in minor additions to their plants while others completely shut down.

Youngstown families, including the Andrewses, Bonnells, Wicks, Armses, Tods, and Stambaughs, each having solidified their positions among the Valley's social and industrial elite, had interests in five of the Valley's eight merchant furnace companies and six of the ten rolling mill firms. These numbers would only increase in the mid-1880s, as impending economic depressions would once again leave mills not owned by Youngstown's social elite in bankruptcy. The social construct of other major industrial centers, such as Pittsburgh and even Cleveland, was much more diverse. Pittsburgh's population became more dispersed than in other cities, its topography playing a major role in developing the city's economic, political, and social structure.[10] Youngstown and the Mahoning Valley, however, were not as geographically varied as Pittsburgh, so there were no significant separations among the Valley's leading social families. In addition, lower population and the comparatively slow rate of industrial growth in the Mahoning Valley left much of the region's iron mills and financial institutions in the hands a select few entrepreneurs by the end of the nineteenth century.

Some struggling industrial families sought to continue their social dominance. The Ward family in Niles, which stood atop the Valley's industrial hierarchy for much of the nineteenth century, attempted to regain their elite status after James Ward's financial implosion in 1874 by reinvesting in their former mills. Coming out of the 1870s, the industrial integrity of Trumbull County's industrial hub was uncertain. However, as the economy recovered, Niles's iron industry began to show signs of life. In October 1878, the *Western Reserve Chronicle* reported, "The revival of the iron business at Niles has given all branches of trade in that place more activity and cheerful outlook than has been felt for several months past."[11] After the company's bankruptcy, the extensive mills of James Ward & Co. were individually parceled out to three different companies, one under control of the wealthy Arms family of Youngstown, the others under Lizzie B. Ward, James Ward's wife, who helped her husband partially restructure Niles's iron industry.

Several families of Niles laborers moved back into town after years of trying to find work elsewhere. The Russia Sheet Iron Mill reported running "double turn," with orders for iron piling up, while the original Old Ward Mill, as it became known, operated with the same urgency.

These two mills, formerly under the James Ward & Co. interests, had been the economic backbone of Niles since the Civil War. Amidst Niles's industrial reorganization in the four-year period following James Ward & Co.'s failure, the Old Ward Mill and the Russia Sheet Iron Mill were restructured under the financial backing of Ward's wife and mother. Ward's wife, Lizzie, inherited a portion of her wealthy father's estate after his death, purchased the Russia mill, and operated it under the Lizzie Brown Ward Co., otherwise known as L. B. Ward Co. In 1882 and 1883, Ward invested $50,000 worth of her inherited money into the Russia mill.[12] In addition, James Ward's mother, also named Elizabeth, purchased the Old Ward Mill and formed the Ward Iron Co. Accordingly, both Ward's wife and mother were the first women to run iron mills in the Mahoning Valley and the state of Ohio, a rare occurrence that usually happened only after the unexpected death of a company president or major shareholder.[13] However, both mills were operated under the fragile management of James Ward Jr.

Ward's family connections gave him a second chance to clear his name and reputation. Thus, he began his dubious second tenure by removing all of the nail machines from the Old Ward Mill to focus primarily on the manufacture of bar and sheet iron rather than cut nails, the latter a staple of the mill's production since the 1850s. By the mid- to late 1880s, the iron cut nail industry had fallen in production to steel cut nails and wire nails. The production of steel cut nails commenced in the United States in 1884, and by 1889, over two-thirds of all cut nails were made of steel, while wire nails steadily increased in production, from 600,000 kegs in 1886 to 3,135,911 kegs in 1890.[14] Wire nails were cheaper to produce and required little to no manual labor, unlike cut nails, and sheet iron remained a valuable commodity on the market, increasing in production in the late 1880s, while iron cut nails steadily decreased. Hence, sheet iron experienced a considerable increase in production in Ohio alone, increasing by over 20,000 tons between 1884 and 1888, while most other states experienced considerable decline or only marginal growth in its production.[15] Ward followed these market trends closely, and by 1882, the Ward Iron Co. and L. B. Ward Co. had an annual combined capacity of about 14,600 tons.[16]

Perhaps the most important means to rejuvenating Nile's industry was the reorganization of the Falcon Iron and Nail Company. Another one of Ward's former mills, the works was immediately successful after being put back into production in the late 1870s under Youngstown businessman and merchant Freeman O. Arms. As president of the company, Arms reconditioned the mill and continued producing cut nails. However, by 1879, Arms retired from active business and was replaced more than adequately by John Stambaugh, one of Youngstown's leading iron manufacturers, whose long history of running a successful iron business

FIGURE 5.2. A view looking northeast across Mosquito Creek in Niles of the Falcon Iron and Nail Company's mill, c. 1895. The sheet mill appears in the center, and a portion of the puddling mill is at the extreme right of the image. The mill proved successful under the management of Youngstown industrialists. The Thomas furnace can be seen in the back left. Courtesy of Niles Historical Society.

presented an ideal opportunity to keep the mill profitable. Focusing his interests on the Brier Hill Iron and Coal Co. in Youngstown, Stambaugh delegated all of the managerial responsibilities at the Falcon Iron and Nail Co. to Freeman Arms's nephew, Myron I. Arms. Arms was the son of Emeline E. Warner and Myron Israel Arms and grandson of the distinguished Mineral Ridge iron manufacturer Jonathan Warner.[17] After attending Rayen School in Youngstown, Arms spent a year studying at Cornell University from 1871 to 1872. He first worked as a bank clerk before taking his position at the Falcon Iron and Nail Company, which flourished under his management. In 1882, when Arms was only twenty-eight years old, he supervised 250 workers, many of whom had witnessed firsthand the devastating effects of James Ward & Co.'s financial failure. Nevertheless, Arms had the mill producing twenty tons of finished iron per day and working double turn. From June 1882 to June 1883, the nail mill produced 96,800 kegs of nails, the largest amount ever manufactured by the works.[18] Editors of the *Western Reserve Chronicle* praised Arms's ability to manage the company, saying, "If Old Trumbull had a few more young men like Mr. Arms within her borders engaged in business, we would have in this county a big boom in the manufacture of iron."[19] In November 1885, Arms added a large sheet mill to the ironworks, similar to how James Ward converted the product of the Old Ward Mill from cut nails to exclusively sheet iron (figure 5.2). Although the company continued to produce cut nails until the early 1890s, its sheet iron production grew from 2,500 tons to 12,000 tons a year during that period, reflecting the product's production growth in both the Mahoning Valley and the entire country.[20]

FIGURE 5.3. By founding the Niles Fire Brick Co. in 1872 and the Thomas Furnace Co. in 1879, John Rhys Thomas (1834–1898) supplanted James Ward Jr. as Niles's leading manufacturer. He remained prominent in Mahoning Valley industrial circles until his sudden death in January 1898, when his business interests moved into the hands of his five children. From *The Iron Trade Review* 29 (January 2, 1896): 12.

Despite its rejuvenation, Niles's industry continued to fragment. In 1879, Welshman John Rhys Thomas, an experienced mechanic, blacksmith, brick maker, and stonemason, purchased the small blast furnace built by William Ward in 1870.[21] Thomas repaired the stack and put it back into operation in October 1879, establishing the Thomas Furnace Company (figures 5.3 and 5.4). Thomas, however, was already familiar with Niles industry. In 1872, he organized the Niles Fire Brick Company as a partnership with Lizzie B. Ward, building a small plant across from the Old Ward Mill. It supplied the Valley's puddling, heating, and blast furnaces with heat-resistant firebrick.[22] Thomas's endeavors in Niles's industry launched a legacy not unlike that of the Ward family before and during the Civil War. Perhaps ironically, Thomas purchased James Ward's 1862 Italianate mansion built on the south side of the Mahoning River opposite the iron mills but far enough away to elude the constant smoke and noise. Niles was beginning to see a shift not only in its social elite but also in its manufacturing base. The transition from a single dominant, integrated enterprise to an agglomeration of small iron companies was a drastic change in the organization of one of northeast Ohio's largest nineteenth-century industrial hubs.

Although Niles saw clear signs of economic rejuvenation from a number of different investors, it would be limited. In 1881, C. H. Andrews delivered another large economic blow to Niles by removing Niles Iron Co.'s puddling and rolling mill to Haselton, directly adjacent to his two blast furnaces.[23] Two years later, Andrews merged all of his interests under one management, creating an extensive integrated ironworks just outside the Youngstown city limits (figure 5.5). This was one example of

FIGURE 5.4. Workers (center) pose in front of the Thomas furnace in Niles, c. 1880. When purchased by J. R. Thomas in 1879, the furnace was among the smallest in the Mahoning Valley. Nonetheless, Thomas's investment in the plant over the next fifteen years made it one of the region's most efficient. Courtesy of Niles Historical Society.

FIGURE 5.5. Andrews Bros. Co.'s Haselton Iron Works evolved into one of the most extensive in the Mahoning Valley after C. H. Andrews removed the mill from Niles to Haselton in 1881. The rolling mill appears to the right, while smoke and steam largely shroud the puddling department. The photograph dates from 1889. From *Youngstown Illustrated* (Chicago: H. R. Page & Co., 1889).

how Youngstown manufacturers favored the city because of its central location in the Valley and its six railroads that connected with nearly all major lines in the country.[24] Although Andrews and many other Youngstown iron companies had their mills under sound management, some overexpanded in a vulnerable period of industrial growth. With immense capital, Brown, Bonnell & Co. was one of the Valley's most financially stable enterprises throughout its existence, never waning in the wake of economic depressions. It was also one of the few in the Valley to significantly increase the size of its mills after the Panic subsided, with the last major additions and investments to the plant occurring during the Civil War.

Between 1879 and 1882, the company nearly tripled its wrought iron production with the addition of the No. 3 puddling and bar mill. Workers in the new mill manufactured specialty fish plate used in adjoining railroad rails, which supplied the company's wholesale warehouses in Chicago and other major cities. In 1880, the company also constructed the No. 4 puddling mill, containing nineteen double puddling furnaces, parallel to Market Street. Overall, the company added seventy puddling furnaces for a combined 124—the most of any other iron mill in the country—expanding Brown-Bonnell's capacity from 25,000 to about 70,000 tons per year.[25] To supply these new mills, the company in 1882 purchased the Anna furnace in Struthers, which, along with the Phoenix and Falcon furnaces, gave Brown-Bonnell an output of about 64,000 tons of pig iron per year; this could supply all of the company's puddling furnaces with the mills running at full capacity. In time, the Youngstown *Vindicator* characterized the Brown-Bonnell mill as "the greatest, largest iron mill in all the West," a title the mill well deserved for its merit (figure 5.6).[26]

The firm's rapid expansion in the early 1880s was similar to James Ward & Co.'s overzealous growth in Niles after the Civil War. These improvements to the company's mills, however, were not under the sound financial management of local shareholders, but rather Chicago men. The primary catalyst of the company's rapid expansion was new president Herbert C. Ayer. The son of Chicago iron merchant and major Brown, Bonnell & Co. shareholder John V. Ayer, Herbert secured an interest in his family's company after he was granted an honorable discharge while serving for the Confederate Army, thus allowing him to move back to Chicago.[27] After his father's death in 1877, Ayer inherited the entire company in Chicago, accumulating a large fortune and acquiring a significant interest in Brown, Bonnell & Co. He began to alienate other company directors his father had appointed over the years, leaving one of Chicago's oldest and most well-respected merchant iron dealers almost solely in his hands.[28] In addition, Ayer was known as an extravagant spender, "whose personal expenditures were enough to exhaust the profits of many a large concern."[29] In 1881, Ayer arranged to construct

FIGURE 5.6. Herbert C. Ayer's expansion of Brown, Bonnell & Co. created the largest ironworks in the United States on the Mahoning River's north bank near downtown Youngstown. The No. 4 puddling mill appears on the extreme left, and the Phoenix furnace rises above the lone tree just right of center. The No. 1, No. 2, and No. 3 rolling mills are concentrated in the center of this 1889 photograph. From *Youngstown Illustrated* (Chicago: H. R. Page & Co., 1889).

the "largest plate and sheet mill in the world" at the Brown-Bonnell mill, but the project never materialized.[30] Such a spendthrift mentality would not bode well for Ohio's largest iron company, particularly after Ayer controversially bought out the interests of all of Brown, Bonnell & Co.'s Youngstown-based shareholders in 1879, including the company's original founders, Richard and Joseph H. Brown.[31]

The outcome of the buyout engendered not only the extreme expansion of the Brown-Bonnell mill but also the creation of a new company from many of Brown, Bonnell & Co.'s former local stockholders. In September 1879, Richard Brown, Henry O. Bonnell, W. Scott Bonnell, Charles D. Arms, and Peter M. Hitchcock, a Cleveland industrialist who also had interests in Brown, Bonnell & Co., formed the Mahoning Valley Iron Company from the defunct iron rail mill originally built by the Valley Iron Co. on Youngstown's east side. Conforming to market conditions, company directors rehabilitated the mill and converted it to produce merchant bar iron and other wrought iron products rather than iron rail. In 1882, the company diversified its product line by adding a mill for making iron shafting.[32] The shafting plant was essentially wire-drawing on a large scale that created circular stock up to three inches in diameter, principally used as axles for transmitting power. Shafting was initially made of wrought iron until open-hearth steel containing 0.15% carbon was made available from other sources in the Pittsburgh and Wheeling districts in the mid- to late 1890s. The company added pig

FIGURE 5.7. The outgrowth of the Brown-Bonnell buyout was the formation of the Mahoning Valley Iron Company, which transformed an unsuccessful iron rail mill into a thriving integrated ironworks. In this 1889 image, the Hannah furnace, named after the wife of company director Charles D. Arms, rises from the left background, while the shafting works appears to the left and the nail factory to the right. The puddling and rolling mills extend from the left to the right behind the Hannah furnace. From *Youngstown Illustrated* (Chicago: H. R. Page & Co., 1889).

iron production by buying James Ward & Co.'s long-idle blast furnace in Niles and removing it to their Youngstown plant, completing the old rail mill's transformation into a fully integrated ironworks (figure 5.7).[33] This industrial renewal in Niles and Youngstown echoed the country's overall growth in the early 1880s.

Likewise, the rejuvenation of the Valley mill and others throughout the region following the 1873 Panic served to create thriving ethnic working-class communities reliant on the mills in which they worked. The neighborhood surrounding the Mahoning Valley Iron Company quickly earned the fitting title of "Smoky Hollow," as it was often saturated with smoke from the nearby mill. Irish, Germans, and, by the turn of the twentieth century, Italians dominated the ethnic makeup of the neighborhood. Outside of the mill, new immigrants whose family members came to work for the Mahoning Valley Iron Co. opened small groceries and dry goods stores from the front of their small tenant houses. This trend propagated throughout the Mahoning Valley's industrial communities, including in Niles, which by the turn of the twentieth century saw its Italian-born population grow considerably. Such large immigrant communities became the norm as the iron mills grew from meager workforces of only a few dozen to hundreds following the Civil War. Consequently, the Mahoning Valley's working-class immigrants forged strong community bonds through shared ethnic values amidst the shadows of the mills and their largely paternalistic workshop culture.[34]

The growth and expansion of the Mahoning Valley Iron Co. and others in the region initially seemed positive to the Valley in terms of employment and fiscal development. However, the prosperity enjoyed by many of the Valley's rolling mills and their workers would be short-lived, as the country's great economic growth after the 1873 Panic resulted in overproduction. On a visit to the United States in 1882, English sheet iron manufacturer William Molineaux inspected many of the country's iron- and steelworks, leading him to believe there was a dangerous production surplus. He wrote his thoughts in a letter to the American Iron and Steel Association:

> I think that your iron manufactures are fully five years ahead of the demand. Take, for instance, the country between Cleveland and Pittsburgh. The capacity of the mills is fully three times as great as it was eight years ago, when I was here last. Brown, Bonnell & Co., at Youngstown, have doubled, and perhaps trebled, their capacity. At Sharon, where there was one mill before, now there are two or three. At Newburgh the extensions have been very considerable. . . . In building new mills and remodeling old ones you are making iron and steel faster than you are able to use it. The natural consequences will be lower prices and less work to do.[35]

The production of pig iron, steel of all kinds, and rolled iron (excluding rails) in the United States in 1882 was the largest in the country's history.[36] Despite the country's massive industrial growth, most Mahoning Valley industrialists were hesitant to build entirely new mills when incorporating new iron companies. George Summers, a former superintendent at James Ward's ironworks, and his brother James Summers, a skilled roller at Brown, Bonnell & Co., were the only individuals to build an enterprise from scratch without inherited wealth or a foothold in the Valley's industrial elite. Summers Bros. & Co. Ironworks, although small, was situated across from the Anna furnace in Struthers and operated primarily by members of the Summers family, producing only nine tons per day. The firm evolved into one of the more profitable iron enterprises in the Valley, remaining a small niche producer of sheet iron for stamping works and mower and reaper manufactories (figure 5.8).[37] After production commenced in August 1882, the company never expanded beyond its financial limitations during the 1880s, a time when many firms, such as Brown-Bonnell, were pressed for orders and believed that the only solution was to build higher-capacity mills rather than produce specialized products.[38] Although some firms such as Summers Bros. remained small and financially secure, the industry itself was highly unstable, being what historian David Brody described as either "panic-stricken" or "strained to utmost capacity."[39] Thus, there was a

FIGURE 5.8. The workforce at the Summers' sheet iron mill in Struthers poses near the puddling furnaces in this photo taken in 1883. A pile of scrap iron sits in the foreground, which was reworked into sheet iron. The mill was small and remained one of the lowest-capacity iron mills in the Mahoning Valley until the turn of the twentieth century. Photo courtesy of the Struthers Historical Society and Museum; Marian Kutlesa, curator.

marked increase in union activity and labor strife, which, along with economic decline stimulated by overproduction, helped to dismantle some of the Mahoning Valley's largest firms, leaving only the efficient producers behind.

LABOR STRUGGLE AND FINANCIAL TURMOIL

Large-scale organized strikes in the Mahoning Valley were not commonplace in the 1860s and 1870s outside of coal miners' strikes. In fact, there are few important documented strikes in the region, and some had little to do with wages. For example, in 1870, puddlers at Brown, Bonnell & Co. struck largely because of principles and "incompetent workmanship."[40] The puddlers described one incident:

> One of the boilers put in the squeezers a ball which he considered too cold to roll; he therefore notified the roller not to roll it. The latter thoughtlessly put it in the rolls, and broke a box and spindle, for which

the boiler was wrongfully, as we believed, discharged, the employers holding him responsible for the roller's neglect.[41]

They called a strike shortly thereafter.

Less important strikes occurred at various Valley furnace plants and rolling mills during the 1870s, but they had little effect owing to the depressed economic conditions.[42] Organized activity among puddlers and other rolling mill workers was greater in Pittsburgh, where, in 1858, puddlers formed an underground union called the Sons of Vulcan with the goal of improving wage rates. Union membership among ironworkers increased from 600 in 1868 to over 3,300 by 1876, the same year the Amalgamated Association of Iron and Steel Workers formed.[43] This new organization merged three skilled labor unions, including the Sons of Vulcan, and represented puddlers, heaters, rollers, and roll hands (figure 5.9).[44]

Almost immediately following the establishment of the Amalgamated Association, local members of the union formed five different lodges in and around Youngstown.[45] By 1880, there were over 150 lodges in the prominent iron and steel districts in the country, while membership grew to over 16,000 by 1882.[46] Members of the union were composed of all classes of labor in the iron and steel mills. Consequently, when one department had a dispute with the wages being paid to a certain class of worker, all other departments in the mills gathered to resolve the situation; this often led to one department refusing to work and being joined by all of the others, resulting in a large-scale strike throughout the major iron and steel manufacturing centers.[47] After the harsh summer of 1881, which limited the amount of crops produced, there was an increase in the cost of living in the spring of 1882. When iron companies failed to increase workers' pay, the Amalgamated Association threatened to call a general strike if there was no revision in the sliding scale for wages by June 1, 1882. The scale, devised by Pittsburgh ironmaster Benjamin Franklin Jones in 1863 and described by him as "a means to create a community of interests between capital and labor," tied skilled ironworkers' wages to the selling price of iron.[48] The Amalgamated Association called for higher rates for various kinds of labor and a new scale adopted by manufacturers that included an advance of fifty cents per ton to puddlers, extra pay for rollers during the hot summer months, a 15% to 30% increase for workers in sheet mills, and an advance of ten cents for those working scrap iron into new products.[49] Iron manufacturers from the Mahoning Valley, as well as those from all major manufacturing centers, met in Pittsburgh to discuss the new scale, but ultimately rejected the Amalgamated Association's proposed wage increases.

The result was an immense strike that affected 116 rolling mills in western Pennsylvania, Ohio, Indiana, West Virginia, Illinois, and Wisconsin and involved 35,000 ironworkers.[50] In Youngstown, *The Vindicator*

FIGURE 5.9. Rollers and heaters pose with their tools in this undated photograph at an unknown Youngstown rolling mill. Most skilled ironworkers such as these were a part of the Amalgamated Association of Iron and Steel Workers, which became a highly active union in the Mahoning Valley in the early to mid-1880s. Courtesy of Tom Molocea.

screamed: "Ten Thousand Men Go Out. Every Rolling Mill in Youngstown, Warren, and Girard Shuts Down—Furnaces Banked Up—Mines to Shut Down."[51] Despite operations in the Valley coming to a halt, Youngstown was a small theater in the strike compared to Cleveland and the Pittsburgh area. The National Tube Works Company in McKeesport successfully started their mill by employing nonunion men. However, six Amalgamated Association strikers were arrested for conspiracy in an attempt to keep the mill from starting.[52] In Cleveland, two small iron companies signed the scale within a week of the strike being called, likely because of fear of losing extended work and profit. The Cleveland Rolling Mill Company also attempted to employ nonunion men, only to have them turned back by striking workers and their wives who blocked entrances to the mills, persuading most from working. According to reports, only 300 men worked the company's mill, which normally required a workforce of 5,000. "The workmen were very crude and awkward and in the rail mill it took 76 men four and a half hours to run out six rails. Such work will not redound to the financial advantage of the company," reported *The Vindicator*.[53] Attempts to run mills in Youngstown with nonunion men failed entirely.[54]

The strike dragged on for sixteen weeks. Finally, at a September 19 meeting in Pittsburgh, Amalgamated Association president John Jarrett announced, "Gentlemen of the Amalgamated Association, the strike is ended."[55] The union could no longer hold out against manufacturers and accepted the old wage scale. Most mill owners admitted that they resisted signing the new scale not because of the increase in wages, but rather because they wanted to disorganize the Amalgamated Association, thereby discouraging others from joining the union.[56] However, this mind-set was detrimental to some Youngstown iron manufacturers, namely Brown, Bonnell & Co. The company's large-scale expansion in the three years before the strike and the resultant outlay of capital did not bode well during periods of extended idleness. The company began making repairs to their mills in early September in the hopes that an agreement would be reached and secured a number of employees to work their furnaces. Brown-Bonnell president Herbert Ayer agreed to house their employees on the mill grounds, and put out a help-wanted ad looking for puddlers, heaters, and rollers working for $5.50 per ton with "steady work guaranteed" and "McDonald's watershields to every furnace."[57] This was a desperate plea from Ayer. Hugh McDonald's patented shield for puddling furnaces was a hollow piece of iron through which water passed that was placed in front of the furnace, which kept puddlers cooler and made for a more tolerable working environment. Advertising such "luxuries" helped sway nonunion workers to come work for the company.

During the strike, Ayer remained in Chicago, where he feared his striking workforce, asking for protection from the local militia.[58] He was unsympathetic to his workers' demands, but the end of the strike was only a marginal victory, as the combination of four months' worth of lost profit and a declining iron market weighed heavily on his companies in Chicago and Youngstown. The weakening of the iron market, however, was so "gradual and tranquil" that many iron manufacturers were not worried about an impending crash. The price of rolled iron did not increase during the strike, indicating that supply exceeded demand. As a result, the demand for pig iron from merchant furnaces decreased, and by December 1882, the market for both pig iron and rolled iron was the lowest since 1878.[59] In 1883, the iron market witnessed a severe decline not unlike that of the early 1870s. In a review of the year 1883, James Moore Swank stated, "We have had a natural reaction from extraordinary activity, a great shrinkage in the building of railroads, and a capacity for the production of iron and steel that has been in excess of the country's wants and that has brought low and still lower prices."[60]

The inactivity of the iron market severely affected Ayer and his business in Chicago. In February 1883, Ayer filed for bankruptcy because his warehouses in Chicago had taken a drastic hit when the city's Union Iron and Steel Company failed with over $3 million in liabilities. This coupled with the extreme loss of profit from the 1882 strike and Ayer's personal spending of over $100,000 in 1881 left the future of Brown,

Bonnell & Co. in a state of uncertainty.[61] To make up his losses during the strike, Ayer sold iron for less money than it cost to make, but he later expressed a belief that running the mills even at half capacity during the summer of 1882 would have been a better choice than shutting down entirely, "even at sacrifice of principle."[62] His massive failure did not sit well with those in the Mahoning Valley. An observer from Youngstown exclaimed,

> There is not little excitement in the Mahoning Valley, not so much because of apprehended danger as from a dislike that the pride of the valley—the iron business—which has always stood so high in the commercial iron world, should be involved in the speculative difficulties of a Chicago concern.[63]

Luckily for Brown, Bonnell & Co.'s 2,000 workers, a number of Cleveland industrialists, including Amasa Stone and Charles A. Otis, were large shareholders in the company. Although Stone experienced other major financial setbacks during the spring of 1883 that resulted in him taking his own life in May of the same year, Otis, H. P. Eels, and W. H. Harris incorporated a new company, keeping the name Brown, Bonnell & Co. The company appointed Fayette Brown as receiver to pay off its liabilities of nearly $1.4 million over a planned four-year period.[64] Brown, a Cleveland banker associated with the Stewart Iron Co. in Sharon, Pennsylvania, and the Brown Hoisting and Conveying Machinery Company in Cleveland, agreed to sell the $2 million property containing Brown-Bonnell's mills, coal mines, coke ovens, and limestone quarries to its creditors once the debt was paid off. Brown's receivership took double the proposed time, and it would be eight years before the company was free from debt and reorganized as the Brown-Bonnell Iron Company in 1892.[65] Consequently, the company made sacrifices in production capacity, leasing their Anna furnace in Struthers to other Cleveland industrialists and largely sitting idle in terms of physical expansion and output growth. Technological development also stagnated. By 1890, Brown-Bonnell's Phoenix and Falcon blast furnaces were among the oldest in the state.[66] Furnaces of similar size and output, such as the Eagle furnace in Youngstown, were readily abandoned in the 1880s because of their inefficiency.

Five months after Ayer's failure, the collapse of the iron market had yet another devastating affect on Niles and James Ward's newly organized companies. Ward stood his ground for much of the boom period and did not expand his mills, operating at a profit and "turning out larger and more valuable products than ever before."[67] However, in March 1882, Ward agreed to construct a new rolling mill in New Philadelphia, Ohio, promising to employ 250 men as long as the town donated $40,000 and twenty acres of land for its development.[68] The $150,000 mill began operation in March 1883, just as demand for iron leveled off. The mill operated for less than four months before operations came to a roaring halt; an announcement in Warren's *Western Reserve Chronicle* read: "GONE UNDER! Serious Financial Trouble in Trumbull's Iron City."[69] Once again, Ward overextended his reach in the industry, almost mirroring his financial collapse in 1874. He failed to uphold his part of the contract with the town of New Philadelphia and did not receive his $40,000. Ward's other company, L. B. Ward Co., which operated the Russia mill, also filed for bankruptcy, amounting to liabilities of over $1 million, leaving any hope for industrial rebirth in Niles in shambles. Two massive financial embarrassments within ten years struck a massive blow to James Ward's ego. After the latest fiasco, Ward moved to Pittsburgh, where he lived out his days as attorney to the William H. Brown estate while completely disassociating himself from the iron industry (figure 5.10). By 1884, Niles's ironworks were entirely disconnected with the Ward name for the first time in the town's history.

Niles would not return to its old form. Another two-month-long strike in the summer of 1885 crippled the Valley's rolling mills, leaving many families of ironworkers without food. When the strike ended in August 1885, merchants and millworkers throughout the Valley actually celebrated with fireworks.[70] Three months after the end of the strike, however, Ward's former Russia mill, operated by Struthers iron manufacturer George Summers Sr. under a $3,000 loan from Niles's ironworkers and other capitalists, crumbled once again.[71] For much of 1885, Summers ran the Russia mill steadily, employed about 150 men, and, as was supposed, turned a profit. Although some difficulties arose between Summers and the wage scale presented by the Amalgamated Association, these complications were amicably settled until a disagreement arose between Summers and the mill's shearers, who were responsible for shearing the ends of muck bar for rerolling. Summers claimed that it was also the duty of the shearers to tie and bundle the iron for reheating, they refused, and Summers hired an extra man to do this work at one cent per bundle—one cent that was deducted from the pay of the shearers. The shearers called district Amalgamated Association vice president James Nutt to remedy the situation, and he gave Summers a week to pay the men "what was owed to them."[72] Summers did not yield, asserting that he would rather shut down the entire mill than give in to the Amalgamated Association.

After this disagreement with the Amalgamated, employees in the Russia mill detected something was amiss when no ore or pig iron were coming into the mill for the following week's production. In the week leading to Summers's financial disaster, he was seen giving checks to those who had loaned him the money to originally start the mill, but

FIGURE 5.10. James Ward Jr. (left) and his wife, Lizzie (second from left), visit friends in New Castle, Pennsylvania, in this photograph taken in the early 1900s. After his second major financial collapse in 1883, Ward gave up on the iron industry and spent much of his remaining years in Pittsburgh, where he died in 1919. Courtesy of Niles Historical Society.

when an employee asked Summers when they were to receive their wages, Summers answered, "It is hard to say; but do not expect pay today."[73] Summers's words spread like wildfire through the rest of the mill; the din of the machinery ceased, work stopped, and explanations were sought. Little was it known that Summers had just filed for bankruptcy and owed his employees between two and three months' worth of wages, about $18,000 altogether. Editors of Warren's *Western Reserve Chronicle* reported on the frightening scene that took place in Niles on the night of October 31:

> The men grew wild with excitement. Mr. S[ummers] was closeted in his office and the building surrounded with a determined crowd of men, women and children, who emphatically refused to let Mr. S. leave until some satisfactory arrangements could be had. They demanded money or blood; many threats of life were expressed. Some of the men had indulged to drink, and it is thought nothing would have saved the unfortunate man from death at the hands of these men.[74]

Women went to Summers's house and pleaded with him to share his Sunday dinner. One poorly clad woman with two small children, barefoot and holding on to their mother, forced her way into Summers's office and begged him to give her some of her husband's money to buy clothing and food for her family, who were cold and starving. This enraged the adjacent crowd and resulted in an angry mob forcing their way into Summers's office to try to get hold of him, breaking windows and office furniture in the process.[75] Russia mill employees kept a watchful eye on Summers's house, assuring he did not attempt an escape out of town, while others overthrew a car of pig iron consigned to the Russia mill from Andrews Bros. Co. Negotiations between Summers and the workers never transpired, however, and the Russia mill eventually fell into the hands of Myron Arms, becoming a part of the Falcon Iron and Nail Company, Niles's only stable iron firm.

The decade of the 1880s was a turbulent period for many of the Valley's leading iron concerns. Several other firms were caught in financial trouble, including one of Youngstown's leading merchant pig iron firms, Himrod Furnace Company. The New York–based company had long flirted with financial disaster, and finally filed for bankruptcy in August 1884. President R. A. Wight attempted to secure certain creditors over others by delivering several mortgage deeds upon other premises and the company's personal property to several corporations.[76] The company's three furnaces were leased to other successful Valley companies, but they ultimately deteriorated over the years, only to be dismantled in 1895. One of the Valley's largest rolling mill enterprises, Cartwright McCurdy & Co., also fell into bankruptcy in 1884 and was subsequently taken over and revitalized by a number of Youngstown financiers and industrialists, including members of the Wick family. Joseph Butler described the Wick name as being synonymous with "fiscal integrity and unusual ability, for high character, and for public spiritedness."[77] Despite being one of Youngstown's oldest and wealthiest families, the Wicks did not really begin to use their elite social and financial status to advance the region's iron industry until the late nineteenth century. In 1880, the Wicks purchased a controlling interest in the Girard rolling mill, which was put back into operation in 1878 (figure 5.11).[78] By 1883, along with the Girard mill, the Wick family held interests in the Falcon Iron and Nail Co. and the Youngstown Rolling Mill Co., both of which endured the depressed economic conditions of 1882 through 1885. However, the rolling mills would not be the focal point of the Wick family or most other local industrialists during the 1880s.

ADAPTING TO A NEW MARKET

Despite many of the Mahoning Valley's rolling mills being in financial limbo during the 1880s, the region's primary product was still pig iron, resulting in the Valley branding itself as a famous merchant furnace center. Wealthy Youngstown financiers invested in the rolling mills to limit

FIGURE 5.11. The Girard rolling mill, photographed here in the 1890s, was one of several iron mills in the Mahoning Valley that ended up under the Wick banking interests in the late nineteenth century. The Mahoning Valley Historical Society, Youngstown, Ohio.

financial collapses and impose their jurisdiction over the industry, but more they sought to revive the region's blast furnaces in order to control both the primary and secondary processes of iron manufacturing. Making pig iron was also a slightly more stable branch of the industry, with fewer labor troubles, and it usually promised a better monetary return. The Wicks were instrumental in this. The most ambitious member of the family was Henry Kirtland Wick (figure 5.12), one of ten children of Colonel Caleb B. Wick.[79] As part of the second generation of the family born in Youngstown, Henry Wick assisted in developing his family's banking interests during the Civil War and later became involved with coal shipping in Youngstown, Pittsburgh, and Buffalo, serving as president of the large coal firm of H. K. Wick & Co. He had several interests in coal and iron ore ports along Lake Erie and owned additional coal and timberlands in the south.[80] Wick invested in numerous Mahoning Valley industries and played a key role in merging the area's rolling mills in the 1890s. His brother, John C. Wick, also a member of the Wick Bros. banking firm in Youngstown, had a strong financial mind, was involved with coal mining and railroad interests, and organized the Wick National Bank in 1894.[81] He joined his brother Henry on many of his industrial ventures, primarily for fiscal reasons. Henry and John's cousin, Myron C. Wick (figure 5.13), was another one of Youngstown's leading financiers and business minds. Along with banking and industrial interests, he invested in real estate and gas and electric companies in Youngstown, as well as silver mines in Idaho.[82]

FIGURE 5.12. Henry K. Wick (1846–1915) was one of Youngstown's principal financial and industrial leaders in the late nineteenth century. His shrewd business techniques would give him firm control over the Valley's iron and steel industry in the 1890s. From *The Iron Trade Review* 44 (February 4, 1909): 266.

FIGURE 5.13. Henry Wick's cousin, Myron C. Wick (1848–1910), was also a banker and a visionary in the iron and steel industry. Often working with his cousin Henry, he aided in restructuring many of the Mahoning Valley's iron mills after economic depressions in the 1870s and 1880s. From *The Iron Trade Review* 44 (January 28, 1909): 228.

Henry, John, and Myron C. Wick rejuvenated the Valley's merchant pig iron industry by restarting operations at one of its old, well-established blast furnaces. Together, they formed one of northeast Ohio's most successful merchant iron firms: the Ohio Iron and Steel Company. The old Mahoning furnace in Lowellville, despite being rebuilt to modern standards in 1872, was an ambitious project. The Wick family had

FIGURE 5.14. Ohio Iron and Steel Co.'s Mary furnace, seen here c. 1909, was the oldest in the Mahoning Valley when the Wicks took over operations. Further modernization of the plant was inhibited by the hillside behind the furnace, once an attractive feature for its original builders in 1845, but a nuisance as the industry changed geographically and technologically. Author's collection.

little to no experience in operating a blast furnace plant. In addition, the village of Lowellville failed to attract industrial growth in the latter half of the nineteenth century because of its narrow topography and the furnace's location against the steep valley wall, preventing any plant expansion (figure 5.14). This resulted in the furnace being isolated from the widespread industrial development that took place in Youngstown. Other businessmen did not intend to invest in the furnace, which was "in good repair" but required an extensive overhaul after sitting idle for six years when its previous owners filed for bankruptcy in 1874.[83] However, Lowellville's most attractive feature to furnace operators was its limestone quarries, which still supplied most Mahoning Valley pig iron manufacturers. The Wicks owned vast amounts of these limestone deposits. Thus, they took a gamble on the furnace property; after evaluating its worth, the Wicks organized the first merchant pig iron company under their management. They clearly understood the financial risks involved. Incorporated in February 1880, the capital of the company was a mere $35,000, suggesting that the Wicks and other directors were initially skeptical about the furnace's potential to produce a profit.[84] The company's board of directors appointed as its president sixty-six-year-old iron and coal veteran Thomas H. Wells, who had worked alongside Governor David Tod in managing his coal mines in the late 1850s and 1860s. After a month of repairs, Ohio Iron and Steel Co. put their furnace into operation on April 21, 1880, shipping the first pig iron made to Henry Wick's Youngstown Rolling Mill Company.

Despite their inhibitions, the furnace was immediately profitable, if only because of the financial prowess and shrewd business ability of the Wick family. However, another member of the company, Robert Bentley, proved imperative to the company's success, becoming one of the finest young business managers in the Mahoning Valley. A Youngstown native, Bentley grew up under the guidance of the Wick family and began working for Youngstown's First National Bank when he was only nineteen years old.[85] Throughout his life, Bentley was an ambitious businessman, being involved in numerous enterprises, including the Carbon Limestone Co., Republic Rubber Co., Ohio Leather Co., Wick Oil & Gas Co., and Dollar Savings and Trust Company, among countless others.[86] The Ohio Iron and Steel Company was Bentley's first experience in managing a high-stakes iron firm, but he excelled from the start, leading it to become arguably one of the Valley's most celebrated and historic iron smelters.[87] Bentley had the honor of renaming the company's blast furnace, which was known by most in the Valley simply as the "Mahoning" since its construction in 1845, and he christened it Mary, in honor of his mother, Mary McCurdy. Despite being old and outdated, the furnace's product reached customers all around the country. In 1882, Ohio Iron and Steel shipped 54,820 tons of foundry iron to 286 customers, from the Atlantic to the Pacific coast.[88]

The Wicks were also innovative in their branding and marketing. Ohio Iron and Steel was the first merchant pig iron company in the Mahoning Valley and the state of Ohio to include the word *steel* in their corporate title, despite never actually manufacturing the metal. This was indicative of the growing popularity of steel in the late 1870s and early 1880s. It was a flashy term that some iron manufacturers began to use to entice potential clients, particularly as new Bessemer steelworks were built in major manufacturing centers in the Midwest. In 1884, Abner C. Harding, a mechanical engineer, wrote, "The manufacture of iron by puddling seems doomed; steel is taking its place rapidly," despite the United States having a record number of puddling furnaces the same year with 5,265.[89] Although there was still a demand for pig iron from Bessemer and open-hearth steel companies (and a shrinking demand from puddling mills outside the Mahoning Valley), many of those same steel companies began to reduce their reliance on buying pig iron off the market. A director of Youngstown's Brier Hill Iron and Coal Co. commented on the situation:

> Our market is mainly in Pittsburgh and the West. We send but little iron East, as we are met with competition of Eastern made irons. . . . Our business prior to 1883 was mainly in making Bessemer pig-iron for various

steel works west of the Alleghanies [*sic*]. This outlet has been lost to us by reason of the Bessemer works making their own pig metal and converting it in its molten state as it comes from the blast furnace.[90]

In Pittsburgh, Carnegie's Edgar Thomson Steel Works began integrating its Bessemer converters with blast furnaces after years of securing iron from outside firms, including the Struthers Iron Co., which provided the great steelworks with Bessemer iron in the late 1870s.[91] The E. T. Steel Works built five large stacks between 1879 and 1881 to make the company self-sufficient in its pig iron supply. The Lucy and Isabella furnaces continued their remarkable production rates, providing Pittsburgh's rolling mills and Bessemer steelworks with another 250,000 tons of pig iron per year. Steel companies in Cleveland began to follow Carnegie's methods of integration when the Cleveland Rolling Mill Company built another large furnace in 1879 and put the first of the two Central furnaces into production in 1882. The second, eighty-foot-tall Central furnace began making iron in 1887, giving Cleveland a collective annual pig iron capacity of about 245,000 tons that year.[92]

Valley pig iron manufacturers had to diversify their product and began to develop their own brands of pig iron to accommodate a niche market. Until more advanced methods of analyzing the chemical composition of pig iron came about in the late nineteenth century, most ironmasters spent little time evaluating their iron for various refining processes. In the late nineteenth and early twentieth centuries, metallurgists classified pig iron in two different grades: those that underwent transformation into other forms of ferrous products, such as wrought iron and steel, and those that were not "materially changed in composition or nature" and were used for foundries and castings (table 5.1).[93]

TABLE 5.1. Characteristics and Uses of Various Grades of Pig Iron

Grade of Pig Iron	Class	Percent of Silicon	Percentage of Phosphorous	Use
No. 1 foundry	2nd	< 0.035	0.5–1.0	Castings
No. 2 foundry	2nd	< 0.045	0.5–1.0	Castings
No. 3 foundry	2nd	< 0.055	0.5–1.0	Castings
Gray forge/Mill	1st	< 0.100	< 1.0	Wrought iron
Bessemer	1st	< 0.050	< 0.1	Bessemer steel
Low phosphorous	1st	< 0.030	< 0.03	Bessemer steel
Basic	1st	< 0.050	< 1.0	Open-hearth steel

Source: Robert Forsythe, *The Blast Furnace and the Manufacture of Pig Iron*, 2nd edition (New York: David Williams Company, 1909), 287.

After the Civil War, ironmasters experimented with several grades of iron ore, mixes of ores, and amounts of fuel to manufacture the distinct grades of iron required by their consumers. In the early 1860s, Jonathan Warner experimented with using local black-band iron ore in his Mineral Ridge furnaces without mixing them with Lake ores. Consequently, he was the first in the United States to successfully produce "American Scotch," a soft, fluid pig iron popular among foundries throughout the country.[94] Scotch iron made from similar black-band ore mined in Scotland and Wales was a major import into the United States for much of the nineteenth century, but many considered authentic American Scotch iron to be "better than the imported" type.[95] As more blast furnace plants developed their own chemical laboratories for analyzing pig iron in the late nineteenth century, the composition of their product became extremely important and often defined a company's brand. Before 1890, Lowellville's Ohio Iron and Steel Co. sent their pig iron to Youngstown for analysis. After establishing a laboratory on furnace grounds around 1891, the company reported, "Experience taught the greatest care in the manufacture of our product by the association of daily analysis. We are able to preserve uniformity that sustains its demand."[96] Ultimately, most pig iron manufacturers in the Valley and across the country adopted on-site chemical analysis of their product, thus meeting customers' needs as the industry became more diverse and competitive.

Certain types of iron ore prohibited ironmasters from making particular grades of pig iron. For example, Valley blast furnace managers could not produce Bessemer-grade pig iron with black-band ore because its phosphorous content was too high.[97] In addition, founders believed that they could not make soft castings without Scotch pig, and after the Civil War, the pig iron had a large market for casting stove plate and hollowware. Edward Kirk, a foundry owner who wrote several treatises on the quality and use of foundry irons, stated that he "melted many tons of these [Mahoning Valley Scotch] irons for light as well as heavy castings and found them to run as soft as any brand of Scotch pig . . . ever melted."[98] However, the production of Scotch iron in the Mahoning Valley after the Civil War largely gave way to the manufacture of mill and Bessemer iron for steelworks and Pittsburgh rolling mills.

As steel companies rebounded financially after the Panic and built their own blast furnaces, there was an increase in demand for Scotch and foundry iron for construction and fabrication purposes. Thus, some Valley iron manufacturers once again started to produce Scotch iron using only black-band ores along with their standard foundry and mill irons made with Lake ores. Black-band iron ore was still mined in significant enough quantities in Mineral Ridge and other areas in the Valley. In 1885, for example, Trumbull and Mahoning Counties produced 32,685 tons of the ore.[99] Using the black-band iron ore that remained, Valley merchant furnace companies advertised "American Scotch Foundry Iron" to cater to the thousands of foundries that sought pig iron to make castings in the early to mid-1880s. James M. Swank summarized this newfound consumer base:

> The explanation is found in a variety of concurrent influences. Undoubt-edly most importance should be attached to the fact . . . that we have an immense population, which must be supplied with all kinds of machinery, and with stoves, ranges, and heaters, domestic utensils, wagons and carriages, plows and other agricultural implements, mechanics' and labor-ers' tools, cast and wrought iron water and gas pipe, bridges for country roads, and many other articles, all of which require iron and steel. . . . Lastly, may be mentioned the increased demand in recent years for pig iron due to the construction of new iron and steel works, particularly in the Southern States. All these confluences have contributed toward the maintenance . . . of a large demand for pig iron, despite the falling off in the demand for steel rails and for some other finished products. Many of the articles are produced directly from pig iron in foundries.[100]

One of the largest foundries in the Youngstown area was William Tod & Co. In 1881, *American Machinist* reported the company making "rail presses, and a large amount of rolling mill and steel works' machin-ery."[101] William Browning Pollock, a Pittsburgh native who became a successful blast furnace operator in the Mahoning and Shenango Val-leys before the Civil War, established the Mahoning Boiler Works of William B. Pollock Co. in Youngstown in 1863. The company's small foundry and machine shop along Basin Street in Youngstown fabricated iron and steel castings for the construction of blast furnaces, receiving so much business that after the Panic ended, the company built a larger plant on South Market Street in 1881 (figure 5.15).[102] The Pollock Co. constructed thirteen blast furnaces around the country between 1866 and 1873. Foundries and machine shops such as these were major consum-ers of the product of local blast furnaces, particularly the famous Scotch iron. Writing in 1900, Edward Kirk stated,

> When the Scotch pig craze was at its height, a number of furnacemen in this country, with the usual American inventive genius, conceived the idea of imitating Scotch pig, and a number of brands of iron were made and put on the market . . . but as Scotch pig is not in demand to the extent it was some years ago, furnacemen have generally dropped the term American-Scotch, and the irons are known only by their local or furnace names.[103]

FIGURE 5.15. Foundry and machine shops provided the Mahoning Valley's merchant furnace companies with their primary market in the 1880s. This 1898 image, taken during the reconstruction of the Market Street bridge, shows two of Youngstown's largest such enterprises side by side; William Tod & Co. is on the left and William B. Pollock Co. is on the right. ©*The Vindicator.*

In 1898, there were 228 different brands of foundry and Scotch pig iron made by various furnace companies around the country.[104]

Lowellville's Ohio Iron and Steel Company was among the first furnace companies in the Mahoning Valley to remake their brand to meet demand. After rebuilding their furnace in 1883, the company began to advertise two different varieties of pig iron: the "Mary," made with Lake ores, and "Ohio Scotch," made with black-band. In 1893, an ad put out by the company advertising their brand of "No. 1 Mary Ohio Scotch" (figure 5.16) challenged other American Scotch pig iron manufacturers, as well as Scottish ironworks who exported their product to the United States: "Many inferior Irons are to-day being put on the market and called 'Ohio Scotch' Foundry, and in many cases have been sold to our customers with intent to deceive."[105] Ohio Iron and Steel claimed their product equaled the chemical composition of Scotch irons imported from ironworks in Langloan, Coltness, Glengarnock, and Carnbroe in Scotland, and they demanded that customers get an analysis of the iron if they doubted its authenticity. Indeed, many other merchant furnace

THOS. H. WELLS, President.
JOHN C. WICK, Vice-President.

F. H. WICK, Treasurer.
R. BENTLEY, Sec'y and Gen'l Mgr.

MARY FURNACE

THE

Ohio Iron & Steel Co.

LOWELLVILLE, OHIO,

MANUFACTURERS OF PIG IRON.

SPECIALTY:

AMERICAN SCOTCH FOUNDRY IRON.
BRAND, MARY OHIO SCOTCH.

OFFICE OF

THE OHIO IRON & STEEL CO.

Believing the trade will be interested in the great progress made in producing in the United States a Foundry Iron in every respect equal to the Imported Scotch, we give herewith comparative analyses of four well-known brands of Imported Scotch and our No. 1 Mary Ohio Scotch Foundry Iron. We challenge comparison of these analyses. Many inferior Irons are to-day being put on the market and called "Ohio Scotch" Foundry, and in many cases have been sold to our customers with intent to deceive.

Please ask for "Mary Ohio Scotch," and see that you get it; and demand an analysis with every order, if you are in doubt.

Respectfully,

THE OHIO IRON & STEEL CO.,

LOWELLVILLE, OHIO.

No. 1 MARY OHIO SCOTCH.

Metallic Iron	92.01
Silicon	3.15
Graphite	2.97
Combined Carbon	.25
Phosphorus	.425
Sulphur	.018
Manganese	1.20
	100.023

IMPORTED SCOTCH.

	Coltness.	Glengarnock.	Curnbroe.	Langloan.
Metallic Iron	91.34	91.800	90.65	92.177
Silicon	2.93	2 021	2.93	1.68
Graphite	3.14	2.147	2.90	2.99
Comb. Carbon	.40	.880	.76	.75
Phosphorus	.628	1.121	1.12	.642
Sulphur	.048	.037	.03	.021
Manganese	1.08	1.915	1.51	1.74
Copper			.04	
Titanium			.06	
	99.566	99.921	100.00	100.00

AGENTS:

Pickands, Brown & Co., Chicago, Ill.

Pickands, Mather & Co., Cleveland, Ohio.

N. S. Bartlett & Co., Boston and New York.

(106)

FIGURE 5.16. Ohio Iron and Steel Co. was one of the Valley's most vocal producers of American Scotch Iron. This 1893 ad challenges consumers to find a product equal to that made at Lowellville's Mary furnace. From Chas. W. Sisson, *The ABC of Iron* (Louisville: Press of the Courier-Journal Job Printing Co., 1893), 106.

companies in the Mahoning Valley began branding their product as Scotch iron. Like Ohio Iron and Steel Co., some firms produced genuine American Scotch iron from Mineral Ridge black-band ore, including the Thomas Furnace Company in Niles, the Haselton furnaces, and Brier Hill Iron and Coal Company's furnaces. Managers at Andrews & Hitchcock's Hubbard furnaces mixed three-quarters Trumbull County black-band ore and one-quarter Lake ores to produce "Hubbard Scotch," which sold in place of Scotch pig iron.[106] By 1884, only two furnace companies in the Mahoning Valley did not make some type of Scotch foundry iron, while, perhaps surprisingly, only one company produced Bessemer-grade iron. This was a significant shift from the 1870s, when five different Valley furnace companies made Bessemer iron. Such a shift in product was a clear indication of Valley pig iron manufacturers' intent to reinvent their industrial trademarks according to market trends.

The foundry pig iron craze was not limited to the Mahoning Valley, as many other merchant furnace companies in Ohio began producing only foundry iron in the early 1880s. There was a significant divide between iron companies needing mill pig iron for puddling furnaces and other firms requiring iron for making large castings. Like the Mahoning Valley, markets for pig iron used in puddling furnaces began to shrink in other Ohio districts. Counties along Lake Erie contained only nine rolling mills, several of which produced less than 5,000 tons annually. Central Ohio and the Hocking Valley, including five different counties, contained only seven rolling mills, mostly being smaller works. Along with the Mahoning Valley, ironworks located along the Ohio River consumed the highest amount of pig iron. However, each of the twenty-eight charcoal furnaces in the Hanging Rock Iron Region in southern Ohio made only foundry iron for railroad car wheels and other machinery. The fourteen furnaces in the same region using coke made both mill pig iron for rolling mills and foundry iron, while some such as the Milton Furnace and Coal Company in Wellston, Ohio, reportedly produced American Scotch from ore found in the Hanging Rock region.[107] The fourteen furnaces in Ohio's Hocking Valley also only produced variations of foundry pig iron. In western Pennsylvania, several Shenango Valley furnaces in New Castle, Sharon, and Sharpsville made Bessemer-grade iron, and two, the Douglas furnaces of Pierce, Kelley & Co., even made pig iron suitable for using in open-hearth steel works.[108] Perhaps surprisingly, the 1882 *Directory to the Iron and Steel Works of the United States* lists Youngstown's Brier Hill Iron and Coal Company as the only merchant pig iron company in Ohio to produce Bessemer-grade iron.

Throughout the 1870s, the Brier Hill Iron and Coal Co. operated on a limited basis, like most iron companies. In 1873, the company rebuilt one of their Grace furnaces, tore down the other with the intentions to rebuild, and banked the Brier Hill furnace, the company's oldest stack, by the end of the year. Brier Hill operated just one furnace for much of the depression, and, despite the other's foundation being finished and the site ready for reconstruction in the winter of 1874, it did not make any financial sense to spend huge amounts of capital on a new furnace when prices were low.[109] Coming out of the panic, Brier Hill Iron and Coal Co. had large stocks of unsold pig iron lying in the stockyards. "The grass had grown around it until some of it was nearly hidden," wrote Joseph Butler, who left the Girard Iron Co. in 1878 for the position of general manager at the Brier Hill Iron and Coal Co.'s furnaces.[110] Brier Hill president John Stambaugh actively sought out Joseph Butler because of his management skills at the Girard furnace. Butler had no money to invest in the company, but Stambaugh and his associates, George, Henry, and William Tod, agreed to pay Butler a larger salary than he received at the Girard Iron Co., along with stock in the company.[111] His first task with the company was to sell all of the pig iron in the yards. "It was our custom to run the furnaces continuously. When the price was low and the market inactive, we piled the iron, and sometimes had acres of it," Butler wrote.[112] Indeed, large stocks of unsold pig iron were common among most furnace companies during economic downturns. In 1876, stocks of unsold pig iron in the United States totaled 760,908 tons; in 1878, it decreased to 574,565 tons, and in 1879, it reached its lowest total since the early 1870s, at 141,674 tons.[113] When pig iron prices increased, Brier Hill Iron and Coal would market their unsold stocks at a "good profit," whereas other furnace companies ceased production (figure 5.17).[114]

Unlike most other established Mahoning Valley merchant iron companies, Brier Hill Iron and Coal asserted itself as one of the region's preeminent enterprises of its kind. Its directors and officers began investing in Lake Superior iron ore mines. Other than Jonathan Warner, few Valley iron manufacturers had invested in raw materials outside of the region, with most only engaging in the mining of coal in Trumbull and Mahoning Counties. The Himrod Furnace Company, whose manager A. B. Cornell invested in coke ovens in the Connellsville region with Brier Hill president John Stambaugh in the late 1870s, also helped organize the Himrod Hematite Company in 1872. The company leased eighty acres of land in the Lake Superior iron ore region, built docks, and shipped roughly 50,000 tons of iron ore a year to various enterprises, primarily the Himrod Furnace Company.[115] The economic panic prevented other Valley iron firms from purchasing their own iron ore lands until the late 1870s, when shareholders of the Brier Hill Iron and Coal Co. heavily invested in Lake ores, which accounted for one-half of the total ore mined in the United States by 1880, largely used by furnace companies in western Pennsylvania, Ohio, and Chicago.[116] Stambaugh, his son Henry H. Stambaugh, Butler, and Henry Tod formed the Brier

FIGURE 5.17.
Brier Hill Iron
and Coal Co. fur-
nace crew, c. 1885.
German and Irish
immigrants dom-
inated the com-
pany's workforce
in the late nine-
teenth century.
Courtesy of Tom
Molocea.

Hill Mining Co., opening the Brier Hill mine in Menominee, Michigan, on about eighty acres of land with a shaft 190 feet deep. Stambaugh, Butler, and H. H. Stambaugh also formed the Youngstown Iron Mining Company, buying a forty-seven-acre lot in Crystal Falls, Michigan, for $80,000.[117] The company's Youngstown mine consisted of an eighty-foot shaft on the west bank of the Paint River. Lake ores were not only an abundant resource but were generally lower in phosphorous content than other ores, making them much more desirable for Bessemer pig iron manufacturing, a staple of Brier Hill's product.[118] In the early 1880s, other Valley pig iron manufacturers, including Andrews & Hitchcock, sent workforces to the Lake Superior area with prospects of developing their own mining concerns.[119]

Along with being one of the Valley's first major investors in Lake ores after the panic, Brier Hill Iron and Coal was also among the best managed and most progressive iron firms in the state under Joseph Butler's supervision. After several years of successful management at the Brier Hill Iron and Coal Company, editors of *The Iron Age* recognized Butler as "one of the best blast furnace managers in the country."[120] His efforts to develop Brier Hill's product to meet market demands, and do so profitably, was rare among Valley blast furnace men, particularly in an unstable economy. Butler was a major proponent of bringing the manufacture of Spiegeleisen (German for "mirror-iron") to Brier Hill, a form of pig iron that contained from 10% to 30% of manganese orig-inally produced in Germany and Hungary by smelting manganiferous iron ore with charcoal. Spiegeleisen was a difficult form of pig iron to manufacture. Writing in 1875, George J. Snelus, a member of the Associate Royal School of Mines in Workington, England, stated, "Although spiegeleisen has been long known as a special product of certain iron-making districts, its real nature was for a long time but partially understood, and its manufacture, in England at least, was, to a great extent, a secret even so lately as two years ago."[121] Spiegeleisen was initially praised for its purity but later was used specifically in making Bessemer steel. Toward the end of the refining process in the Bessemer converter, a small amount of oxidized iron dissolved back into the steel, and did not turn into slag, as it did in the puddling process.[122] The manganese in the Spiegeleisen extracted the oxygen from the ferrous oxide, transferring it to the slag and leaving carbon behind, thus converting the iron into steel.

Making Spiegeleisen in the blast furnace required iron ore containing a large amount of manganese, which were primarily found in the Siegen District of Germany and in smaller quantities in Brendon Hill, England.[123] In the early 1880s, new sources of iron ore with high manganese content were found in Michigan, Virginia, Georgia, Alabama, and Arkansas, but ironworks in Europe produced the majority of Spiegeleisen and imported it to steelworks in the United States at a cheaper rate than could be produced domestically.[124] However, a select number of American blast furnace companies produced Spiegel iron in the late 1870s, including a few in New Jersey, Pennsylvania, Georgia, and Alabama. The New Jersey Zinc Company was among the first in the United States to produce the product in 1870, using two experimental twenty-foot by seven-foot blast furnaces; the Bethlehem Iron Company commenced production of the product in 1875 using iron ore imported from Spain; the New Jersey Iron Company made several thousand tons in 1877; and the Lucy furnaces in Pittsburgh also produced the iron.[125] However, there was a drastic contrast between the total production of Spiegeleisen made in the United States compared with the amount converted into Bessemer steel. In 1877, 8,845 tons of Spiegeleisen were produced domestically, while Bessemer steelworks in the United States converted 562,227 tons of Spiegeleisen the same year.[126] Despite this, production of domestic Spiegeleisen rose to over 33,500 tons in 1884, as the country experienced a "marked diminution in the use of foreign ore."[127] The demand for the product was there, and Butler and the Brier Hill Iron and Coal Company knew they could corner the market if they produced Spiegeleisen economically, as no other furnace companies in Ohio produced the iron.

To do this, the company built a small experimental blast furnace at their works in 1880 to smelt manganiferous ores. The furnace only measured forty-five feet by ten-and-a-half feet and sat next to the Brier

Hill furnace (the two-furnace complex was renamed the Tod furnaces). Brier Hill Iron and Coal's first contract for Spiegeleisen came from the Joliet Steel Company, one of the largest Bessemer steel companies in the Midwest, which did not begin making iron from its own blast furnaces until June 1880 and January 1882. Brier Hill purchased a cargo of high-manganese Spanish iron ore in order to manufacture the 2,000 tons of Spiegeleisen required by the steel company.[128] The furnace managers at Brier Hill did not have the knowledge to properly smelt these ores, and the company could not furnish the amount needed by Joliet Steel. To alleviate the situation, Butler called upon the aid of Julian Kennedy, former head blower at the Brier Hill furnaces and one of Pittsburgh's leading blast furnace engineers. Kennedy grew up in the Mahoning Valley and spent his early years overlooking the Mahoning furnace in Lowellville, watching and marveling at the furnace operations from the nearby hillside.[129] In 1879, Kennedy left his position at Brier Hill for the prestigious job of superintendent at Carnegie's Edgar Thomson blast furnace plant, in charge of constructing one of the country's most ambitious and advanced works of its kind. Butler and Kennedy had always been friends, and Butler knew that the Pittsburgh engineer "understood blast furnace practice thoroughly."[130] Kennedy's schedule did not allow him to travel to Youngstown, but instead he recommended his protégé and fellow Edgar Thomson engineer, Edward L. Ford (figure 5.18).

At only twenty-six years old, Ford was already well traveled and highly educated in the processes of making iron and steel. He was born in Troy, New York, in 1856 and nine years later witnessed the first successful commercial Bessemer steel produced in the United States in his hometown. After graduating from Yale University in 1876, Ford worked at the Bessemer steelworks in Troy for a brief period before accepting a position at the Cambria Iron Company's Bessemer plant.[131] He traveled extensively across the country in 1879, examining various iron- and steelworks, noting their efficiency and characteristics of the processes. In 1880, Ford traveled to Europe, where he devoted an entire year to studying steel in its various forms at works in England, France, and Germany before taking his position at the Edgar Thomson Steel Works. After Kennedy's recommendation, Ford left Pittsburgh for the Mahoning Valley, where the Brier Hill Iron and Coal Co. immediately placed him in charge of the small furnace producing Spiegeleisen to fill the order for the Joliet Steel Company. According to Butler, Ford "took hold and ran the furnace successfully, enabling us to complete our contract and preventing serious loss."[132]

Butler and other members of the Brier Hill Iron and Coal Co. had high hopes that Spiegeleisen would become a staple product of both their works and other manufactories consuming the metal:

FIGURE 5.18. A former engineer at Andrew Carnegie's Edgar Thomson Steel Works, Edward L. Ford (1856–1927) came to the Mahoning Valley in 1882. He perfected the process of "washing iron" at Brier Hill and helped organize the Valley's first open-hearth steel plant used for making specialty steel castings. From *The Iron Trade Review* 29 (January 2, 1896): 12.

It is the intention of our company to resume its manufacture. . . . If the metal is properly protected in ten years time every ton of spiegel used in the United States can be made here from American-mined manganiferous ores, and at prices which will in time be much cheaper to the consumers.[133]

This statement from Butler indicated the company's determination to grow their already specialized market thanks to Ford's achievements. If the market for Bessemer iron continued to decline, the company would have been forced to give into consumer demand, to which Butler stated, "To utilize our plant in making Bessemer metal would involve building steel works convenient to the blast furnaces. This we decided was unwise to do."[134] Although Brier Hill was exclusive in its Spiegeleisen production—being one of only five firms in the United States to produce the metal in 1882—the company jumped on the Scotch iron bandwagon like other Valley iron manufacturers and began producing "Brier Hill Scotch" in September 1882, selling their entire stock in less than a year. Brier Hill Scotch iron began to take hold in Eastern markets by the mid-1880s, where it had strict competition from anthracite pig iron manufacturers.[135]

After securing a larger market for the company, Ford would not be a temporary addition at Brier Hill Iron and Coal. Not only did he achieve success in producing Spiegeleisen, Ford became one of the Valley's leading civic and industrial leaders. He remained in Youngstown and bought a large house on Wick Avenue with his wife, Blanche Butler Ford, the eldest daughter of Joseph Butler, whom he married in 1887. During this period, he contributed a distinct innovation to the iron and steel industry and was instrumental in establishing the Youngstown Steel Company, one of the region's first new iron and steel mills built since large-scale industrial construction in the Valley ceased at the onset of the 1873 Panic. As its name implies, the company was the first in the Mahoning Valley to make steel by using the open-hearth process. Ford organized the firm in 1882 with John Stambaugh, backed financially by other Valley manufacturers, including Joseph H. Brown and William Tod & Co. They put the plant, described by the American Iron and Steel Association as "one of the largest concerns in the United States for making steel castings," into operation in March 1883.[136] As the mill's superintendent, Ford used his knowledge from studying steelmaking furnaces to make steel ingots, billets, and other castings using a single twenty-ton open-hearth furnace with an output of about 8,000 to 10,000 tons a year.[137]

Many firms in the United States built small, experimental open-hearth furnaces in the early 1880s due to the process's ability to make grades of high-quality steels used for numerous products. With few exceptions, established companies producing wrought iron, along with new steel castings enterprises, utilized open-hearth plants for small-scale production, usually containing only one or two furnaces ranging from five to ten tons.[138] Similarly, Youngstown Steel Company was not meant for large-scale production. It was, rather, an experimental mill that later gave way to a more lucrative product: "washed" iron. Ford's long-term research interests focused on developing methods to further refine pig and wrought iron and, particularly, creating a mechanical puddling process for making wrought iron without the traditional skilled labor of the puddler. While in Europe, he studied trial-and-error approaches of refining pig iron that never made a commercial impact because of flaws in their design. This resulted in Ford experimenting with the Krupp and Bell processes, invented by English ironmaster Sir Lowthian Bell and Friedrich Alfred Krupp, who was part of the Krupp Steel dynasty in Germany. In 1878, Bell found that running molten iron from the blast furnace into a bath of molten oxide removed about 95% of the silicon, 84% of the phosphorus, and only 11% of the carbon.[139] Krupp's process, which he patented shortly after Bell, differed slightly because it required a certain percentage of oxide of manganese. Combining these processes created washed metal with anywhere from 70% to 95% of the phosphorus removed to make a nearly "pure" cast iron, while retaining the carbon content traditionally found in pig iron.[140]

While experimenting with a variation of the Krupp and Bell processes, Ford found that making a commercially pure pig iron was more profitable than making open-hearth steel castings. There was an emerging market for washed metal, particularly as companies built open-hearth furnaces to make high-grade steel castings, boiler plates, and tool steel that required iron with low amounts of phosphorus. In addition,

Brier Hill Iron and Coal could not maintain profitable manufacture of Spiegeleisen using foreign ores due to cheaper European imports, and they attempted to make Spiegel using high-phosphorus Lake ores. According to Butler, Ford's process for dephosphorizing pig iron would consume "time and a large outlay of money," but if successful, it would open a large market for both Brier Hill Iron and Coal Co. and the Youngstown Steel Company, as no other manufacturers in the United States made washed metal.[141] Thus, in 1884, Ford installed his washed metal plant for dephosphorizing pig iron adjacent to Brier Hill's Tod furnaces, using pig iron from these stacks. The process was a success, making Youngstown Steel Company and Brier Hill Iron and Coal one of the most specialized pig iron duos in Ohio.

Although Ford could not profitably make Spiegeleisen on a consistent basis, his experiments in refined iron became an integral part of Butler and Stambaugh's ironworks at Brier Hill. In fact, his efforts were so successful that by 1885, Brier Hill Iron and Coal abandoned the small furnace built to produce Spiegeleisen exclusively, dismantling it in 1887, and the Youngstown Steel Company entirely discontinued making open-hearth steel castings and blooms by the early 1890s. Taking the place of these two products, Ford's washed metal plant manufactured forty-five to fifty tons a day of what Stambaugh, Ford, and Butler believed to be "the best steel material ever produced in the United States," as many large iron and steel companies preferred the metal for manufacturing high-powered steel guns.[142] As a result, the Mahoning Valley's marginal attempt at steel manufacture was short-lived, leaving Youngstown Steel Co. a misnomer, as it produced only washed iron and pig iron for outside consumption (figure 5.19).

As Brier Hill tried to vary their product and market largely through trial and error, they also began to drastically increase their capacity of pig iron through technological improvements, which reflected the progress of most other Valley pig iron manufacturers. By 1884, the company's capacity reached 100,000 tons per year, producing at least 75,000 tons of pig iron annually between 1881 and 1883 with four blast furnaces.[143] The company eclipsed all other merchant iron manufacturers in the Valley by about 40,000 tons, as most other furnace plants in the region consisted of one or two stacks, although many improved their production as smelting increased rapidly during this period. In 1884, Mahoning Valley blast furnaces produced 246,288 tons of pig iron with eighteen furnaces—an increase of nearly 110,000 tons since 1873, when the Valley peaked with twenty-one furnaces.[144] This incredible increase was a result of a number of factors; the most significant was that furnace managers transitioned to using coke as fuel. Connellsville coke steadily decreased in price by the early 1880s, dropping sixty-five cents in three years to

FIGURE 5.19. Brier Hill Iron and Coal sold their Tod furnaces to the Youngstown Steel Company in 1890, giving the latter a steady supply of pig iron for its washed metal plant. Shown here is the Tod furnace complex in 1895. The washed metal plant appears in the extreme right background of the image, partially obscured by the furnace's cast house. Its product was known internationally, as Youngstown Steel would ship thousands of tons of washed metal to England for armor plate manufacturing in the 1890s. From *The Iron Trade Review* 29 (January 2, 1896): 10.

$1.14 per ton by 1883, resulting in Mahoning Valley iron manufacturers preferring the fuel to raw coal.[145] An October 1883 letter from a prominent Valley blast furnace manager stated that "most of the furnaces of the Mahoning Valley use no block coal whatever, running entirely on Connellsville coke, and none use block coal entirely."[146] Those that did mix coal and coke typically increased the coke ratio as the block coal supply decreased in the late 1880s.

The result was larger, higher-capacity blast furnaces. In general, Mahoning Valley furnace companies lagged nearly a decade behind other major pig iron–producing centers in terms of technological advancements. Valley companies did not build new, larger furnaces like those in Cleveland and Pittsburgh during the 1870s. This was largely due to the unwillingness of Valley iron companies to invest in their plants because they relied on an unpredictable market and, in general, did not consume their own pig iron. However, in the early 1880s, most Valley furnace companies rebuilt their furnaces to a size ranging from seventy-five to eighty feet tall, finally equaling the furnaces in Pittsburgh and other prominent districts that used coke. According to the 1886 *Directory to the Iron and Steel Works of the United States*, eight of the eighteen blast furnaces in the Mahoning Valley were rebuilt or remodeled between 1879 and 1884. Some of the largest furnaces were Ohio Iron and Steel Co.'s Mary furnace in Lowellville, which was rebuilt in 1883 to a height of seventy-five feet; Brier Hill Iron and Coal Co.'s Grace furnace, rebuilt in 1882, which measured eighty feet tall; and C. H. Andrews's Haselton furnace, rebuilt in 1880, which stood seventy-five feet in height. Other stacks, including those in Hubbard and Girard, were seventy-five to eighty feet tall, while only eight measured less than seventy feet. Larger furnace construction coincided with many companies transitioning to using Connellsville coke and installing more powerful blowing engines, thus instituting hard-driving and pushing outputs to record numbers.[147] As discussed in chapter 4, managers at several Pittsburgh blast furnaces began implementing hard-driving at their stacks in the early 1870s. In 1874, the *Engineering and Mining Journal* reported that most of the country's top blast furnaces, none of which were in the Mahoning Valley, achieved blast pressures anywhere from three to seven pounds per square inch.[148] By 1880, most Valley furnaces used low blast pressures ranging from three to five pounds per square inch, as documented at the Mary furnace in Lowellville before it was rebuilt.[149] After remodeling their stacks in the early 1880s, Brier Hill Iron and Coal's Grace furnace and Andrews & Hitchcock's Hubbard furnaces achieved blast pressures of eight to nine pounds per square inch; managers of Pittsburgh's Lucy furnace attained this in 1872.[150]

Although many Valley furnaces were on par with the largest stacks in Pittsburgh and Cleveland by the mid-1880s, Youngstown iron manufacturers were still far behind in other industry innovations, namely application of the hot blast. All furnaces in the Mahoning Valley used inefficient pipe stoves heated by coal or excess blast furnace gas. By 1880, the "highest classes" of furnaces in England, Wales, Germany, Austria, France, Belgium, and the United States used the cylindrical firebrick stove patented by Englishmen Thomas Whitwell and Edward Cowper to preheat the hot-blast air.[151] Whitwell patented his stove in 1869 at the Thornby Iron Works in Stockton, England. He had long admired Andrew Carnegie's Lucy furnaces and made frequent visits to the plant in the 1870s, often discussing methods to make operations more efficient.[152] The long-term economic advantages of firebrick stoves were clear, and Carnegie had them installed at the furnaces of the Edgar Thomson Steel Works in 1879 and 1880, which decreased the company's fuel consumption and increased weekly output to 600 tons per furnace.[153] In 1882, Carnegie installed Whitwell's stoves at his Lucy furnaces, while managers of the Isabella furnaces and the Cleveland Rolling Mill Company followed suit. *Scientific American* reported that by July 1880, there were 130 such stoves built at various blast furnace plants in Europe and the United States, usually about three to a furnace.[154] Most Mahoning Valley furnace managers were hesitant to install these stoves because of

FIGURE 5.20. Andrews Bros. Co. invested in modern hot-blast technology at their Haselton furnace, one of the largest in the Mahoning Valley at the time. This 1889 image shows one of three cylindrical firebrick stoves partially hidden behind the hoisting house and to the right of the smokestack. The stoves and larger furnace replaced the second, smaller stack constructed in 1868. From *Youngstown Illustrated* (Chicago: H. R. Page & Co., 1889).

their initial construction costs. In 1880, Brown, Bonnell & Co. contemplated building firebrick stoves at the Phoenix and Falcon furnaces in Youngstown:

> These brick stoves or hot blasts are about sixteen feet in diameter and about sixty feet high. We shall need three at each furnace, the whole cost amounting to from $65,000 to $85,000 somewhere. With them the cost of making pig metal will be greatly reduced. It will take with them only 2,200 pounds of coke to make a ton of iron, whereas now it takes nearly two tons of coal and coke to make a ton. Besides this, we can make nearly twice as much iron—double the capacity of our furnaces. Where we now make fifty-five tons a day, we can then make 100.[155]

The company never installed the stoves, likely due to their high cost, along with the age and comparatively small size of the two furnaces, neither of which had been significantly rebuilt or modernized since after the Civil War ended. Nevertheless, in 1882, the company's Phoenix furnace made ninety-four tons of pig iron in a single day using 2,700 pounds of fuel consisting of a mix of three quarters coke and one quarter raw coal.[156] These numbers suggest that the company had the ability to achieve a 100-ton output per day without expensive firebrick stoves, despite using more fuel than originally desired.

TABLE 5.2. Pig Iron Production in Various Districts in Ohio and Western Pennsylvania, 1895

Region	Total Production (tons)
Ohio	
Mahoning Valley	620,526
Hocking Valley	40,700
Lake Counties (Cleveland district)	286,861
Hanging Rock Iron Region	140,769 (130,312 tons coke; 10,457 tons charcoal)
Western Pennsylvania	
Shenango Valley	820,037
Allegheny County (Pittsburgh district)	2,054,585
Western Pennsylvania (not including Allegheny County)	454,773

Source: *Statistics of the American and Foreign Iron Trades for 1895* (Philadelphia: The American Iron and Steel Association, 1896), 38.

Once again, it would be another ten years after Pittsburgh and Cleveland iron and steel manufacturers began installing firebrick stoves that Valley blast furnace companies would upgrade their hot-blast technology. Three companies, Brier Hill Iron and Coal, Andrews Bros. Co., and the Thomas Furnace Co. in Niles, installed firebrick stoves at their furnaces in 1889 (figure 5.20). In 1890, the three stoves at Brier Hill's Tod furnace (see figure 5.19), which cost about $17,500, reported "doing excellent work . . . making 185 gross tons per day on a very low consumption of coke," with hot-blast temperatures ranging between 1,300 and 1,500 degrees Fahrenheit.[157] In contrast, Brown, Bonnell & Co. recorded hot-blast temperatures of about 850 degrees Fahrenheit using old pipe stoves.[158] Most Valley companies waited until the mid-1890s to install these expensive firebrick stoves, just as the economy fell into the thick of another depression. Despite this, improved technology and better blast furnace practice resulted in Mahoning Valley furnaces producing a total of 620,526 tons of pig iron in 1895, which continued to lead the state by a significant margin; however, pig iron production in Pittsburgh was over three times that of the Valley (table 5.2).[159] Thus, as Mahoning Valley blast furnaces increased in size and capacity and iron- and steelworks in Pittsburgh rapidly became self-sufficient in their iron production, Youngstown's largest iron manufacturers considered a new market for their product when the Scotch and foundry pig iron craze and Pittsburgh's need for iron continuously declined. This market would come by way of locally produced steel, as the pressure to make the metal from consumers and other rival manufacturing centers became increasingly evident.

RESTRUCTURING WITH STEEL

The capital of steelworks in the United States and England generally dwarfed that of the largest iron companies in the Mahoning Valley. The major English steel companies had immense capitalizations of between $5 million and $12 million by 1885, whereas the average iron and steel firm in Pittsburgh was capitalized at $805,000 (this number was not representative of Carnegie's interests, which had a capital of $25 million in 1892).[160] In the Mahoning Valley, the average capital of all manufacturing concerns in 1888 was $270,000, with iron firms rounding out the highest capitalized enterprises.[161] The largest capitalized iron companies were Brown-Bonnell Iron Co., with $1.2 million, and Mahoning Valley Iron Company, at $1 million.[162] Other enterprises had half the capitalization. Andrews Bros. Co. and Andrews & Hitchcock each had a capital of $500,000, Brier Hill Iron and Coal Co. was capitalized at $432,000, and Cartwright, McCurdy & Co. incorporated in 1877 with a capital of $320,000.[163] Unsurprisingly, three of the top four capitalized companies in the Valley were integrated iron firms. However, just as the Panic of 1873 served to eliminate weaker iron companies, the decline of the iron market and the economic depression of the mid-1880s did the same to those struggling companies that survived the 1870s. As a result, Youngstown's social elite gained almost total control over the region's iron mills, having interests in eight of the Valley's ten rolling mill firms and five of the seven merchant furnace companies.

Despite their dominance, leading Youngstown iron companies faced increasing difficulties in the market. One growing handicap for mill owners was the cost of production, particularly for pig iron. The Mahoning Valley no longer had a local source of raw materials, excluding small amounts of black-band iron ore used by select merchant furnace companies. Blast furnaces were also landlocked and sat in between the iron ore docks of Cleveland and the Connellsville coking district south of Pittsburgh. Each required rail transportation to the Mahoning Valley, and this often put iron manufacturers at a disadvantage in terms of freight charges and higher overall production costs per ton of pig iron. In 1887, it cost Mahoning Valley furnaces an average of $4.47 to produce a ton of pig iron, compared to $3.07 for Cleveland furnaces and $3.06 for those in Pittsburgh.[164] In addition, Pittsburgh iron manufacturers fared much better in dull market conditions than those in the Mahoning Valley because of the former's higher volume of iron and steel consumers. To alleviate this, Valley pig iron firms could manipulate the railroads for a reduction in freight charges, find closer consumers of their product, reduce labor and increase hours, or enter into steelmaking.[165] Unlike steel companies in Pittsburgh, Mahoning Valley iron manufacturers did not have the power or profit margins to influence the mighty railroad companies. Wage reduction often proved counterproductive and only increased the threat of costly strikes.

Luckily for Valley iron manufacturers, Bessemer steel production continually grew, and with it came the need for pig iron. In 1884, the United States had twenty Bessemer plants with forty-five converters; in 1894, there were forty-three plants and ninety-five converters.[166] The year 1895 saw the United States produce 4,909,128 tons of Bessemer steel, the largest amount in the country's history.[167] One of the most significant new Bessemer plants for Mahoning Valley pig iron manufacturers was the Shenango Valley Steel Company, which in 1892 erected two eight-ton converters in New Castle with an annual steel billet capacity of 140,000 tons. Not only was it the first large-scale steel plant built in either the Mahoning or Shenango Valleys, but it also gave a new, relatively local market to merchant pig iron firms in the region. Thus, Mahoning Valley blast furnaces began to refocus their production on Bessemer iron, including Brier Hill Iron and Coal, which leased one of the defunct Himrod furnaces specifically for that purpose.[168] Accordingly, in the late 1880s and early 1890s, nearly a quarter of the total amount of pig iron produced in the Mahoning Valley was Bessemer grade. Open-hearth steel also began to gain more popularity for use in producing structural steel and other castings, as the country witnessed a 15% increase in its production from 1891 to 1892. Although the manufacture of rolled iron products such as sheets, plates, bar, hoop, and skelp still outnumbered their steel counterparts, the latter began to increase significantly in the late 1880s, while iron products experienced a much slower rate of growth.[169] It was clear that steel's economy and popularity began to push it to the forefront for use in all products, while iron had taken a back seat for the first time in the country's history.

Succumbing to this market demand, veterans of Mahoning Valley industry pushed for consolidations of the region's iron mills and, finally, for steel production. Some mill owners, however, had long anticipated the growth and popularity of steel. As early as 1884, the Wicks began purchasing Bessemer steel blooms from Pittsburgh to roll both steel and wrought iron hoops and bands at the Youngstown Rolling Mill Company, a practice that gained popularity among most other Wick-owned rolling mill firms in the late 1880s and early 1890s.[170] By 1892, nearly all rolling mills in the Valley reported producing both wrought iron and steel products. However, the process of shipping steel blooms from Pittsburgh to Youngstown became expensive, and some companies, such as

Myron C. Wick's Cartwright, McCurdy & Co., continued to invest in puddling and wrought iron production. Nevertheless, the product of Valley rolling mills began to change with the market. As consumption of structural iron and steel increased, most iron firms began to shift their product to specialty shapes, including angle splices, channels, I-beams, bridge rivets, shafting, and angle iron and steel. Proprietors of the Falcon Iron and Nail Co. in Niles successfully produced the Valley's first tin plate in May 1893 with the construction of the Falcon Tin Plate & Sheet Co. on land just north of the old Russia mill. Locals affectionately called the plant the McKinley Mill, named after Republican representative and Niles native William McKinley, who framed the Tariff Act of 1890 that raised duties on imports by 50% in order to protect domestic industry from foreign competition.[171] The tariff's protective tin plate clauses commenced the large-scale production of tin plate in the country after decades of relying on European imports.

In addition, the Old Ward Mill in Niles—one of the oldest in the region—found a market niche to compete with steel and stay relevant. The mill, taken over in 1887 by Cleveland industrialist J. Morgan Coleman and Girard blast furnace manager Henry B. Shields, largely followed the wrought iron pipe market cornered by Pittsburgh iron baron and Girard furnace owner A. M. Byers.[172] Under the influence of Byers, Shields manufactured tube iron and pipe casing at the Old Ward Mill, a product that Byers and many others believed was superior to the steel variety because of its better welding, threading, and resistance to vibration and corrosion.[173] Others in Pittsburgh, however, thought differently. In 1892, an officer for the Carnegie interests stated,

> Five or six years ago iron received its deathblow from Bessemer steel. Then a dozen Pittsburgh mills had to shut down. The natural gas boom which came then created such a demand for pipe iron that these mills again started to make iron for the manufacture of pipe. The pipe industry has gone down, however. These mills are trying to run in face of this fact and it is no wonder they are not able to keep going. Bessemer steel is the material of the future.[174]

Nevertheless, Shields and other Mahoning Valley iron manufacturers had opposing opinions on the pending transition from iron to steel in the Youngstown region.

Shields's position as an iron pipe and pig iron manufacturer perhaps skewed his attitude toward the steel question. He made his view on the situation clear in a March 1892 interview with Youngstown's *Vindicator*:

> While steel is taking the place of iron for many purposes, I do not think that the iron is to be entirely eliminated. The talk about all the puddlers

soon being out of a job, is idle. There were more puddlers at work the last year than ever before, and a great number of puddling furnaces were erected. It is also time that a great many mills have changed their puddling furnaces into steel converters, but all the puddlers will have to do is to learn to convert steel and their jobs remain. It is a question whether steel pipe will ever prove a success. Their hardness, which makes them so hard to cut, will prove an impediment to their use.[175]

Judging from his interview and background in iron production, Shields likely had little knowledge of how steel production affected ironworkers. The transition of puddlers from such a skilled position in the rolling mill to a largely semiskilled position in the steel mill was not the same class of work. Between 1892 (when Shields made his statement) and 1907, the number of puddling furnaces in the United States decreased by nearly 50%. In most cases, steel companies did away with such skilled positions, as the goal for large-scale steel manufacturers was to make the steelmaking process as mechanical as possible. For example, in 1910, of the 23,000 men employed by Carnegie Steel Co. in Allegheny County, 17% were skilled, 21% semiskilled, and 62% unskilled.[176] Mechanized steel production was the primary culprit. In contrast to Shields's statement, Mahoning Valley Iron Company president Henry O. Bonnell, a major advocate for a steel mill in Youngstown, believed that a Bessemer steelworks would increase production while drastically reducing skilled labor:

> With the most improved and best machinery and appliances, 1000 tons steel billets can be produced daily with the labor of not over 200 men. There are now in the Mahoning Valley 477 puddling furnaces (counting each double furnace as two.) These furnaces now employ 954 puddlers, 954 helpers, 236 muck roll hands; total, 2144, without counting standing turn men. These furnaces and men produce 1050 tons of muck iron daily, equal to the product of [the] steel plant proposed. On the basis of four persons being dependent on each adult workman, the puddling furnaces of the valley support directly a population of 10,720. . . . This being the case, the inevitable conclusion will be that unless the cost of producing puddled iron is reduced, steel will take the place of muck bar, and the puddler's occupation will be gone.[177]

Bonnell, like most of the Valley's leading iron manufacturers, was new to large-scale steel production. Despite the potential loss of skilled jobs in the puddling mills, he understood that changes needed to occur to keep Youngstown relevant in the overall progress of the iron and steel industry.

Along with Bonnell, the Wick family was among the largest advocates for a steel plant. By 1891, Henry Wick and his cousin Myron C.

Wick controlled almost half of the Valley's finished iron and steel production, while Henry also held interests in the massive Brown-Bonnell Iron Company. In the following year, the Wicks combined their rolling mill firms under the management of one immense company named Union Iron and Steel Co. The $1.5 million enterprise controlled four Valley rolling mills, including the Girard mill, the Warren mill, Cartwright, McCurdy & Co.'s mill (known as the "Lower Mill"), and the Youngstown Rolling Mill Company's mill (known as the "Upper Mill").[178] Together, the company had an annual capacity of nearly 140,000 tons, making it the largest in the Mahoning Valley.[179] Shortly after this consolidation, Henry Wick immediately began to heavily promote the idea of a steel plant to Valley iron manufacturers. This was not new, as many mill owners had mulled the idea for a number of years, but nothing had ever materialized. According to Joseph Butler, "The movement originated in a desire to avoid the necessity for shipping pig iron from Youngstown to Pittsburgh and again shipping the steel ingots made from it there over the same distance to Youngstown rolling mills."[180]

A steel plant was not a small endeavor. Wick and other Valley iron manufacturers understood the massive outlay of capital needed to construct a large and efficient Bessemer plant. George D. Wick, vice president of the Union Iron and Steel Co., was also in favor of a steel plant but was wary of the capital costs. He observed,

> With a capital stock of $500,000, we could erect a very desirable steel plant and leave us a working capital of $100,000. Such a plant is not only an imperative necessity by reason of the inroads that steel [is] making in the iron business, but would prove an excellent investment to all who put their capital into it. . . . The sooner we have a steel plant here the better.[181]

However, the interested parties did not have the financial backing of Brown-Bonnell Iron Co., whose Cleveland shareholders were content with buying steel from other sources. In addition, raising the capital stock proved difficult, as some iron manufacturers believed a steel plant would reduce demand for their product and throw those that did not produce steel out of business.[182]

Nevertheless, the Wicks' rolling mill companies were not the only enterprises interested in the proposed steel plant. In February 1893, Henry Wick estimated that the steelworks' capacity would be between 1,000 and 1,400 steel billets per day, which required the output of four to five blast furnaces "as large as the largest furnace now in operation in this city."[183] Consequently, merchant blast furnace companies became major contributors of capital to the steelworks. Joseph G. Butler and other furnace operators felt the new plant was a prime opportunity to

FIGURE 5.21. In New Castle, merchant pig iron firms and rolling mills faced a similar problem in the market as Mahoning Valley iron companies. As a result, two New Castle blast furnace firms—the Crawford Iron & Steel Co. and Raney & Berger Iron Co.—joined interests to construct a Bessemer steel plant adjacent to their furnace plants. This c. 1896 photo shows the Shenango Valley Steel Company's plant in the center background and Raney & Berger's blast furnaces to the right. From *Lake Erie and Ohio River Ship Canal* (Baltimore: The Friedenwald Company, 1897), 178.

establish a closer and cheaper consumer base for the region's merchant pig iron firms. This was the same ideology employed by merchant furnace companies in New Castle when organizing the Shenango Valley Steel Co. in 1891 (figure 5.21).[184] Thus, in July 1892, Youngstown's leading iron manufacturers contributed to the formation of the Ohio Steel Company, initially capitalized at $750,000 but later increased to $1 million in the spring of 1893.[185] H. O. Bonnell and the Mahoning Valley Iron Company gave $100,000, insuring Bonnell's position as vice president of the new steelworks until his death in January 1893, when Butler took the vice president position, with Henry Wick as president.[186] Youngstown's iron companies estimated that their rolling mills would roll about 600 to 700 tons of steel from the plant per day, or almost 75% of its total daily production.[187] Indeed, towns throughout the Mahoning Valley clamored for the opportunity to have the steel plant built within their municipality. The towns of Girard and Struthers offered free sites for the plant along the Mahoning River and near the Girard and Anna blast furnaces, respectively.[188] The city of Youngstown offered the company $25,000 and tax exemption. An unknown official from an unknown Mahoning Valley town with a population of 5,000—likely Warren or Niles—granted

$75,000 and 150 acres of land that included access to three different rail-roads.[189] Ultimately, Ohio Steel Co. decided to construct their massive Bessemer works on the Hawkins farm in Youngstown located on the south side of the Mahoning River opposite the Brier Hill Iron and Coal Company. Youngstown not only had the highest concentration of rolling mills in close proximity to the acquired farmland but also had immediate access to several large blast furnaces just across the river, whereas Niles and Warren together only had one furnace. The site also fit into the plan of Youngstown iron manufacturers to centralize iron and steel production.

In the fall of 1892, *The Vindicator* reported that Youngstown's steel plant would be the "third largest in the United States," with only the Edgar Thomson and South Chicago mills exceeding its "mammoth proportions."[190] *The Vindicator*'s words were promising but perhaps exaggerated statements; however, they exhibited the hope for a grand economic boost that steel manufacture could bring to the Mahoning Valley, which for too long had relied upon iron production. The Panic of 1893 delayed construction of the plant until 1894, as the economic depression threw 10,000 Valley ironworkers into unemployment in the fall of 1893.[191] Although the market for iron and steel remained dull, mill activity revived slightly in August 1894 after the House of Representatives passed the tariff bill the same month, despite iron and steel prices steadily falling until January 1, 1895. One major investor in Ohio Steel Co. sold his stock well before the mill was complete to "not be involved in the trouble" that he believed inevitably awaited the company.[192]

Ohio Steel Co. pressed on, making significant progress on their mill regardless of the economic circumstances. The company painted the entire plant, including the fences, a dark red color, which looked "no more attractive than most of such structures," according to editors of *Scientific American*.[193] The efficiency and technological ingenuity of the steel plant is what mattered most to Ohio Steel Co.'s investors. With the plant near completion in August 1894, industry observers believed that "the possibilities of the new enterprise suggest that Youngstown's future as an influential factor in the iron and steel world may fairly be expected to eclipse the record of her most prosperous days."[194] In other words, the opening of the steel plant invoked a sort of industrial renaissance for the Mahoning Valley. Ohio Steel Co. complemented the region's aging iron mills, but more importantly initiated a thriving merchant pig iron industry and long-term investments in big steel production after the turn of the twentieth century.

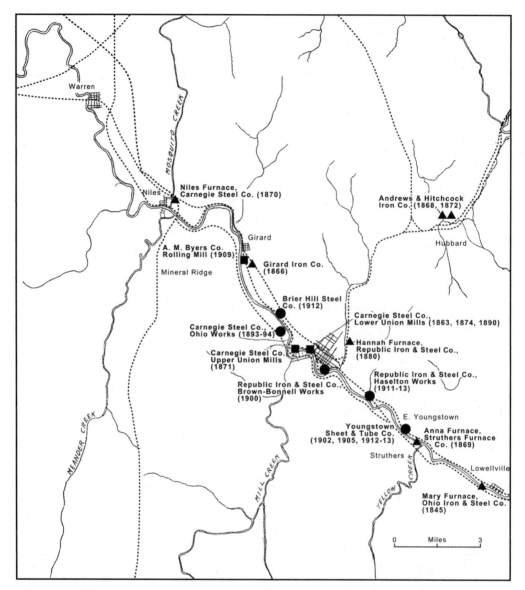

FIGURE 6.1. This map illustrates the Mahoning Valley's industrial transformation through the year 1913. Coal mining all but ceased and coal railroads (dashed lines) were removed. The coming of large-scale steel production changed the industrial structure of the region, as Bessemer and open-hearth steel plants (circles) overtook the region's iron mills.* Some merchant furnaces (triangles) remained, while steel production concentrated around Youngstown.

*Note: Squares represent rolling mills, though two of the three shown stopped producing wrought iron in favor of finished steel products. A. M. Byers Rolling Mill, however, produced wrought iron.

CHAPTER 6

STEEL, CONSOLIDATION, AND THE FALL OF IRON, 1894–1913

The function of the merchant blast furnace has been changing from period to period. . . . The time was when the making of pig iron was a trade, or an art, by itself. . . . In the days when the puddling furnace held chief sway there were many blast furnaces and many iron mills and the former sold to the latter. . . . As steel making developed and integration became the vogue, steel works built blast furnaces of their own, and in some cases even bought blast furnaces.
    —*The Iron Age* 105 (April 29, 1920): 1252–1253.

ON A FREEZING –4 DEGREE DAY on February 4, 1895, a large crowd of spectators gathered around the Bessemer converters of the Ohio Steel Company in Youngstown. The eager gathering impatiently waited for the first-ever blow of Bessemer steel in the history of the Mahoning Valley. However, just as the company experienced troubles in completing the plant during the 1893 Panic, there was a near disaster when making its first ten tons of steel. The intense cold broke one of the main steam pressure lines while the first heat of steel was being blown. The blower in charge of maintaining and turning the vessel nearly let the molten contents of the converter drop onto the damp, frozen ground because of the lack of pressure in the steam valve. The result would have been disastrous. The crowd and other nearby observers came within one bad decision on the blower's part of "being blown into eternity" by an explosion of molten iron and steel.[1] Luckily, he let the massive nine-and-a-half-foot diameter converter rest as it stood rather than trying to turn it downward. The crowd instead saw a magnificent show of pyrotechnics in one of the most remarkable and dramatic processes in manufacturing iron or steel:

The hiss and roar of the escaping gas is full of evil forebodings, happily unfulfilled, as though some bad spirit were escaping from imprisonment. At the beginning of the blast the flame is short and not highly luminous. . . . Soon the flame increases in size and brilliancy; it becomes deep yellow or orange, with bluish streaks, intermingled with sparks from the

FIGURE 6.2. Ohio Steel Company's Bessemer converters, 1895. The steelworks, co-owned by a number of Youngstown rolling mill and pig iron firms, relied upon other Valley merchant furnaces for its iron. From *Scientific American Supplement* 40, no. 1018 (July 6, 1895): 16,265.

metal. . . . Next comes the violent period, when the noise is deafening. The molten mass gurgles and bubbles and seems on the point of boiling over. The flame becomes dazzling in its intensity. . . . Then comes the last and peaceful period, when the flame becomes intensely hot and purple or violet in color. The noise becomes less and less, the flame grows dimmer, and then, a puff of brown, dusty smoke announces that the blast is finished.[2]

The entire process took from eight to ten minutes. At capacity production, the company's two ten-ton converters had the ability to make 125 to 150 heats per day, or nearly 2,500 tons of steel every twenty-four hours (figure 6.2).[3]

The new steel plant was a point of pride for all of those in the Mahoning Valley. With a workforce of about 2,000, a new worker community named Steelton materialized just south of the works, which saw 300 to 400 houses erected in a matter of months, just as Youngstown reached a population of nearly 35,000. The plant itself was truly a product of local ingenuity and capital. Much of the iron and steel, engines, and structural work used in building the steelworks came from Valley manufacturers. Youngstown Bridge Company constructed the cupola house, Youngstown founders and machinists Lloyd, Booth & Co. built the large blooming mill, and William Tod & Co. fabricated several massive engines for driving the mill.[4] In addition, the plan and construction of the entire plant was under the direction of Pittsburgh engineer and Mahoning Valley native Julian Kennedy. One of the best in the business, Kennedy in the early 1890s began a general consulting and contracting firm in Pittsburgh for designing furnaces and steelworks throughout the country. Kennedy's experience and his personal and professional relationships with Mahoning Valley iron manufacturers permitted the construction of Youngstown's first steelworks without "miscalculation" or "encountering of the unexpected."[5] The long-term success of the mill was in the hands of forty-seven-year-old Thomas McDonald, former manager of the Allegheny Bessemer Steel Company's Bessemer plant in Duquesne, Pennsylvania.

McDonald was no stranger to Bessemer steel mills. He began his prominent career working alongside Kennedy in the Bessemer department at Carnegie's Edgar Thomson Works. The company promoted him to turn foreman in 1880 and to night superintendent in 1884.[6] After McDonald had taken charge of the Duquesne Bessemer plant for a number of years, Henry Wick and other Ohio Steel Co. officers believed he had the knowledge to successfully manage their Youngstown steelworks. The company's original plans under Kennedy were for two eight-ton Bessemer converters rather than ten-ton vessels. McDonald advised against this, as many of the largest steel companies, including his former employers, Carnegie Steel, used converters that were ten to fifteen tons in capacity.[7] This resulted in Ohio Steel Co. increasing their machinery and overall size, which, in turn, raised the amount of capital needed. Despite this, McDonald's management was key to the steelworks' success, particularly in a dull market for iron and steel. After the company manufactured its first heat of steel in February 1895, the mill produced well under its rated capacity for almost two years. During the last week of January 1896, the company made a single day's output of 1,252 tons, the highest amount reached since operations began.[8] Production did not increase that year. By August 1896, the company's output for the entire month was only 3,373 tons. However, after July 1897, Ohio Steel Co.'s monthly output never fell below 30,000 tons.[9]

Although McDonald's original plan for the mill was to produce rail, the market for the product dwindled; however, there was still demand for sheet bars and billets. The price of steel rails in Pennsylvania dropped from $28 per ton in 1896 to $18.75 per ton in 1897. The price of steel bars and billets did not drop as dramatically, only decreasing by $3.75 during the same period.[10] The overall unremitting fall of iron and steel prices in the mid- to late 1890s was largely due to high domestic competition rather than foreign. Competition among American tin plate manufacturers lowered the price of the product to such an extreme that it was cheaper to import it.[11] Although financial failures among iron and steel companies in the Mahoning Valley were miniscule during this period, some of the largest such companies in the United States felt the effects of the national depression. Among those that filed for bankruptcy were the Maryland Steel Company, Philadelphia and Reading Coal and Iron Company, and the Pennsylvania Steel Company, to name only a few.[12] In 1898, a report from the American Iron and Steel Association read, "We have not built any new Bessemer steel-rail mills for a number of years because the fact has been made only too plain that we had enough mills of this character and more than enough."[13] The association also made a farsighted declaration regarding the country's iron and steel mills, saying that those that were "neither well situated nor equipped with the best appliances for cheap production, or which can make iron only and not steel, must give up the struggle for existence."[14] This statement defined the Mahoning Valley's old iron mills. Although the Ohio Steel Company achieved increasing success after a few years of operation, most iron mills in the Valley wavered and failed to streamline to any extent, as economic conditions halted most modernization.[15] As a result, the Valley's largest iron rolling mills were financially and technologically limited. Most Valley mill owners believed steel from the Ohio Steel Company was enough to keep their iron mills afloat in a highly competitive economy. However, this was only a temporary and short-sighted solution.

IRON MILLS IN THE STEEL ERA

On March 21, 1895, Joseph Butler appeared as the keynote speaker at the Montgomery Republican Club's banquet in Youngstown. He summarized the importance of the Mahoning Valley and its iron and steel industry in a toast partially reproduced by the *Iron Trade Review*:

> The industrial activity of the valley is represented by 12 blast furnaces, 12 rolling mills, one steel plant and about 100 other manufacturing establishments with invested capital of $20,000,000 with 15,000 employees. The present daily output of the furnaces is 2,500 tons, of the mills 2,000 tons and the steel plant 500 tons, which will shortly be doubled. . . . The Mahoning valley is also noted for the manufacture of products other than bar and pig iron, such as tubing made in large quantities at Warren and Youngstown, cold rolled shafting, railroad splice bars, coupling pins, cotton ties and horseshoe bar. It has been estimated that the horses of the United States carry around with them 100,000 tons of iron and steel in the shape of horseshoes. One-fourth of this is produced in the Mahoning

valley, and one-half of the cotton crop of the South has been baled for the past quarter of a century with cotton ties made by Mahoning valley manufacturers.[16]

In addition to these remarkable numbers, the Valley increased its iron and steel product and output from 2,465,400 tons in 1888 to 4,000,000 in 1892.[17] However, there began a transition in leadership among the Mahoning Valley's industrial elite. With the death of several of the region's most prominent iron manufacturers in the late 1880s and early 1890s, younger business partners and family members took over operations of the Valley's largest iron companies.

Despite this new leadership, these modern mill managers were cognizant of the fact that their iron plants needed substantial upgrades to compete with other rival companies and districts. Nevertheless, in most cases the Ohio Steel Company took precedence and only served to aid the outdated rolling mills in the Valley because of limited funds resulting from depressed market conditions. Although most rolling mills increased their annual output because of the rolling of steel billets made at the Ohio Steel Co., many mill owners retained their puddling capacity. After 1890, Valley iron companies did not build any new puddling mills, and most either maintained their already extensive puddling departments or removed some to make room for different types of mills to increase their finishing capacity. Thus, most companies focused on varying their products. Companies, particularly smaller iron and steel firms, chose this strategy because it isolated and protected them from the actions of competitors and helped fight the rampant fluctuation of the economy.[18] Branding products also helped build a good reputation with these companies' clients.

As mentioned in chapter 5, Valley iron companies began concentrating on manufacturing wrought iron and steel structural shapes for the construction, ship, and bridge-building industries. By the late 1890s, Mahoning Valley Iron Co. and its new president, George D. Wick, developed brands for their special types of products, just as many merchant pig iron companies did in the 1880s. The company used the brand "M. V. I." for their shafting, "I. X. L." for iron used in making horseshoes, "B. Q." for bridge iron, and "M. S. B." for staybolt iron, a bolt and short rod used to connect opposite iron or steel plates subjected to pressure, such as steam.[19]

Other Valley iron companies tried their hand at product differentiation and centralization by investing in cost-effective mills. Brown-Bonnell, for example, significantly reduced its puddling capacity for preference of rolling bigger classes of iron and steel. In April 1895, the company produced the largest iron bridge sections ever rolled in Youngstown, shipping orders for angle iron that measured seven-eighths

FIGURE 6.3. Youngstown's Lower Union Mill, built during the Civil War and shown here in 1895, became one of the few to undergo modernization in the 1890s. In the left foreground, houses along North West Street sit in the shadow of the old No. 1 mill, which stretches from left to right in the center of the image. The 1874 addition to the plant can be seen in the right background. From *The Iron Trade Review* 29 (January 2, 1896): 11.

of an inch thick by seventy feet long and weighed 2,600 pounds to Detroit for bridge construction.[20] Brown-Bonnell also abandoned its old cut nail mill in 1893—one of the oldest in the Mahoning Valley, dating back to 1846—to focus on more profitable products. However, after the construction in 1892 and 1893 of a new $100,000 bar mill, Brown-Bonnell made little technological progress and largely stagnated in production and output for the remainder of the decade.[21] The Wicks' Union Iron and Steel Co., on the other hand, differentiated their products at each of their four mills, with a focus on making smaller classes of iron and steel. The Girard rolling mill produced specialty iron for chains, bolts, nuts, and agricultural implements, and the company's Warren mill focused primarily on the manufacture of skelp iron for piping and tubing. In addition, Union Iron and Steel was one of the few rolling mill companies in the Valley to put money into their mills throughout the 1890s. Myron C. Wick primarily invested in newer, more efficient machinery for his company's larger Upper and Lower Union mills in Youngstown (figure 6.3).[22] These two mills served as the primary consumer of steel from Ohio Steel Co., leaving the Girard and Warren mills largely outdated.

The Wicks configured their Union Mills to roll larger amounts of iron and steel while at the same time decreasing the skilled labor needed in the traditional method of rolling. There, they introduced the first continuous and semicontinuous rolling mills ever constructed in the Mahoning Valley. Iron and steel manufacturers in the United States used continuous mills since the mid-1880s, but most were limited to rolling wire rod.[23] Contemporary observers of the industry believed that "old fashioned works may find it difficult to compete with" continuous mills.[24] As capacity increased in large-scale iron and steel mills, the old method of hand rolling metal became inadequate to meet demand. Indeed, several larger iron and steel companies added the new technology to their works in the 1890s. This included Jones & Laughlins (J&L) in Pittsburgh, who in the mid-1890s completely removed their puddling mills in favor of producing only open-hearth and Bessemer steel. As a result, J&L installed a continuous billet mill in 1894, while Myron C. Wick added a continuous mill to the Union Iron and Steel Co.'s works in Youngstown two years later.[25] Wick's mill was able to handle a full-length steel billet made at the Ohio Steel Company. The billet entered a thirty-foot continuous heating furnace and continued down a declined hearth, where an automatic push rod pushed it into the first of six sets of roughing rolls. The roughing rolls were set in a straight line and connected by gears so that each set of rolls had a fixed speed greater than that of the previous set of rolls.[26] The billet increased in length as it passed through each set of rolls without the aid of workers. Additional experimentation on the flying shear, which cut the hot billet or hoop as it passed at high speed through the finishing set, allowed the removal of hand shearing and the installation of an automatic device. The company installed another continuous mill in December 1897.[27] Like most new technology, those running these mills initially had difficulty and resorted to trial and error until reaching efficient operations:

> It took some time to get the tonnage up to a point where it exceeded the hand mills, but eventually the old time limit of three hundred and ten bundles of cotton tie per twelve hour turn was left far behind and the cost greatly lowered. The hoop trade was then able to secure coils of hoop containing hundreds of feet of continuous strands instead of the old time "scroll" bundles with strands of thirty or forty feet.[28]

To make room for their increased output of finished iron and steel, Union Iron and Steel removed the puddling furnaces from their No. 1 mill, converting it into a warehouse. In addition, the company dismantled its entire puddling department at the Upper Mill in 1898, converting it to an iron and steel finishing mill exclusively.[29]

Myron C. Wick was one of the few mill owners in the Mahoning Valley who had the foresight to adapt his mills to the rolling of steel. Despite many mill owners introducing product differentiation, most in the region did not have the same mind-set of steel over wrought iron, and some, like the Summers Iron Works in Struthers, remained small niche producers. Between 1895 and 1899, the owners of the small ironworks transitioned to an iron and steel finishing works without using the puddler's labor, thus lowering the overall costs of production.

As Valley rolling mill companies attempted to adjust to the new era of steel production in the region, local merchant furnace companies thrived. Ohio Steel Co.'s need for pig iron prompted many pig iron producers in the region to again make Bessemer pig iron in high quantities after a decade of producing mostly foundry and Scotch iron. In 1895, Butler stated, "The Mahoning valley has been noted for a quarter of a century for the special good quality of its Bessemer grade of pig iron, and the product has been shipped to all the prominent steel plants in the country. It will, however, be largely used at home."[30] Furnaces such as Mahoning Valley Iron Co.'s Hannah stack—for years an exclusive producer of pig iron for puddling furnaces—began producing Bessemer-grade pig iron almost entirely for Ohio Steel Co. This was true of nearly all blast furnaces in the Mahoning Valley, despite the price of Bessemer pig iron dropping nearly nine dollars per ton from 1890 to 1898.[31] Of the thirteen active blast furnaces in the Mahoning Valley in 1892, only five produced Bessemer iron, while the other primary product was mill iron for puddling furnaces and foundry iron. By 1898, ten of the eleven active blast furnaces in the Valley produced Bessemer iron, and some began to produce basic pig iron suitable for use in open-hearth steelmaking.[32] In addition, the output of Youngstown's steelworks grew enough to merit the use of additional pig iron from merchant furnace companies in the Shenango Valley.[33]

The Mahoning Valley's most prolific merchant furnace companies in Youngstown, Hubbard, Struthers, and Lowellville, expended from $100,000 to $500,000 in improvements to their plants in the 1890s.[34] This ensured that pig iron production reached high enough levels to fully supply the needs of the steel plant. As a result, many blast furnaces companies increased their annual capacity by twenty to forty thousand tons in this period. Modernizations such as new firebrick stoves and larger blowing engines at Andrews & Hitchcock's Hubbard furnaces increased the company's capacity from 73,000 to 130,000 tons of pig iron a year (figure 6.4).[35] In Lowellville, Ohio Iron and Steel Company manager Robert Bentley spent $40,000 remodeling the Mary furnace in 1894 and installed three firebrick stoves. This resulted in an increase in overall production by one-third, to nearly 250 tons of iron per day, while decreasing fuel consumption.[36] Four years later, the company embarked on an extensive overhaul of the entire furnace, spending over $150,000. Bentley's investments paid off. Ohio Iron and Steel deemed the 1899–1900 business year "the most prosperous of the entire 21 years' existence of the company."[37] In addition, the company praised Bentley for his "excellent executive ability" and management, which "earned money for both himself and the other stockholders."[38]

FIGURE 6.4. Andrews & Hitchcock Iron Company's two blast furnaces in Hubbard, seen here around 1905, remained one of the most successful merchant iron companies in the Mahoning Valley since their construction in 1868 and 1872. The company endured as a factor in the region's iron and steel industry until the First World War, and the plant itself stayed operational until 1960. This photograph dates from 1905. Courtesy of David Madeline.

Brown-Bonnell Iron Company, however, failed to modernize their furnaces, likely because they were not involved with Ohio Steel Co. Therefore, expending money on modernizing and enlarging their Phoenix and Falcon stacks for small-scale puddling mills proved redundant and unnecessary. These two furnaces were among the most advanced of their kind before and during the Civil War, but technological stagnation proved their downfall. In 1891, the company described the Phoenix and Falcon as "old-fashioned shells, very small at the top, very little room in the shell . . . anything but models, but they have done good work."[39] The company dismantled the smaller Falcon furnace in August 1893 with plans to construct one of the largest furnaces in the Mahoning Valley on its site.[40] However, the Panic of 1893 prevented any such construction. The Phoenix remained in operation, producing a modest 125 tons of pig iron per day. Other companies who modernized their furnaces for maximum output usually produced 250 to 300 tons daily, thus attracting outside investors who sought to take advantage of the Valley's renewed affinity for pig iron production.

In 1896, Cleveland industrialists purchased Struthers's Anna furnace and incorporated the Struthers Furnace Company, led by the management of Mahoning Valley native Samuel Allen Richards, son of prolific Youngstown ironmaster William Richards. Under Richards, the company transformed into one of the most progressive and successful merchant

FIGURE 6.5. Samuel Allen Richards (standing, top left) poses in front of Anna furnace with the Struthers Furnace Company's baseball team, c. 1905. Richards successfully managed the company until 1909, when he retired from active business. Courtesy of Shelley Richards.

pig iron firms in state (figure 6.5).[41] Richards instituted labor-saving technology at the Anna furnace with the installation of a mechanical pig-casting machine in 1898. Invented by Sloss Iron and Steel Co. blast furnace superintendent Edward A. Uehling in the early 1890s, the mechanical pig caster eliminated the harsh labor of carrying pig iron performed by the "lowest ranks of unskilled labor."[42] Uehling did not mechanize pig casting solely for profit or improved output but for moral reasons. He stated,

> The task of breaking and carrying out the iron from the casting beds of even a moderate sized furnace is not a fit one for human beings. If it were possible to employ horses, mules, or oxen to perform this work, the Society for the Prevention of Cruelty to Dumb Beasts would have interfered long ago, and rightfully so.[43]

Though Uehling's own company rejected his ingenious device, Andrew Carnegie purchased the patent for Uehling's machine in 1895 due to the large capacity of the Lucy furnaces in Pittsburgh.[44] The machine consisted of a series of cast iron molds that rotated on an endless chain and conveyor (figure 6.6). Molten iron was poured from a ladle into the molds and circulated on the conveyor while being cooled by

FIGURE 6.6. S. A. Richards installed only the second pig-casting machine in the country at the Anna furnace (the first was installed at Andrew Carnegie's Lucy furnaces in Pittsburgh). In this 1898 image, the casting machine extends from an iron shed near the furnace's casting house. From *American Manufacturer and Iron World* 63 (November 18, 1898): 727.

air and water sprays. Under Uehling's patent, a single-strand machine could handle 400 tons of iron in one day, while double-strand pig casters handled up to 2,400 tons in twenty-four hours.[45] The new device not only saved the excruciating labor involved with casting iron in traditional sand beds, but it allowed for the production of basic (or sand-free) pig iron used for basic open-hearth steelmaking. As open-hearth steel grew in output and practice in the late 1890s, the acid silica-based pig iron cast in sand on the furnace floor became unsuited for the basic open-hearth process. The acid base quickly ate away the open-hearth furnace's refractory brick lining; casting in "chills" on a pig-casting machine eliminated this silica base. Thus, Struthers's pig-casting machine—only the second installed at any blast furnace plant in the United States—not only allowed Richards to increase the Anna furnace's output but also gave him the ability to reduce labor and expand the company's product for another growing pig iron consumer.[46]

Large integrated steel companies in the country followed suit. Managers of the Carnegie Steel Company had pig-casting machines installed at each of the corporation's blast furnace plants in the late 1890s. The National Tube Works Company in McKeesport, Pennsylvania, installed pig casters at their Monongahela furnaces; the Tennessee Coal, Iron and Railroad Company added the device at their massive four-furnace plant

FIGURE 6.7. Brier Hill Iron and Coal Co.'s Grace furnace was one of the primary suppliers of Bessemer-grade pig iron to Ohio Steel Co. Providing iron for a large-scale steel plant prompted the company to greatly enlarge the Grace furnace's capacity in the 1890s, as seen in this c. 1899 photo. Courtesy of the Youngstown Historical Center of Industry and Labor (MSS 109).

in Ensley, Alabama; and Jones & Laughlins did the same with their four furnaces in Pittsburgh.[47] Other furnace companies in the Mahoning Valley were slower to adopt the technology, largely because their output was not yet high enough to merit mechanization of the casting process. Youngstown's Brier Hill Iron and Coal Co. was the second merchant furnace company in the Valley to install a $50,000 pig-casting machine in August 1900 to take care of the Grace furnace's 100,000 ton per year capacity (figure 6.7).[48] Remaining Valley merchant furnace companies would not adopt these machines for another ten to twenty years. The use of casting with pig-casting machines was necessary for increased productivity and could not be built on a small scale. Such devices were only profitable when the daily output of the furnace was high, usually 500 tons or more.[49]

Despite the significant investment in making Mahoning Valley blast furnaces efficient enough to supply Ohio Steel Co.'s Bessemer converters, officers of the company turned to the idea of integrated operations rather than open-market pig iron. Pittsburgh's Carnegie Steel Company largely pioneered the modern integrated steelworks that became the norm among major steel companies in Pittsburgh, Youngstown, Gary, Chicago, and Cleveland in the twentieth century. In 1896, the new four-blast furnace plant of Carnegie Steel's Duquesne Works caused a "veritable sensation" in the industry when put into production.[50] Each furnace stood at a towering 100 feet and were served by the industry's first fully automated raw materials handling system. Designed by plant engineer Marvin Neeland, such modern contrivances included an iron ore storage yard, stock bins for raw materials storage, and an automatic skip hoist bucket system used to convey raw materials to the top of the furnace, thus foregoing manual labor such as top fillers and bottom fillers.[51] Large-scale Bessemer and open-hearth steel companies needed such mechanization to increase their production. As a result, the Duquesne furnaces achieved world record production of over 600 tons of pig iron per day. Furnace operators around the country, including Henry Wick and the Ohio Steel Company, began to emulate this model.[52]

After the construction of the Ohio Steel Co.'s Bessemer facilities, the firm left enough space for a blast furnace plant, although other additions took precedence. As business for the steel plant grew, company officials found that a great deal of money could be saved with the addition of their own blast furnaces to directly supply molten iron, rather than remelting cold pig iron. Butler, Wick and other members of Ohio Steel Co. calculated that a two-furnace plant similar to Carnegie Steel's Duquesne Works would cost about $1.2 million and save the company seventy-five cents per ton of iron produced by eliminating the remelting process.[53] However, Ohio Steel Co.'s board of directors were not initially enthused by these numbers. They eventually approved the plan and issued $1 million in bonds for the project in August 1898 and March 1899.

Once finished, the Ohio Steel Company's furnaces were the largest in the world (figure 6.8). At 106 feet tall, they exceeded the height of the massive Duquesne furnaces and the two furnaces built in 1898 and 1899 by the Lorain Steel Company in Lorain, Ohio.[54] Duquesne engineer Marvin Neeland served as the project's chief engineer, responsible for the construction of the Mahoning Valley's first entirely new blast furnaces since before 1880. Neeland designed the furnaces to closely resemble the efficiency obtained by the Duquesne plant. Ohio Steel's furnaces utilized automatic charging equipment, or skip hoists; two steel ore bridges that spanned the iron ore stockyard; and a bin system for conveying raw materials to the skip hoists. Most importantly, Ohio Steel Co. could transfer molten iron directly from the blast furnace in ladle cars to the Bessemer converters without re-melting pig iron.[55] The two furnaces produced about 600 tons of iron per day, which, together, was still not enough to wholly supply the Bessemer converters. Despite this, these furnaces' daily output still doubled that of the largest merchant furnaces

FIGURE 6.8. Ohio Steel Company's two massive blast furnaces and eight firebrick stoves shown under construction in 1899. In December 1901, the No. 2 furnace (right) set a world record for pig iron production in a single month, with 19,645 tons produced, an average of 633 tons per day. Courtesy of the Youngstown Historical Center of Industry and Labor (MSS 109).

in the Mahoning Valley. In addition, no other blast furnace in the region utilized the modern, labor-saving equipment installed at Ohio Steel's plant. Editors of *The Iron Trade Review* stated that "the new furnace plant promises special economies in the handling of material" and "that the minimum had been reached in the number of men required for the laborious and diversified service called for in connection with blast furnace work."[56]

The new No. 1 furnace first produced iron in February 1900 and the second furnace did so in June, under the management of the newly formed National Steel Company. The $59 million company formed in February 1899 to control the supply of crude steel and tin plate bars, purchasing Ohio Steel Co.'s works along with several mills in the Mahoning Valley, Shenango Valley, and others in the "Central West." Directors of National Steel included two former Ohio Steel Co. officers, Henry Wick and William H. Baldwin. National Steel's principle organizers recognized Wick as "one of the best informed men in the manufacture of the heavier forms of iron and steel in the United States."[57] As a result, Wick took the position of first vice president, while William E. Reis, former president of the Shenango Valley Steel Company in New Castle, served as president of the concern. Along with Ohio Steel, National Steel absorbed the Shenango Valley Steel Co.; Buhl Steel Co. and Sharon Iron Co. in Sharon, Pennsylvania; King, Gilbert and Warner Company in Columbus; Bellaire Steel Company in Bellaire, Ohio; and Aetna-Standard Iron and Steel Company in Bridgeport, Ohio.[58] National Steel nonetheless believed that the Ohio Steel Company's Works—purchased for $3.4 million—was their "largest and most complete Bessemer plant."[59] In March 1899, National Steel also acquired the property and assets of the Thomas Furnace Company in Niles for $175,000.[60] Along with the blast furnace, Thomas Furnace Co. owned the Aetna Iron Company in Minnesota, which consisted of iron ore mines in the Mesabi Range, opened in the early 1890s.[61] Indeed, National Steel sought to become independent of all outside markets, such as pig iron, coke, iron ore, and limestone, thus combining its steel plants with sources of raw materials. The creation of National Steel was the result of a number of previous corporate consolidations and competing holding companies formed in 1898 and early 1899. Such combinations were formed to prevent competition among independent manufacturers, limit production and centralize modern mills, stabilize the industry in periods of economic turmoil, and, most importantly, maximize profits for shareholders. Ultimately, these mergers served to eliminate the Mahoning Valley's iron industry, ushering in the era of steel.

MERGERS IN THE VALLEY

Historian Naomi Lamoreaux aptly referred to this period in United States history as the "great merger movement in American business," which also serves as the title of her important 1985 book on the subject. This movement played an indispensable role in changing the industrial makeup of the Mahoning Valley. Unlike Pittsburgh, the Valley's largest iron and steel mills were nearly all individually owned, with the exception of the Union Iron and Steel Co. merger under the Wick family. In the 1880s and 1890s, Carnegie consolidated several of Allegheny County's largest iron and steel concerns, thereby removing his immediate competition. By the mid-1890s, the $25 million Carnegie Steel Co., Limited controlled five steelworks in the Pittsburgh area, including the Edgar Thomson Works, the Duquesne Works, the Homestead Steel Works, and the Upper and Lower Union Mills in Pittsburgh. The latter two mills, built during the Civil War, saw their puddling furnaces removed in late 1892 largely due to the fallout of the Homestead Steel Strike. Smaller iron and steel firms found it increasingly difficult to compete with such a massive company. Rockefeller's Standard Oil pioneered this method of monopolistic consolidation by acquiring nearly all of its competing companies in the 1880s.[62] The idea was to control and establish minimum prices by adopting the "dominant-firm strategy."[63] Hogan stated that "many small rolling mills and blast furnaces, which were managed and financed locally, soon found their resources and markets sharply restricted."[64] Indeed, the contemporary business mentality favored the corporate organization of iron and steel firms:

> As for the smaller industrial concerns, which may be forced out of existence, or to adapt their business to new conditions, their sacrifice is only a part of the price which must be paid for progress. They must go the way of the stage coach with the advent of the locomotive. . . . Whatever is for the greatest good must prevail in the end, and if a concern capitalized at millions can produce at a lower cost than one with only thousands invested, no amount of sympathy for the "little fellow" will save him from being crowded out.[65]

Nevertheless, Mahoning Valley iron manufacturers did attempt to adapt to the monopolistic conditions that prevailed in the late 1890s.

In late 1896, Brown-Bonnell Iron Co. president Samuel Mather proposed a merger of Youngstown's largest iron and steel companies, a plan that *The Vindicator* appropriately called a "Gigantic Deal."[66] By mid-January, 1897 stockholders of Brown-Bonnell Iron Co., Union Iron and Steel Co., and Mahoning Valley Iron Co. (the Ohio Steel Company also became involved later in the month) drew up a near-definite plan to consolidate these companies under single management with a capital of $5 million.[67] In response to the proposed deal, *Iron Age* stated that "the . . . four concerns, under one management, would be a power in the trade, and would be able to manufacture and market material at prices which under individual management, they could not hope to do."[68] If the deal materialized, stockholders stood to earn more profit than if the companies remained separately operated. In addition, the deal had the ability to make the projected company the largest iron and steel firm in the state of Ohio. "The consolidation of these concerns means that they will be able to not only handle a large amount of business, but be in a position to meet on equal terms any iron and steel corporation in the United States," boasted *The Vindicator*.[69]

By the beginning of February 1897, however, the push for these companies' consolidation slowed considerably. One of stockholders involved with the deal cited the Wicks' motivation of potentially controlling all of Youngstown's major iron and steel concerns as the primary culprit: "There is some feeling that the move to consolidate has been made for the purpose of enabling the Wick interest to secure a controlling interest and then run the three plants to suit themselves."[70] Because the Wicks controlled Union Iron and Steel Co. and Ohio Steel Co., a controlling interest in Mahoning Valley Iron Co., and a "block of stock" in Brown-Bonnell Iron Co., other stockholders believed they could "pool together and control everything."[71] Ultimately, the Cleveland stockholders of Brown-Bonnell opposed the merger, and the massive deal, as well as any other local consolidations, failed to materialize. The Mahoning Valley's iron companies thus remained independent. This weakened their position in the market, just as the country's leading financiers and industrialists sought widespread national consolidation.

In the late 1890s, over 1,800 companies dissolved into the mergers and consolidations that swept the country.[72] In the iron and steel industry alone, eleven large mergers occurred between 1898 and 1900 that saw nearly 200 previously independent companies consolidated with the likes of National Steel Co., Carnegie Steel Co., American Steel and Wire Co., and Federal Steel Co., among many others.[73] The primary architects of the merger movement in iron and steel were Andrew Carnegie, William H. Moore, and J. P. Morgan. Moore, a prominent New York attorney and financier, controlled National Steel and three other concerns that dissolved many of the Valley's independent iron mills; very few area companies fell under the Carnegie or Morgan interests, which primarily absorbed the large steel plants in Pittsburgh, Cleveland, and Chicago, as well as railroad, iron ore, coke, and bridge-building companies. Apart from National Steel, which produced 12% of the country's steel output, one of Moore's other concerns was the American Tin Plate Company. Formed in December 1898 with a capital of $50 million, American Tin Plate included nearly all of the tin plate manufacturers in the United States, with thirty-nine plants in Pennsylvania, Ohio, Indiana, Maryland, West Virginia, Illinois, and New York. By May 1899, editors of the *Iron Trade Review* reported that the company enjoyed "a complete monopoly" of the tin plate business in the United States.[74] The company consolidated the Falcon Tin Plate and Sheet Company in Niles, the only one of its kind in the Mahoning Valley. Moore also controlled the $52 million American Sheet Steel Company, formed in March 1900. This concern consolidated the highly competitive western sheet iron and steel industry. Among the company's 164 sheet mills were the Falcon Iron and Nail Company and Russia Sheet Iron Mills in Niles and the Summers Iron Works in Struthers.

The Moore concern with perhaps the most significant impact on the Valley was the American Steel Hoop Company, incorporated in April 1899. Although not as large as Moore's other concerns, the $33 million company consolidated rolling mills that produced hoops, cotton ties, billets, and bars. American Steel Hoop combined seven mills in Pennsylvania (including the three Isabella furnaces in Pittsburgh), seven mills in Ohio, and one in Georgia. The consolidations in Ohio included the Wicks' Union Iron and Steel Company, which National Steel originally purchased for $1.4 million, until the Moore interests transferred the mills to American Steel Hoop in April 1899.[75] Despite its large capital, American Steel Hoop did little in the way of modernizing Youngstown's Upper and Lower Union mills or the Girard and Warren mills. However, the Wicks' modernization of the Youngstown mills in the late 1890s proved valuable at the time of the buyout. These mills had the highest annual capacity of any of American Steel Hoop's mills in Ohio or Pennsylvania, with a combined production of 250,000 gross tons.[76] The company's Girard and Warren mills, which still heavily relied on wrought iron production, only had an annual capacity of 35,000 tons each. In addition, American Steel Hoop's rolling mills were dependent on the crude steel manufactured by Moore's National Steel Company. Thus, the symbiotic relationship between Youngstown's Bessemer steelworks (known as the Ohio Works) and the Upper and Lower Union mills remained just as it had under the Wick interests. In an October 1899 interview, American Steel Hoop president Charles S. Guthrie listed the overall advantages of such massive consolidations as being "so many I can hardly give them all."[77] "There is great economy in having all the management in one office," he said. "The economy goes through every department and every mill from one end to the other. Then we have the benefit of the good work of one mill to compare with that of another."[78] Guthrie also stated that the money saved from shipping finished products was a major benefit: "We ship from the mill nearest the consumer. For instance, we ship from Youngstown west to Cincinnati; from Pittsburgh and the Eastern mills we ship east or south."[79]

The economic advantages of these consolidations were clear. American Steel Hoop earned a profit of $720,000 in its first nine months, and by March 31, 1901, earned over $1.6 million. Likewise, American Tin Plate accumulated a surplus of $6.3 million in the same period.[80] Despite these earnings, the creation of Moore's mostly interdependent group of companies seldom added to the physical facilities for producing steel in the Mahoning Valley. Besides completing Ohio Steel's two blast furnaces (with a third built during 1900 and put into production in March 1901), one of the few but significant changes implemented by National Steel at their Ohio Works was the commencement of steel rail production. The mill's "first-class" rail equipment produced its first steel rail at 12:30 P.M. on May 14, 1900, a moment that many in the Valley had eagerly anticipated since the early 1870s.[81] It was the first steel rail ever produced in the Mahoning Valley. Operations at the Valley's largely outdated rolling mills in Girard, Warren, and Niles continued without any significant additions, but many workers under American Steel Hoop actually saw their wages increase between 15% and 25% by 1900.[82] In 1898, the United States finally emerged from the six-year economic depression, and iron and steel prices rose by the close of the year. In addition, the production of pig iron, Bessemer steel, and open-hearth steel in 1898 was the largest in the country's history to that point.[83] One of the biggest factors for this economic boom was the Spanish-American War. Swank noted that the war had a "stimulating effect upon the business activity of the country by creating a demand for supplies for the army and navy."[84] Shipments of Lake Superior iron ore and Connellsville coke, along with production of all leading articles of iron and steel in the United States, increased from 1897 to 1898, with the exception of iron and steel cut nails. In the same period, the Mahoning Valley increased its total rolled iron and steel production from 252,512 tons in 1897 to 358,943 tons in 1898.[85] This substantial increase in production and in iron and steel prices prompted even more large-scale mergers.

In May 1899, the Republic Iron and Steel Company formed independent of the Moore, Carnegie, and Morgan interests. Described as a "rolling mill trust," Republic fundamentally defined the direction of the Valley's remaining independent iron mills in the first decade of the twentieth century. The company focused its consolidations on the western and southern bar iron mills outside of the Pittsburgh district. With a capital of $55 million, Republic consolidated thirty-six rolling mills, six blast furnaces, mining properties on Lake Superior, coal lands in the Connellsville region, ore and coal land near Birmingham, Alabama, and interests in limestone properties in Lowellville, New Castle, and Birmingham.[86] In the South, Republic owned an open-hearth steel plant in Birmingham, the two Thomas blast furnaces, and the Birmingham Rolling Mills. The company's other acquired steelworks consisted of only open-hearth furnaces at its Duluth, Minnesota, plant. In the immediate Midwest, Republic owned a number of rolling mills in Illinois, Indiana, and Ohio. Its Ohio mills were concentrated in Leetonia and Youngstown, the latter the site of Republic's corporate headquarters. In the Mahoning Valley, the company essentially absorbed the remaining independent mills that Moore's concerns overlooked or were unable to secure. Brown-Bonnell Iron Co., Andrews Bros. Co., and Mahoning Valley Iron Co. each sold their mills and assets to the new trust (figure 6.9). Republic also acquired these companies' blast furnaces, as well as two small furnaces in Sharon and New Castle, Pennsylvania.

Several former officers of companies taken by the Moore interests joined with Republic. Former King, Gilbert and Warner Co. president Randolph S. Warner helped organize Republic after National Steel absorbed his Columbus-based steel company. Warner served as Republic's first president. Myron C. Wick signed on with Republic as chairman after National Steel and American Steel Hoop absorbed Youngstown's Union Iron and Steel and Ohio Steel Company. George D. Wick agreed

FIGURE 6.9. The smoke-ridden iron mills of Brown-Bonnell Iron Co. in Youngstown, as seen looking northeast across the Market Street viaduct in the mid-1890s. Westward winds often blew thick smoke from the mill's No. 4 puddling plant (left) across the bridge, enveloping those trying to cross. The immense plant was the most important of Republic Iron and Steel's acquisitions. Library of Congress, Prints & Photographs Division, Detroit Publishing Company Collection (LC-DIG-det-4a28742).

to serve as first vice president after Republic purchased the Mahoning Valley Iron Company in which he served as president. Several other Valley iron manufacturers also held positions with Republic. Joseph Butler took charge of the department of distribution, while James A. Campbell became district manager.[87] Despite the massive size of the company, Warner, the Wicks, and other directors had great concern over Republic's ability to produce steel over wrought iron. In January and February 1899, the mills acquired by Republic produced 96,120 tons of wrought iron products. During the same period, these mills combined to produce only 21,221 tons of steel bars and shapes.[88] Modernization of Republic's mills to compete with the other major consolidations required millions of dollars of investment. Fortunately, Republic recognized the profitable mills from the costly and obsolete plants. The latter type consisted primarily of puddling and bar iron mills in outlying regions detached from Republic's planned centralization of operations in Youngstown. By 1901, the company dismantled eight plants in Alabama, Indiana, and Illinois. The towns that lost their ironworks—often considered the financial backbone of a small community—were not so thrilled. In one instance, such economizing by Republic was, as one newspaper put it, "the story of another prosperous Ohio town absolutely ruined by the trusts."[89] However, in reference to Republic's mills in the Mahoning Valley, one stockholder promised the company would modernize: "You can say that the mills here [Youngstown] will not only be kept in steady operation but that many important improvements will be made. Further than that a large steel plant will be erected to supply the mills with steel billets."[90]

This promise was only true for one of the company's Youngstown mills. In late June 1899, Warner, Myron C. Wick, and James Campbell inspected all of Republic's mills in Youngstown to determine the repairs and investments needed. Before leaving for a Chicago business trip, Campbell authorized the company to proceed with upgrading the Brown-Bonnell mill, in which Republic invested the "bulk of the money."[91] Amidst these planned investments, Republic decided to abandon the mill's old Phoenix furnace, as "it was the opinion of the officials that it was not worth the money it would cost to refit it, and that the money could be invested to better advantage elsewhere."[92] Republic scrapped the historic furnace in the fall of 1899.[93]

The demolition of the Phoenix stack symbolized the company's expanding agenda for steel production. Unlike other major steel combines running the Mahoning Valley's mills, Republic spent a significant amount of money updating and adding to their mills. Between May 1899 and December 1901, the company invested just over $1.9 million of its income modernizing their mills.[94] This was much more of necessity for Republic than for National Steel or any other of Moore's concerns, as little modernization occurred at Republic's three Youngstown plants in the 1890s and none had steel production. Thus, Republic spent much of this money in building Youngstown's second Bessemer steel plant on the site of Brown-Bonnell's former Phoenix and Falcon furnaces (figure 6.10). On September 27, 1900, hundreds of people gathered around Republic's new Bessemer plant of two five-ton vessels.[95] The crowd "was perched on the elevated pulpits and platforms and every other place of vantage" to watch the converters in action. After the first blow of Bessemer steel at 8:02 A.M., the crowd celebrated the event, described as "worthy of commemoration in the town's history of industrials."[96] As evidence of Republic's intention to centralize operations in Youngstown, the company removed the Union Steel Company's rolling mill and Bessemer converters in Alexandria, Indiana, to Youngstown as part of the latter's Bessemer plant. The new Youngstown plant also consisted of one of the Bessemer converters and other equipment from the Springfield Iron Company in Springfield, Illinois.[97] Republic invested money into repairing and upgrading all of their blast furnaces in Sharon, New Castle, and Youngstown to render the company as independent as possible from buying pig iron off the market (figure 6.11). Still, large

FIGURE 6.10. Republic Iron and Steel officially entered the steel industry in the Youngstown district with the installation of two five-ton Bessemer converters at the Brown-Bonnell Works in 1900. This photo shows the converters shortly before operations began. Courtesy of the Youngstown Historical Center of Industry and Labor (MSS 12).

steel companies aimed to have their blast furnace capacity somewhat less than their steel-producing capacity. If steel production was pushed to its limit, merchant furnaces provided the remaining pig iron. In 1901, for example, the combined capacity of Republic's Youngstown area furnaces was 330,000 tons a year, while the company's Bessemer converters had an annual capacity of 350,000 tons of steel ingots.[98]

Republic only just began to modernize its Brown-Bonnell plant despite James Campbell and George D. Wick resigning from their positions at the company in the fall of 1900. Both men believed the Mahoning Valley's mills should stay locally owned and operated (figures 6.12 and 6.13). In addition, Campbell, described as a "stubborn, persistent man," may have been disappointed with Republic's move to Bessemer steel production, as he believed the lasting qualities of wrought iron to be greater than that of steel.[99] Thus, perhaps Republic's biggest advocate for the large-scale continuation of wrought iron production left to organize his own company with Wick, called the Youngstown Iron Sheet & Tube Company. As an indirect result, other Youngstown mills under Republic received less investment, much to the chagrin of the company's skilled puddlers and rollers. Even after the commencement of the

FIGURE 6.11. The Hannah furnace on Youngstown's east side, once a part of the Mahoning Valley Iron Company and located between present-day Valley Street and Lansing Ave., was one of Republic's several detached blast furnace plants. The company initially relied on isolated furnace plants such as the Hannah to supply its Bessemer converters and later on an open-hearth plant. The Mahoning Valley Historical Society, Youngstown, Ohio.

Bessemer plant, the small Haselton plant—formerly owned by Andrews Bros. Co.—remained shut down. Republic intended to use the plant as a "merchant mill for the fulfillmen[t] of small orders."[100] Company officials tried to find the workers of the Haselton mill other positions at the Brown-Bonnell or Mahoning Valley Works, both of which ran "in all the departments which are in shape to be operated."[101] In January 1902, editors of the *Iron Trade Review* reported of the Haselton mill that "it is an old plant and may possibly never be operated again."[102] However, a year and a half after Republic's formation, the company finally placed the Haselton mill into operation, though it removed twelve puddling furnaces in the process. After a significant decline in rolled iron production in 1900 and 1902, an increase in prices from $1.84 per hundred pounds of rolled bar iron to $2.13 in the same period prompted Republic officials to start operations at all of their Youngstown bar iron mills.[103] Regardless, the company reduced the majority of its iron production to the smaller Haselton and Mahoning Valley Works.

Republic continued its extensive modernizations to its Brown-Bonnell plant. In 1902, the company rebuilt the Bessemer plant and enlarged the converters from five to ten tons each, increasing annual

FIGURES 6.12 AND 6.13. George D. Wick (1854–1912), left, and James A. Campbell (1854–1933), right, were among the most notable local industrialists that emerged from the era of consolidations. Their expertise in managing companies such as Union Iron and Steel and Mahoning Valley Iron Co. helped create Republic Iron and Steel Co. Unhappy with their roles in the trust, the two men established one of Youngstown's most iconic iron (and later steel) companies: Youngstown Sheet & Tube. Courtesy of the Youngstown Historical Center of Industry and Labor (MSS 140 and MSS 134).

production of steel billets from 350,000 tons to 500,000 tons.[104] To handle this increase in steel production, Republic implemented the labor-saving technology of the continuous mill. Such modernization reduced skilled jobs throughout the mill but allowed the company to stay ahead of their competitors. It also left sixty men unemployed.[105] *The Vindicator* reported: "Men who have been employed at the Brown-Bonnell mill for many years are included in the sixty who will be dropped and they are now looking for employment, having been given an intimation that they will not be needed any more."[106] By October, the company reduced their skilled workforce even more with the installation of a new billet mill that included a twenty-six-inch semicontinuous and an eighteen-inch continuous mill.[107] In addition, between 1902 and 1904, the company drastically reduced the Brown-Bonnell plant's puddling department from eighty-two to thirty-eight furnaces. On March 29, 1902, the company's old No. 1 puddle mill—the longtime nucleus of the plant since its construction in 1846—was finally abandoned. The employees of the mill saw this as both a melancholy and a significant moment in the industrial history of the Mahoning Valley. After workers formed the last piece of iron, they presented their long-respected puddle boss, John D. Hodge, with a "costly" cigar holder and a "box of fine cigars."[108] Republic would

FIGURE 6.14. The once great iron mill of Brown-Bonnell transformed into a steel finishing plant under Republic Iron and Steel in only five years. In this 1905 photo, only the No. 4 puddling mill (far left) remained from the major additions implemented to the plant by Brown, Bonnell & Co. between 1879 and 1881. The Mahoning Valley Historical Society, Youngstown, Ohio.

build a "decidedly modern and up to date" continuous finishing mill on the grounds of what observers at the time called the "oldest [mill] west of the Allegheny mountains."[109]

By the end of 1902, Republic Iron and Steel had all but eliminated iron production at its Brown-Bonnell mill, transforming it into a first-rate steelmaking and steel finishing plant. The company even constructed "a convertible sheet bar and rail mill," rolling its first Bessemer steel rail on April 22, 1905.[110] Soon enough, the name "Brown-Bonnell" was the only similarity between the once massive nineteenth-century iron mill and Republic's hastily streamlined plant. After 1902, only one old puddling mill remained in the complex (figure 6.14). Republic made moderate repairs to its old Mahoning Valley Works, leaving the plant solely to produce bar iron, which remained a strong product on the market. In January 1902, Republic reported that its entire output of bar iron was sold for the next six months.[111] Ultimately, the company agreed to keep the old plant in production for these purposes, leaving puddlers and other skilled ironworkers with a temporary safe haven of employment. In fact, Republic's Mahoning Valley Works remained one of the country's few producers of iron and steel cut nails. In 1907, the mill was one of only fourteen in the United States that still produced the product, as all other cut nail production in the Valley ceased by 1895. However, cut nail manufacture in 1909 only amounted to 1.2 million kegs, whereas wire nail producers turned out nearly 14 million kegs the same year.[112] Expensive and outdated production of cut nails resulted in Republic dismantling its fifty-five-machine nail mill in 1908. Despite leaving its other Youngstown plants virtually "as is," Republic continued investment in other important districts. Aside from its Brown-Bonnell mill, the company placed a considerable amount of money in its Birmingham plants, adding a new blast furnace in 1902 and greatly expanding its coking plant the same year. By June 30, 1902, Republic spent over $4.3 million in new construction at its most important plants in the South and Midwest.[113]

Republic's primary competitor, particularly in the Mahoning Valley, arrived with the creation of the United States Steel Corporation in 1901. Hogan described the organization of the company as "perhaps the most daring venture in corporate history up to that time. Never before had so many large companies been drawn into one unified operation."[114] The world's first billion-dollar corporation merged all of the combines formed under Carnegie, Moore, and Morgan into one massive trust. Spearheaded by J. P. Morgan and Judge Elbert H. Gary, U.S. Steel absorbed National Steel, National Tube, American Steel Hoop, American Tin Plate, American Sheet Steel, Carnegie Steel, American Steel and Wire, and Federal Steel for $492 million.[115] As of April 1, 1901, U.S. Steel owned and operated seventy-seven blast furnaces, 110 open-hearth furnaces, and thirty-seven Bessemer converters.[116] In 1902, the corporation produced 44.8% of the country's pig iron, 65.7% of its Bessemer and open-hearth steel, and 65% of its Bessemer steel rails.[117] U.S. Steel controlled all of the Mahoning Valley's mills formerly under the Moore syndicate and, like Republic, dismantled unprofitable plants, constructed new mills, and continued adding other companies to their already extensive list of properties.

In the Mahoning Valley, U.S. Steel invested in and centralized operations to their Youngstown plants. The corporation increased the Ohio Works' annual capacity by 50,000 tons in two years and underwent sweeping modernization at the Union Mills in Youngstown.[118] The latter included the addition of steam cranes to unload steel billets, fans mounted throughout the mill, small engines to transport product around the mill yard, hospitals, sanitary toilets, and even police systems.[119] However, the Lower Union Mill's puddling department remained intact and the corporation's old rolling mills in Girard, Warren, and Niles received little to no investment. The ardently nonunion policies of U.S. Steel and its subsidiary companies shifted to the managers of each of the Mahoning Valley's remaining iron mills, including those under Republic. Although Carnegie Steel (a subsidiary of U.S. Steel that managed Youngstown's Ohio Works and the Union, Girard, and Warren mills after 1903) was among the country's leaders in bar iron production, strikes as a result of antiunion agendas and other labor struggles put the work of the Valley's skilled ironworkers in significant danger.[120]

THE END OF IRON

In 1901, the *Pittsburgh Dispatch* wrote a story on the municipal and industrial growth of Youngstown. "The sale of local plants to the trusts a couple of years ago by local capital resulted in a large sum of money coming into the town, which overloaded the banks until even yet they have more money on deposit than they can well handle," reported the newspaper's editors.[121] Indeed, the growth in the iron and steel market in the late 1890s and the following years marked a period of great prosperity for the Mahoning Valley, and Youngstown especially. Banks such as the Dollar Savings and Trust Company constructed an eight-story, $200,000 building at Wick Avenue and Central Square. In 1903, the bank merged with the Peoples Savings and Banking Company, while financiers established a number of other banking institutions in the early years of the twentieth century. Real estate values and property sales for manufacturing sites and residences steadily increased in Youngstown, although they still did not equal the figures reached in Pittsburgh. Nevertheless, the *Pittsburgh Dispatch* noted that there were not nearly enough houses in Youngstown to accommodate its population, even after 600 homes were built between 1900 and 1901.[122] Despite its population stagnating in the 1890s due to the depression, the Valley witnessed a sharp increase in immigrants coming to the area from southern and central Europe after 1899, pushing Youngstown to nearly 45,000 residents. However, this number was still nowhere near the population of Cleveland or Pittsburgh, both being among the top eleven most populated U.S. cities in 1900. Nevertheless, growing manufacturing cities such as Detroit, Buffalo, and Youngstown would continue to see a population surge in the early twentieth century. Joseph Butler recognized that "modern Youngstown . . . might be said to have had its beginning about the year 1900."[123]

The iron and steel combinations that occurred throughout the country between 1898 and 1901 accomplished several things in the Mahoning Valley. They served to modernize the Valley's industry and largely rid the region of its reliance on iron. Steel was a much larger industry in both profits and general employment. In 1901, a prominent Youngstown banker said that the city had enough money to invest in and buy out any industrial enterprise. At the same time, local companies had invested $4 million in new enterprises and enlargements to old establishments.[124] Wick and Campbell's Youngstown Iron Sheet & Tube, incorporated in November 1900, eventually proved profitable despite many observers saying there was "little chance of success nowadays for independents."[125] Sheet & Tube purchased land along the Mahoning River in East Youngstown for $100 an acre; this was the location of the company's first plant despite several setbacks caused by flooding, which washed away the plant's foundations on more than one occasion. Nonetheless, the land intrigued company officials due to the Pennsylvania and Pittsburgh & Lake Erie Railroads along each side of the river, though several towns presented offers for the mill's construction, including land in Lowellville and even a free site in Niles.[126] Enlargement of the plant and the rebuilding of its foundations required additional money. Thus, Sheet & Tube increased its working capital to $4 million in 1902—seven times its original capitalization—and purchased three joint iron ore properties on the Mesabi Range.[127] Campbell was close friends with U.S. Steel organizer E. H. Gary and agreed with his ideology of vertical integration. Owning sources of raw materials stabilized overall costs for the company and limited reliance on an unstable outside market.[128] Sheet & Tube entered the wrought iron pipe trade, a portion of the industry that the trusts largely neglected and that Pittsburgh's A. M. Byers Co. had succeeded in for nearly forty years.

Other former Mahoning Valley iron manufacturers invested their new money in local industry. Lucious E. Cochran, a major stockholder in Andrews Bros. Co., helped organize the Youngstown Iron & Steel Roofing Company in Haselton after Republic bought out the former company. Cochran authorized the construction of a new sheet mill in 1901 and increased its capital stock to $300,000.[129] Numerous other Youngstown companies, such as Republic Rubber, Youngstown China, and the General Fireproofing Company—formed in 1902 with a capital of $500,000 to manufacture fireproof building materials—began to supplement the region's growing steel industry.[130]

Niles experienced similar industrial reform after a complete dismantling of its nineteenth-century iron mills, largely at the hands of the trusts. Perhaps the most iconic symbol of the Valley's once great iron industry was Niles's Old Ward Mill, which fell under bad management and was ultimately sold to the Pennsylvania Railroad and dismantled in 1901 to make way for a railroad depot.[131] U.S. Steel subsidiary American Sheet Steel Co. promised large-scale investment in its Niles plants, but the opposite proved true. In 1901, U.S. Steel president Charles Schwab visited several of the corporation's mills in the Mahoning, Shenango, and Ohio Valleys. Newspapers labeled his visits a "dooming" tour and observed that if Schwab "decided that a particular steel mill was obsolete and ought to be shut down, then he . . . would cause unemployment, destroy flourishing towns, and ruin the livelihood of farmers and middlemen who supplied those towns."[132] Indeed, between 1901 and 1903, Schwab suggested the closure of thirty plants under U.S. Steel's control.

Among them was Niles's Russia Sheet Iron Mill, which American Sheet Steel dismantled in 1902 without ever putting it into operation.[133] Moreover, the American Sheet & Tin Plate Company (formed in December 1903 by the merger of U.S. Steel subsidiaries American Tin Plate Co. and American Sheet Steel Co.) drastically reduced the capacity of the former Falcon Iron and Nail Co.'s mill by 25%, rarely operating the plant after 1902.[134] In the summer of 1904, company officials received orders to dismantle the plant, a decision that weighed heavily on its former workforce and others in Niles.[135] By the end of the year, Niles's once great nineteenth-century iron mills had all disappeared.

After these serious economic setbacks, however, the Niles Board of Trade helped establish new enterprises within the city that could compete or coexist with larger companies. Several local industrialists organized small-scale sheet steel mills after the turn of the twentieth century. Some of the most prominent were William Aubrey Thomas's Niles Iron and Sheet Co. and Charles Thomas's Empire Iron and Steel Co. in Niles, which produced sheet steel specialties such as shovel and cutlery steel, as well as several other specialty enterprises that produced niche products using steel from larger corporations. Notwithstanding this newfound industrial growth, very few companies invested in the manufacture of iron. Campbell and Wick's Youngstown Iron Sheet & Tube was the only new company in the Valley that built puddling furnaces for manufacturing wrought iron instead of rolling steel exclusively.[136] This trend of moving away from puddling furnaces and wrought iron production was similar across the country. New independent companies formed after 1900 generally did not build large-scale puddling mills, while some added only a few puddling furnaces to produce specialized wrought iron products along with rolling steel.

In spite of Youngstown having a workforce of 12,000 in 1901, skilled workers like puddlers gradually fell victim to the mechanization of the steel industry. Republic Iron and Steel employed 4,500 in the Mahoning Valley, most of which were semiskilled and nonskilled workers. John Williams, secretary and treasurer of the Amalgamated Association, reported that in 1903, Republic only employed a combined total of 1,000 skilled puddlers and finishers at their mills in Youngstown, St. Louis, Muncie, Gate City, and Birmingham.[137] In 1901, the Amalgamated Association estimated that the United States Steel Corporation employed 60,000 skilled workers versus 100,000 nonskilled workers.[138] These numbers would consistently diminish with the eradication of the iron mills and the steel companies' intolerance of union activity.

Although Republic and U.S. Steel's remaining puddling mills held more of a staunch union presence, the region's blast furnaces were not immune to organized labor and strikes. With the exception of the Thomas Furnace Company in Niles, all other Valley merchant blast furnace companies remained independent following the great iron and steel combinations. In Lowellville, for example, Ohio Iron and Steel's Mary furnace proved so valuable that company stockholders proudly announced that the trusts were unable to seize control of its property and assets.[139] Merchant furnace companies still thrived and dominated much of the foundry pig iron market, and they served new steel companies without sufficient pig iron production. In fact, in 1900, the Mahoning Valley produced 1,002,363 tons of pig iron, while the Shenango Valley only produced 800,214 tons.[140] This was the first time since the 1860s that pig iron producers in the Mahoning Valley outproduced their Pennsylvania rivals. By 1904, the Mahoning Valley produced 1,217,186 tons of Ohio's total production of 2,977,929 tons of pig iron that year.[141] Allegheny County, however, still dominated the country's pig iron supply. In 1904, it made nearly half of Pennsylvania's pig iron and over 26% of the country's total production.[142]

Pig iron in the Mahoning Valley still largely came from smaller-capacity merchant furnaces, unlike those in the Pittsburgh region. Furthermore, Bessemer steel production at U.S. Steel and Republic's Youngstown plants grew rapidly. These companies, especially Republic, did not immediately build new, efficient blast furnaces near their steelworks, but rather continued to use former merchant stacks and isolated plants that once provided pig iron to low-capacity puddling mills. Although U.S. Steel's Ohio Works blast furnaces minimized labor by mechanizing nearly every aspect of the pig iron–making process, all other furnaces in the Valley still relied largely on manual labor, such as top fillers, bottom fillers, and iron loaders. This meant that laborers at the smaller furnace plants had some advantage in negotiating their working conditions and wages. In April 1902, the National Association of Blast Furnace Workers held "rousing meetings" in Youngstown, where they argued "without contradiction that the furnace workers are the hardest worked men in the country" and that the "laborious nature of their toil" required an eight-hour workday (figure 6.15).[143] Despite their pleas, blast furnace companies did not concede. On June 1, 1902, the union called a general strike at furnace plants in both the Mahoning and Shenango Valleys. The five-day strike shut down nearly all the blast furnaces in the Mahoning Valley, and workers at furnaces in Sharon and Sharpsville, Pennsylvania, also ceased operations. The strike did not affect U.S. Steel's Ohio Works blast furnaces or the corporation's stacks in Sharon.[144]

Although short, the strike was devastating to the Valley's steel companies and furnace operators. The furnaces involved produced 3,000 tons of pig iron per day and consumed 6,000 tons of iron ore, 3,000 tons of coke, and 750 tons of limestone, thus interfering with railroad operations and steel and iron production at the rolling mills.[145] Furnace

FIGURE 6.15. Iron loaders at Youngstown Steel Co.'s Tod furnace carry 100-plus-pound ingots of pig iron from the hot casting floor and load them onto buggies, c. 1905. The harsh manual labor performed at this furnace and most others in the Valley prompted the large-scale organization of their employees at the turn of the twentieth century. Courtesy of Tom Molocea.

operators did not believe their workers were "organized enough to cause such a complete suspension of work at the stacks."[146] Republic Iron and Steel experienced more difficulty than other companies. At the company's Hannah furnace in Youngstown, superintendent J. W. Deetrick appeared before the furnace's workers at 5 A.M., asking them to line up in front of him. He "told all who wanted to keep working 12 hours, but with an advance in wages to walk up to the pie counter. Only three stepped across the line. They were two colored men and an Irishman."[147] At the company's Haselton furnace, workers expected a settlement between the union and Republic and did not walk out until after their shifts. The men said they would not return without a settlement. Workers at Youngstown Steel Company's Tod furnace quit immediately. Over 2,000 furnace workers walked out, and reports estimated that the strike cost the Mahoning Valley $100,000.[148]

Republic and the other independent furnace companies approached the strike differently. Although most were antiunion, companies like Brier Hill Iron and Coal and Girard Iron Co. attempted to settle with their workers by offering an increase in wages (figure 6.16). Republic, on the other hand, tried to replace their workers with nonunion men. On the second day of the strike, *The Vindicator* reported: "Republic Iron and Steel company is engaged in putting up a number of shacks to be

FIGURE 6.16. Workers at Girard Iron Company's blast furnace pose for a photo in the stock house, 1902. The 1907 Accident Insurance Manual rated the laborious position of working inside the stock house as "hazardous." Courtesy of Shelley Richards.

used in housing negroes who are to be imported here from the south to take the place of strikers at the Haselton furnaces."[149] Eventually, the union struck a deal with Republic Iron and Steel and other furnace companies. These firms agreed to increase wages by 10% to all furnace workers outside of common labor without reducing the twelve-hour workday. In addition, they agreed that no workers lost their jobs.[150]

After the settlement of 1902, labor relations between independent merchant furnaces and their workforces were relatively stable for the remainder of the decade. Those between U.S. Steel and their "problem" plants, however, remained turbulent. Immediately after the formation of U.S. Steel, there was great hostility between the workforce and management at the corporation's plants across the country, including its Mahoning Valley mills. In a 1904 interview, U.S. Steel chairman E. H. Gary stated, "I do not know whether the public will believe it, but a good deal of our time and thought is devoted to a careful and humane study of wages and hours. The welfare of our men can never be lost sight of no matter what happens."[151] Gary was an advocate of positive corporate publicity, whether it was genuine or fabricated for the purposes of dispelling public criticism of U.S. Steel's labor policies. He also believed unions were "un-American."[152] Both before and after Gary's statement in 1904, U.S. Steel handed the blast furnace workers' union and the Amalgamated Association numerous challenges and setbacks.

Despite its reputation in the 1890s as one of the leading merchant stacks in the Valley, the Niles furnace was more or less a perennial problem for U.S. Steel. However, unlike the corporation's other mills in Niles (the Falcon and Russia mills), the old blast furnace was not yet dispensable. U.S. Steel relied on the stack to keep the Ohio Works— one of its most profitable mills—operating at full capacity (figure 6.17). Because of the Niles furnace's age and dependence on manual labor, its organized workforce consistently demanded higher wages and union recognition. Strikes at the Niles furnace in 1901, 1902, and 1907 indirectly prompted Carnegie Steel to authorize the construction of three more high-capacity blast furnaces at the Ohio Works in 1904 and 1908.[153] At the Niles furnace, Carnegie Steel swiftly settled the strikes by bringing in nonunion men and discharging those involved with stopping work. By 1908, the company's six Ohio Works furnaces produced over 1 million tons of pig iron per year, which was a greater output than all twenty-two furnaces in the state of Virginia, and was nearly enough to supply the steelworks.[154] Until Carnegie Steel completed Youngstown's first open-hearth plant of twelve furnaces in August 1909, the company did not need the Niles furnace, which became a backup unit only operated under extreme circumstances until the end of the First World War, when it shut down indefinetely.

After eighteen years of operation under U.S. Steel and its subsidiaries, management failed to strike any settlement with its workforce in Niles. This was a clear indication of the corporation's ironclad nonunion policy and strict agenda of ridding itself of costly plants. However, blast furnaces presented companies with the option of modernization and mechanization. Puddling and hand-rolling mills, on the other hand, did not. As a result, Carnegie Steel was even more intolerant of the Amalgamated Association in the first decade of the twentieth century. Unlike what Republic Iron and Steel did to its Brown-Bonnell plant, U.S. Steel subsidiary American Steel Hoop (and later Carnegie Steel) failed to immediately remove the old puddling departments at the Lower Union Mill in Youngstown, nor did they close their Warren or Girard rolling mills, which were both in a "dilapidated state."[155] It was clear that U.S. Steel officials did not look upon several of the older Youngstown district mills with favor. The corporation thus began to concentrate its efforts on upgrading mills within favorable locations that achieved a high return on investment.[156] Operating uneconomical plants was not a practical business ideology for U.S. Steel or, for that matter, any competing company.

Carnegie Steel's Warren and Girard mills presented the company with efficiency and labor issues, and their positions outside of

Youngstown interrupted the natural production flow of the Ohio Works and Union mills. The Warren and Girard mills also did not have the ability to produce steel or roll high volumes of the metal like the mills in Youngstown. Thus, Carnegie Steel constantly lowered the pay scale for skilled workers such as puddlers, heaters, rollers, and roughers below that of other companies with workers that performed the same class of work. On August 20, 1904, members of the Amalgamated employed at both the Union and Girard mills called a strike after Carnegie Steel's Youngstown general superintendent I. W. Jenks—a former roller never associated with a union—lowered the pay scale of skilled workers by 25% to 71%.[157] An official statement from the Amalgamated Association read, "To accept this meant the annihilation of the iron industry of the Mahoning Valley—a proposition which no sane man or organization could accept."[158] Carnegie Steel defended Jenks's decision by claiming that the company

> had abandoned the former methods of manufacturing iron and steel, and adopted modern appliances and machinery; that its mills are now equipped with continuous furnaces and continuous trains of rolls and other very expensive improvements, all of which increase the output and lessen the labor of the workman; that on account of the improved conditions, the company was entitled to a lower scale of wages.[159]

On August 22, Jenks met with other Carnegie Steel officials and they agreed to employ the open shop principle by importing nonunion men from Pittsburgh to work the Girard, Warren, and Union Mills.[160]

The first men arrived via the Pennsylvania Railroad only three days after the Amalgamated called the strike. They consisted of "13 men, six of whom are colored," while a number of the men inside the plant were cooks who fed the imported workers.[161] Three Polish laborers from Pittsburgh said Carnegie Steel offered them whiskey, a beer, a girl, and four dollars a day to act as strike breakers at the Girard and Upper Union Mill.[162] Paul Hutchinson, a nonunion worker imported from Pittsburgh, made a strong statement regarding Carnegie Steel's questionable strike-breaking strategy:

> Yes, I came here to work in a mill which an agent told me was a new one. After I went through the gate I at once knew that there was something dirty going along. I was to receive $2.50 per day and the agent told me my board was to be free for a time. After that I was to look after myself. I never worked in a rolling mill in my life. My experience as a mill man was a few years I put in in a grist mill in Thoroughfare, W. Va. Yes, I wish I was at home, and many of the other men who were brought here with me wish the same.[163]

FIGURE 6.17. U.S. Steel's Niles furnace was a hotbed of organized labor activity and strikes in the first decade of the twentieth century. U.S. Steel relied on the furnace, originally built in 1870, to supply pig iron to its Ohio Works until after the First World War. The small stack had little in the way of mechanization, unlike the corporation's state-of-the-art Youngstown furnaces. The Mahoning Valley Historical Society, Youngstown, Ohio.

Many nonunion men quickly left their positions in the mills and sympathized with the strikers. Indeed, such reckless policy from Carnegie Steel did not bode well to its Youngstown district workforce. Not long after Hutchinson's statement, a crowd of men and boys attacked four Carnegie policemen in the streets near the Girard mill. The group followed the policemen, said to be Pinkertons hired by Carnegie Steel to escort strikebreakers from Youngstown to Girard, in and out of the mill. The hostile group jumped one of the Pinkertons, who was beaten and received a large gash on the back of his head. In addition, Girard police arrested an eighteen-year-old Pittsburgh strikebreaker for concealing a revolver in his outside pocket. He was "not afraid of the strikers" and was not intimidated to enter the mill alone without Pinkerton protection. Amidst the chaos in the Youngstown district, a U.S. Steel official proudly reported to the press, "We are more than pleased with the way we are operating the mills. At the upper plant the improved mills are working and the tonnage departments are doing as much as we ever expected they would."[164]

However, with their Youngstown mills affected by the strike, Carnegie Steel simultaneously planned to eliminate these costly plants if financially possible. In the spring of 1905, Carnegie Steel president Alva

Dinkey reported that the company had badly oversold its bar iron and steel and asked U.S. Steel president William Corey for money to build an eighteen-inch mill at its newly acquired Clairton Works.[165] If Carnegie Steel approved the new $175,000 mill, they could avoid reopening the Warren mill, "whose best conversion cost would be $7.50 a ton as compared with $4.00 at a new Clairton mill."[166] The new eighteen-inch mill in Clairton more than doubled the capacity of Warren. Dinkey stated, "All the arguments are in favor of building the mill; the finishing mill would be where it belongs, freights would be saved, one more isolated plant would be abandoned, and the organization thus made more compact."[167]

Meanwhile, the strike ended on July 31, 1905, and Carnegie Steel accepted the union's wages, but not without great loss to the company's skilled workforce. Although the company gave in to the union's demands, they delivered a crushing blow to the Amalgamated Association that prevented similar strikes in the immediate future. Carnegie Steel lost thousands of dollars by importing inexperienced strikebreakers who failed to produce any considerable amount of finished iron for the company's customers.[168] After the strike ended, the company decided to remove all of its hand-puddling mills and completely abandon its Warren and Girard mills. Dinkey gained approval for the removal of the eight- and ten-inch mills from Girard to Duquesne, where expectations were that the latter mill's production would double.[169] Another mill from Girard, most likely its twenty-inch mill, was moved to the company's plant in Greenville, Pennsylvania, which operated as a "boom" plant and worked only when other mills were overrun with orders. In addition, modern equipment such as continuous mills replaced all the remaining puddle mills at the Lower Mill in Youngstown, thus ending all wrought iron production in U.S. Steel's Youngstown district. Those who worked at Warren or Girard transferred to Youngstown or were left unemployed. Thus, the corporation successfully defeated the Amalgamated Association, drastically limiting the voice of its skilled labor force and eradicating its costly plants.

Apart from strikes and costly operations, the product of U.S. Steel's Girard, Warren, and Youngstown puddling mills had limited resale on the market. The corporation did not consume its own bar or muck iron and largely sold it off to Pittsburgh's A. M. Byers Co., the country's largest independent producer of wrought iron pipe. The Byers family had connections with the Mellon and Guffey families in the Pennsylvania oil and gas industry, which gave the company a formidable niche in that market as it spread southwest.[170] After Carnegie Steel removed their puddling mills, A. M. Byers looked toward Republic Iron and Steel's remaining puddle mills in Youngstown. Republic sold nearly its entire product made at its Mahoning Valley Works and the No. 4 puddle mill

FIGURE 6.18. View looking north into Girard across the Mahoning River, c. 1900. Heirs to the A. M. Byers Co. believed the empty farmland west of the Girard furnace (right) was a perfect location to build a new puddling mill. When built in 1909, Byers's puddling mill was the last of its kind built in the Mahoning Valley and Ohio. The Mahoning Valley Historical Society, Youngstown, Ohio.

at the Brown-Bonnell plant to Byers.[171] However, these remaining relics of the Valley's nineteenth-century iron industry had limited value beyond what Byers provided for them. In 1905, Republic dismantled its old Andrews rolling mill in Youngstown to build two thoroughly modern blast furnaces to further supply the nearby Bessemer steelworks.[172] The two furnaces each produced 140,000 tons of pig iron per year and served to eliminate reliance on their smaller, detached blast furnaces, thereby progressively centralizing iron and steel production.

Thus, the loss of the old Haselton mill preceded Republic's gradual elimination of its iron plants in favor of integrated steel production. The company's lingering iron mills took a big hit when the A. M. Byers Company, who owned the Girard blast furnace, announced the construction of a modern puddling plant on land adjacent to the Girard stack. The company broke ground on the new $500,000 plant in September 1908 and opened the 800,000-square-foot mill in January 1909 (figure 6.18).[173] The mill, the largest of its kind in the country, with forty-two puddling furnaces, gave employment to the puddlers and rollers who lost their jobs when Carnegie Steel and Republic closed their puddling plants. In March 1910, A. M. Byers Co. added another forty-six puddling furnaces, with all of the plant's muck bar and skelp sent to the company's Pittsburgh plant for finishing into pipe. As a direct result of the Girard mill's opening, Republic ceased operations at the Brown-Bonnell plant's lone puddling

FIGURE 6.19. Rollers in Republic's old Mahoning Valley Works on Youngstown's east side shape bar iron in this September 1911 photo. The old mill was the last of the Valley's nineteenth-century iron mills to remain in operation. The Mahoning Valley Historical Society, Youngstown, Ohio.

mill on July 1, 1909, "a striking example of the policy of the Republic Iron & Steel Co. to become a factor of increasing importance in the production of steel."[174] For years, citizens of Youngstown crossing the Market Street bridge complained about the smoke from the mill, especially "when an east wind blew the black fumes" toward the viaduct.[175] Less than a year after the mill's removal, Republic announced its intent of abandoning most of the company's remaining bar iron mills countrywide. In its 1910 annual report to stockholders, Republic announced, "Owing to the restricted demand for bar iron . . . it was decided to dismantled or permanently close down the following Bar Iron Mills: Corns Works, Toledo Works, Birmingham and Gate City Works, so that Iron Bar production, hereafter, will be confined to the Tudor, Inland, and Mahoning Valley Works."[176] However, three years later, Republic president Thomas Bray stated that the company could not afford to spend much money on the Mahoning Valley Works, noting that "the condition of the iron market today does not justify the company in operating this works" (figure 6.19).[177]

After completing a large open-hearth steelmaking plant of twelve furnaces at its Haselton plant in 1911, Republic solidified its position

TABLE 6.1. Rolled Iron Production in the United States, 1888–1910

	Rolled Iron (gross tons)	Percentage of Total Rolled Iron and Steel
1888	2,153,263	46.5
1889	2,309,272	44.1
1890	2,518,194	42.8
1904	1,760,084	14.7
1905	2,059,990	12.2
1906	2,186,557	11.2
1907	2,200,086	11.1
1908	1,238,449	10.5
1909	1,709,431	8.7
1910	1,740,156	8.1

Source: *The Iron Age* 88 (November 30, 1911): 1169.

Note: Rolled iron includes rerolled scrap iron.

in steel. The new plant produced 575,000 tons of steel ingots per year, which, coupled with 725,000 tons of Bessemer steel annually, required greater pig iron production to supply each steelworks. Thus, Republic allotted nearly all its investment to modern production facilities and built a new blast furnace the same year. The old Mahoning Valley Works proved too expensive and uneconomical for continual operations. Republic abandoned the old plant on September 27, 1913.

With the old Mahoning Valley mill dismantled, Republic completed the total elimination of the Mahoning Valley's nineteenth-century iron mills, which was swift compared to other major industrial centers. Even Allegheny County still boasted fourteen plants with a combined 400 puddling furnaces in 1907 and 1908, while the United States as a whole contained 2,635 puddling furnaces in the same period.[178] Fittingly, the percentage of total rolled iron made in the United States drastically decreased as steel production boomed after the turn of the twentieth century (table 6.1). U.S. Steel's nonunion policies and accompanying steel agenda, along with Republic's drive to compete with other major companies, helped ensure the Mahoning Valley's transformation into one of the region's top steel-producing centers. Still, steel companies in the Valley were not finished. In the years preceding the First World War, the Valley's steel and pig iron production grew almost exponentially. In 1910, observers called the Mahoning Valley "the seat" of Ohio's pig iron industry; it outproduced all other districts in the state by a wide margin, primarily due to steel plants building large batteries of furnaces to supply their own mills.[179] Other once-dominant pig iron regions in Ohio, such as the Hocking Valley and the Hanging Rock Iron Region, faded after steel companies realized their geographical disadvantages. The Mahoning Valley saw the construction of five different steel plants between 1911 and 1918 (four built by local companies), while there were no major steelworks erected in the Shenango Valley during the same period. U.S. Steel, Republic, and the hometown Youngstown Sheet & Tube Company, which dropped the word *iron* from their corporate title in 1905 and added a Bessemer steel plant the same year, became the region's largest employers for much of the twentieth century. The capital invested by steel companies in Youngstown increased from $23 million in 1907 to $375 million in 1920.[180]

Many in Youngstown embraced the rapid changes that took place in the first decade of the twentieth century. Although the puddler's occupation was not yet extinct in the Valley, the men that, according to John Fitch, represented "the oldest, the most picturesque and most self-assertive of the crafts of the iron trade," nonetheless faded into obscurity.[181] Only A. M. Byers Co. in Girard and Youngstown Sheet & Tube employed puddlers, whom in 1909 *The Vindicator* called a "valuable asset to an iron works [during] these times."[182] After the American Iron and Steel Institute officially created the Youngstown District in 1916 by merging the Shenango Valley with the Mahoning, the region became a force among the nation's mightiest steel centers. Youngstown's population grew to 132,358 in 1920, an increase of 200% since 1900. In the same period, industry in the Mahoning Valley transformed from a segmented group of largely outdated iron mills and blast furnaces to the second largest steel center in the United States behind Pittsburgh and Allegheny County. In 1920, the Youngstown District produced 5% of the world's entire steel output and 10% of the output in North America.[183] By 1928, it finally rolled more steel than Allegheny County.[184] Modern steel plants and other mills lined the Mahoning River from Warren to Lowellville. Major newcomers included companies such as Trumbull Steel Co. and Sharon Steel Hoop, as well as a plethora of smaller mills that used steel or its by-product. This transition from iron to steel was extreme and at times challenging, but it pushed the Mahoning Valley into the modern industrial era, in which it remained a formidable powerhouse for much of the twentieth century.

EPILOGUE

THE RISE of the steel industry in Youngstown and the Mahoning Valley is quite atypical compared to other major industrial regions in the United States in the late nineteenth century. Like most of the United States' iron-producing districts, the Mahoning Valley early on followed a general trend of small entrepreneurship to big business practices. Antebellum iron companies consisted largely of skilled ironmasters who operated their mills with help from laborers or business partners. However, the structure and the size of the industry changed after the Civil War, when many well-managed iron companies grew larger and became more technologically advanced, leading to constantly increasing outputs and capital; some companies remained small, while others grew to dominate the industry. Whereas regions like Pittsburgh, Cleveland, Bethlehem, and the Chicago district slowly made the transition to steel production throughout the late nineteenth century, Youngstown's comparatively late move to steel proved hasty—even desperate at times. The region's transformation from iron to steel production occurred largely because of increasing market pressure and the growing preference for steel among the nation's major consumers of the metal. Nevertheless, the tradition of the merchant pig iron industry—a facet of the trade that Youngstown-area iron manufacturers controlled from the 1840s to the 1880s—still survived after the arrival of steel. Indeed, Mahoning Valley iron companies sought to stay relevant in a vanishing industry they had always dominated from the supplier side of the market.

Yet, their initial unwillingness to adopt steel production could have proven disastrous for the region's twentieth-century industrial economy. Iron had a much more limited market by the turn of the twentieth century, when only companies producing niche products thrived. This was particularly true in Pittsburgh, where small companies like Zug & Co.'s Sable Iron Works, originally built in 1845, manufactured special iron used in forging, machine shop work, and railway supplies. Likewise, Pittsburgh's Oliver Iron and Steel Company made a variety of niche iron products, including railway track tools, telegraph and telephone pole equipment, picks and mattocks, and crowbars and wedges.[1] On the other hand, iron firms in the Mahoning Valley produced many of the basic products that most other iron and steel producers began manufacturing, such as iron and steel hoops and bands, nails, spikes, rails, sheets, and plate. Companies such as Brown-Bonnell Iron Co., Andrews Bros., and Mahoning Valley Iron Co. largely produced the same types of products, failed to develop specialty products for a diversifying market, and ultimately fostered a noncompetitive regional market. Thus, without a specific market niche, rolling mill companies in the region had no way of separating from the major iron and steel combines created at the turn of the twentieth century.

Although Mahoning Valley iron manufacturers were some of the largest in Ohio, they never grew to a dominant position among the nation's leading iron and steel companies. Companies such as Jones & Laughlins in Pittsburgh, Bethlehem Iron Co., Cleveland Rolling Mill Co., and Carnegie's iron and steel empire grew to mammoth proportions out of the iron age by innovating with steel. After the Civil War, Mahoning Valley iron manufacturers were content to let others in the industry innovate, a reversal of the moderately widespread innovation that occurred in the region before the war broke out. Valley pig iron manufacturers rarely adopted new technology when other major companies in the industry did so, and this lack of acclimatization to industry trends can largely be attributed to the region's reliance on iron production, specifically pig iron produced for outside consumption. In spite of this, Mahoning Valley pig iron manufacturers were, in fact, very attentive to market conditions. They changed their product from pig iron suitable for Pittsburgh rolling mills and foundries to pig iron for Bessemer steel companies almost immediately after the process's introduction into the United States. When demand for Bessemer iron weakened, they refocused on making pig iron for rolling mills and foundries as the market dictated. Despite local manufacturers' attentiveness to outside market demand, there was never an immediate demand for efficient and high-yield blast furnaces. Thus, merchant furnaces in the Mahoning Valley became victim of extreme variances in supply and demand from outside consumers, and spending large amounts of capital in retrofitting their mills with the latest technology was not practical. Supplying pig iron to an unreliable market and small-scale puddling mills in the region was not grounds for mass production that entailed expensive modernization.

Unlike the Mahoning Valley, some Pittsburgh iron and steel manufacturers—in particular Andrew Carnegie—developed and adopted mass production to supply their massive steel mills. In the late nineteenth century, Pittsburgh and Allegheny County outproduced the region by a huge amount for a number of reasons. Carnegie and other Pittsburgh iron and steel manufacturers wanted to relinquish their reliance on the merchant pig iron market, especially as the need for pig iron in the region steadily increased in the 1870s and 1880s. Thus, the integration of iron and steelmaking proved both efficient and economical, something that low-production iron mills did not require. Adoption of new and modern technology increased the capacity of Pittsburgh furnaces to an average of 75,000 tons by 1890, when Mahoning Valley furnaces averaged around 45,000 tons.[2] Eventually, however, the constant weakening of the iron market forced Mahoning iron manufacturers to rethink their position on steel. It was not until the organization of the Ohio Steel Company in 1892 that merchant furnace companies in the region began to modernize their plants with technology that other manufacturers had adopted twenty years earlier. In addition, rolling mills owned by the wealthy Wick family and directly associated with the Ohio Steel Company saw equipment upgrades due to the arrival of the Bessemer plant. Like other prominent industrialists before them, some Mahoning Valley iron manufacturers modernized with the arrival of the steel industry. Most other mills, however, remained irrelevant, and the region's iron barons opted to sell their plants and retire in luxury.

As a result, outside industrialists and financiers found incredible promise in the Mahoning Valley. Despite the region's antiquated rolling mills and production methods that remained at the turn of the twentieth century, it had established a strong industrial and transportation infrastructure that included railroad connections to all major markets and raw materials. Youngstown's midpoint location between Cleveland and Pittsburgh and New York and Chicago also proved a major selling point to investors. Investment in steel production in the Mahoning Valley ultimately led to a thriving market for merchant iron firms. Not only did the Valley's merchant pig iron firms escape the era of consolidations as independent entities, unlike the outdated rolling mills, but also they were financially stable. These companies and their furnaces constituted the lingering relics of the Valley's vibrant iron era. As was common in the late nineteenth century, merchant furnace companies had to adapt to prevailing market conditions. After 1900, remaining merchant furnaces in the Mahoning Valley had the highest pig iron capacity of all similar companies in the Cleveland, Pittsburgh, and Shenango Valley regions. In addition, the Mahoning Valley's merchant pig iron firms were overall more technologically advanced and produced more iron than most other remaining independent furnace companies in northeast Ohio and western Pennsylvania. In 1910, the Mahoning Valley's seven merchant furnaces had an annual capacity of 991,000 tons, while the capacity of the Shenango Valley's nine merchant stacks was 830,000 tons. Cleveland and Allegheny County together totaled four merchant furnaces in 1910; historically, neither Cleveland nor Pittsburgh were ever prominent merchant pig iron centers, as vertically integrated steel plants quickly took hold in both regions.

In contrast, the Mahoning Valley boasted four of Ohio's most historically successful pig iron firms: Girard Iron Co. in Girard, Andrews & Hitchcock Iron Co. in Hubbard, Ohio Iron and Steel Co. in Lowellville, and Brier Hill Iron and Coal Co. in Youngstown. In Struthers, the Struthers Furnace Company, managed by industry veteran Samuel Allen Richards, transformed into one of the preeminent companies of its kind in the early years of the twentieth century. Like the larger integrated steel companies, control over raw materials was vital for successful operations at the independent blast furnace. Struthers Furnace Co. freed itself of outside fuel dependence by organizing the Struthers Coal & Coke Co. in 1905, which built 160 beehive coke ovens in Fayette County, Pennsylvania.[3] All other merchant pig iron firms held extensive coal and coke properties, as well as iron ore reserves in the Lake Superior region. Andrews & Hitchcock Iron Co. owned 1,123 acres of virgin coal land in Greene County, Pennsylvania, interests in steamship companies in the Great Lakes region, and, most importantly, a 12% interest in the Mahoning Ore Co. in Minnesota's Mesabi Range. The latter had an estimated deposit of 100 million tons of iron ore.[4] Both Brier Hill Iron and Coal and Youngstown Steel Co. owned iron ore interests in the Lake Superior region as well as limestone deposits in Lowellville and coal interests and coke ovens in the Connellsville region.[5]

Proper management and modernization were pivotal for each company's success in the era of big steel. Nonetheless, the industry quickly outgrew its nineteenth-century foundation in the local and regional economy, broadening to a national scale. Most merchant iron companies understood that as their market continued to dwindle, they had to conform and finally produce their own steel, nearly fifty years after the Bessemer process was introduced in the United States. Indeed, some longtime merchant pig iron producers tried to adapt. In January 1910, the stockholders of Lowellville's Ohio Iron and Steel Co. met to vote on increasing the company's capital from $500,000 to $5 million. They wanted to construct an open-hearth steel plant consisting of six sixty-ton open-hearth furnaces, a blooming mill, and finishing mills.[6]

Although company directors voted to approve the increase in its capital stock, cost determinations consistently delayed construction of the new steelworks. By February 1910, Pittsburgh engineer Julian Kennedy drew up tentative plans for the plant, but nothing ever materialized. Robert Bentley explained that "engine builders will not promise delivery for at least eight months, and that prompt delivery of structural material would require premium prices."[7] Ohio Iron and Steel ultimately decided to forego the steel plant and remain a merchant pig iron producer. However, companies that wanted to enter the steel trade were not uncommon. In 1913, rumors circulated that Andrews & Hitchcock Iron Co. planned to build an open-hearth plant in connection with their two Hubbard furnaces. Company president Frank Hitchcock dismissed those rumors, saying that his company "had no thought of making any additions to the furnace at present."[8] Certainly, directors of the Valley's merchant pig iron firms believed that steel was the key to maintaining consistent and profitable operations in the future. Joseph Butler, Edward Ford, and other members of the Brier Hill Iron and Coal Co. and Youngstown Steel Company understood this concept. Hence, the two companies became the first—and only—Mahoning Valley merchant pig iron firms to successfully transition to steelmaking.

In January 1912, officials from Brier Hill Iron and Coal, Youngstown Steel, and several other local independent iron and steel finishing companies discussed the idea of combining forces to create an integrated steelworks. Butler, H. H. Stambaugh, and other members of Brier Hill Iron and Coal first considered expanding their interests after the November 1908 death of company president George Tod. Butler noted,

> Mr. Tod's death was the first break in our circle, as it were, he being the first of the men who started in young with me in the company. I think his passing caused all to realize that we were no longer young and led us to think of the future of our organization more than we had done previously.[9]

The result was the $15 million Brier Hill Steel Company. One source reported that the enlargement of other steelworks' pig iron capacity "forced the steel plant proposition on the blast furnace interests" of Brier Hill Steel.[10] Those involved in the local combine were the merchant furnace plants of Brier Hill Iron and Coal and Youngstown Steel Co., as well as three sheet steel companies in Niles, including the Thomas Steel Company. Brier Hill Steel acquired a large tract of land west of its Grace furnace complex and, as a "legacy from the old Brier Hill Iron & Coal Company," constructed its seven-furnace open-hearth plant there along the Mahoning River.[11] The company produced its first open-hearth steel in February 1914, which marked the fourth such plant in Youngstown, transforming the city into "one of the largest steel centers in the country" and serving to eliminate one of the country's historic pig iron firms.[12]

After Brier Hill Steel's successful launch into Youngstown's pool of steel companies, only a few merchant pig iron firms remained in the Valley. However, the demand for iron during the First World War ultimately undid these independent companies. During the Great War, pig iron prices hit $50 per ton for the first time since the Civil War.[13] Even before the United States entered the war in Europe, there was a fear that the "war demand" for steel would make the metal scarce in the United States. In July 1915, the country's steel mills operated at 85% capacity and by August they reached "substantially full operation."[14] In 1917, Mahoning Valley steel companies rolled "exceptionally large tonnages" of sheet steel for the government war effort and the allies.[15] Naturally, the need for pig iron skyrocketed. By September 1917, all twenty-five blast furnaces in the Youngstown District were in operation. In 1918, Brier Hill Steel supplemented their pig iron production with the construction of the Jeannette furnace in Youngstown. Republic Iron and Steel added to their pig iron output by building a fifth blast furnace at the Haselton complex, which brought the company's total production up to 2,000 tons per day; Carnegie Steel and Youngstown Sheet & Tube produced 3,000 tons of iron a day for the war effort and domestic manufacturers.[16] However, Sheet & Tube's pig iron output was still insufficient for wartime demand. To counter this, the company sought to add existing plants to their properties rather than build new ones.[17] On April 16, 1916, Sheet & Tube acquired all of Andrews & Hitchcock Iron Company's property and assets for $2.5 million.[18] The latter company's failure to convert to steelmaking made it an easy target for larger, more established steel companies. *The Vindicator* hailed the buyout as "the greatest deal ever recorded in industrial circles of the Mahoning and Shenango district."[19] This acquisition was only a precursor to Sheet & Tube's purchase of Brier Hill Steel in 1923.

Andrews & Hitchcock Iron was not the Valley's only merchant pig iron firm absorbed during the war. In 1918, Sharon Steel Hoop Co. purchased Ohio Iron and Steel Co.'s Mary furnace in Lowellville to supply the former's nearby open-hearth facility. Incorporated in 1900, the Sharon, Pennsylvania–based company acquired Youngstown Iron and Steel Co.'s Lowellville steel plant in 1917. Previously, Sharon Steel Hoop did not have the capability or facilities to produce its own pig iron, but it was one of the largest producers of steel hoops and bands used in making barrels for the cotton industry. After the war, however, most steel companies did not need their small, isolated blast furnace plants. In Youngstown, Republic Iron and Steel officials illustrated these economizing measures in their 1925 annual report by stating that their outlying furnaces "were no longer necessary for the Company's

FIGURE E.1. Throughout the country, merchant furnaces fell to the fierce competition of integrated steel companies, including many in the Mahoning and Shenango Valleys. The remains of the Fannie furnace plant in West Middlesex, Pennsylvania, built in 1873 and shown here in 1933, depict a common scene shared by many such plants in the 1920s and 1930s. Shattered windows plague the three-story blowing engine house, while remains of boilers and a stove shell lie in the center of the photograph. Demolition of the plant began in 1929. Courtesy of Hagley Museum and Library, American Iron and Steel Institute Collection photographs and audiovisual materials.

steel operations" because "they could not be economically operated as Merchant Blast Furnaces." Thus, the board of directors decided to "dismantle them and sell the real estate."[20] Those small furnaces that did survive became secondary plants and only operated in times of extremely high demand. In the first nine months of 1920, merchant furnaces in the United States produced 7,067,812 tons of pig iron versus 20,052,458 tons by nonmerchant stacks.[21] Between 1925 and 1929, steel companies and smaller merchant furnace firms abandoned over ninety stacks because of the weakening market for merchant iron and the severe lack of continuous profits (figure E.1).[22] In 1929, editors of *The Iron Trade Review* summed up the old furnace problem:

> Obsolescence and abandonment of smaller blast furnace stacks have been under way for some time. Merchant stacks . . . came to their end as the result of unprofitable conditions, location, high cost or inability to meet competition. . . . At the same time, the modern stack, refined by recent engineering research, is a much larger operating unit than the older stack and produces a heavy tonnage at a much cheaper cost per ton. . . . It is a manifestation of one phase of constant change in the iron and steel industry, change that has been overturning established methods with startling rapidity ever since Bessemer announced his method of producing steel.[23]

Just as the wrought iron industry experienced drastic change because of the world's general preference for steel, merchant furnaces underwent radical transformation from the decentralized state of the old iron industry to the large-scale integrated operations of big steel companies. What the Mahoning Valley lost in small furnace plants it gained in productivity and steel output.

Still, the fall of the merchant furnace went hand in hand with the near-elimination of its nineteenth-century counterpart, the puddling mills. Notwithstanding the efforts of many engineers to mechanize the manufacture of wrought iron in the late nineteenth and early twentieth centuries, few succeeded in inventing a commercially viable process. In the early 1920s, Youngstown's Edward L. Ford developed such a process that A. M. Byers Co. later improved upon at an experimental plant in Warren, Ohio.[24] Nevertheless, most companies naturally favored steel over a mechanical wrought iron process that had yet to prove its economic value. In contrast, the basic principles of producing pig iron in the blast furnace never changed. In many cases, such as those that occurred at several merchant furnaces in the Mahoning Valley, companies could adapt blast furnaces with modern technology, albeit at great cost. Since the late nineteenth century, steel companies and merchant iron firms alike tried to reduce the overall manufacturing cost of pig iron as much as possible while at the same time increasing tonnage rates. In addition, as with puddling, high wage rates at small, hand-fed furnaces forced many plants to close or to convert from manual to mechanical operations. Between 1860 and 1930, the number of blast furnaces in the United States shrunk by 50%, but their total capacity increased tenfold.[25]

It is true that the merchant furnace industry experienced a slower decline in the early twentieth century than puddling and wrought iron production, but with few exceptions, both industries came to a quiet end at the onset of the Great Depression and subsequent years of economic decline. Such a defining moment in the Mahoning Valley occurred in 1929, when A. M. Byers Co. announced their intentions to permanently shut down the Girard furnace after sixty-two years of operation, declaring the plant "obsolete."[26] The furnace had supplied pig iron to the company's Girard puddling plant since 1909. This event signified the end of Byers's reliance on the traditional puddling process in favor of a modern, $12-million mechanical wrought iron plant in Ambridge, Pennsylvania, that produced nearly ten times as much as a single puddler.[27] A. M. Byers president A. H. Beale stated that the new plant meant "the gradual elimination of the hand process." "Puddlers are scarce now," he said, "Young men do not take up the profession any longer. Of late years we have had difficulty maintaining uniform operations, especially through the warm months."[28] In response to Beale's decision, *Vindicator* writer Ernest N. Nementi predicted, "The days of the picturesque old puddlers, artists of the iron and steel industry of years gone by, are numbered."[29] Indeed, by 1930, both A. M. Byers and Youngstown Sheet & Tube closed their puddling plants; the latter was largely in response to Byers's cheaper mechanical method. For A. M. Byers, a company that did not produce steel, mass production and profit gain prevailed over the tradition and romance of the old iron industry, whereas Sheet & Tube focused on steel production exclusively. Almost simultaneously, the puddler's occupation and the merchant furnace, familiar sights in the region since the 1840s, disappeared in the Mahoning Valley. "Iron, the King," as one Youngstown iron manufacturer declared in 1892, had finally been dethroned.

TABLES

Iron and Steel Works in the Mahoning Valley

TABLE A.1. Mahoning Valley Blast Furnaces, 1802–1898

Name(s)	Year Built, Location	Owners (years)	Year Abandoned or Last Operated	Year Dismantled
Hopewell; Heaton's furnace; Eaton furnace	1802, Struthers	James and Daniel Heaton (1802–1805); Daniel Heaton (1805–1808); Montgomery, Clendennin & Co. (1808–1812)	1808	After 1812
Montgomery	1806, Struthers	Montgomery, Clendennin & Co. (1806–1812)	1812	After 1812
Maria; Mosquito Creek furnace	1812, Niles	James Heaton (1812–1830); W. Heaton and J. Robbins (1830–1842); McKinley, Reep & Dempsey (1842–c.1846); J. Robeson & Co. (c.1846–c.1849); Robeson & Bowell (c.1849–c.1852); Robeson & Battles (c.1852–1856)	1856	After 1856
Trumbull; Mill Creek	1833, Youngstown	Isaac Heaton (1833–early 1840s); David Grier (1840s–1855)	1855	After 1855
Mahoning; Ada; Mary	1845, Lowellville	Wilkeson, Wilkes Co. (1845–1853); Alexander Crawford & Co. (1853–1864); Hitchcock, McCreary & Co. (1864–1871); Mahoning Iron Co. (1871–1874); McCreary & Bell (1878–1880); Ohio Iron and Steel Co. (1880–1918); Sharon Steel Hoop Co. (1918–1936); Sharon Steel Corporation (1936–1962)	1960	1963
Eagle; Philpot's furnace	1846, Youngstown	Ohio Iron and Mining Co. (1846–1851); Crawford, Morris & Co. (1853–1856); Crawford & Murray (1856–1862); Eagle Furnace Co. (1862–1873); Cartwright, McCurdy & Co. (1873–1883); Brier Hill Iron and Coal Co. (1888)	1883	1888

TABLE A.1. Mahoning Valley Blast Furnaces, 1802–1898 *(continued)*

Name(s)	Year Built, Location	Owners (years)	Year Abandoned or Last Operated	Year Dismantled
Brier Hill; Tod No. 1	1847, Youngstown	James Wood & Co. (1847–1856); David Tod (1856–1865); Tod Iron Co. (1865–1867); Brier Hill Iron and Coal Co. (1867–1890/1891)	1890/1891	1890/1891
Falcon	1850, Youngstown	Charles T. Howard (1850–1851); James Ward & Co. (1851–1856); Charles T. Howard (1856–1858); James Laughlin & Co. (1858–1862); Canfield & Alford (1862–1864); Brown, Bonnell & Co. (1864–1892); Brown-Bonnell Iron Co. (1892–1893)	1893	1893
Phoenix	1854, Youngstown	Crawford & Howard (1854–1856); Crawford & Son (1856–1863); Brown, Bonnell & Co. (1863–1892); Brown-Bonnell Iron Co. (1892–1899); Republic Iron and Steel Co. (1899)	1899	1899
Meander; Porter	1857, Austintown; moved to Mineral Ridge in 1863/1864	Smith, Porter & Co. (1857–1859); George C. Reis (1862–1863/64); Jonathan Warner (1863/64–1866); Mineral Ridge Iron and Coal Co. (1866–1868); Brown Iron Co. (1868–1871); James and Lizzie Ward (1871); Jonathan Warner (1872–1875); Jonathan Warner (1878–1880)	1873	Around 1880
Himrod No. 1	1859, Youngstown	Himrod Furnace Co. (1859–1884); Brier Hill Iron and Coal Co. (1885–1890); S. Frank Eagle (1890–1893)	1893	1895

TABLE A.1. Mahoning Valley Blast Furnaces, 1802–1898 (*continued*)

Name(s)	Year Built, Location	Owners (years)	Year Abandoned or Last Operated	Year Dismantled
Falcon; Elizabeth; Hannah	1859, Niles; moved to Youngstown in 1880	James Ward & Co. (1859–1874); Mahoning Valley Iron Co. (1880–1899); Republic Iron and Steel Co. (1899–1925)	1925	1925
Grace No. 1	1860, Youngstown	Brier Hill Iron Co. (1860–1867); Brier Hill Iron and Coal Co. (1867–1897)	Torn down in 1873, rebuilt in 1882; abandoned in 1897	1897
Himrod No. 2	1860, Youngstown	Himrod Furnace Co. (1860–1884); Brier Hill Iron and Coal Co. (1885–1890); S. Frank Eagle (1890–1893)	1893	1895
Ashland	1860, Mineral Ridge	Wood, Warner & Co. (1860–1862); Jonathan Warner (1862–1866); Mineral Ridge Iron and Coal Co. (1866–1868); Brown Iron Co. (1868–1871); James and Lizzie Ward (1871); Jonathan Warner (1872–1875); Jonathan Warner (1878–1880)	1875	Around 1880
Grace No. 2	1861, Youngstown	Brier Hill Iron Co. (1861–1867); Brier Hill Iron and Coal Co. (1867–1912); Brier Hill Steel Co. (1912–1923); Youngstown Sheet & Tube Co. (1923–1974)	Torn down in 1887, rebuilt in 1890; abandoned in 1961	1974
Wheatland	1862, Mineral Ridge; moved to Mercer County, Pennsylvania, in 1862	James Wood & Sons (1862–1868); James Wood's Sons & Co. (1868–1875)	1875	Around 1880
Girard; Mattie	1866, Girard	Girard Iron Co. (1866–1916); A.M. Byers Co. (1916–1929)	1929	1929
Haselton No. 1; Laura	1867, Youngstown	Andrews Bros. Co. (1867–1899); Republic Iron and Steel Co. (1899–1926)	1925	1926

TABLE A.1. Mahoning Valley Blast Furnaces, 1802–1898 *(continued)*

Name(s)	Year Built, Location	Owners (years)	Year Abandoned or Last Operated	Year Dismantled
Haselton No. 2	1868, Youngstown	Andrews Bros. Co. (1868–1888)	1884	1888
Hubbard No. 1	1868, Hubbard	Andrews & Hitchcock (1868–1892); Andrews & Hitchcock Iron Co. (1892–1916); Youngstown Sheet & Tube Co. (1916–1961); Valley Mould & Iron Co. (1961–1967)	1960	1967
Himrod No. 3	1868, Youngstown	Himrod Furnace Co. (1868–1884)	1878	1887
Anna	1869, Struthers	Struthers Iron Co. (1869–1875/1876); Struthers Furnace Co. (1878–1882); Brown, Bonnell & Co. (1882–1885); Struthers Furnace Co. (1885–1892); Brown, Bonnell & Co. (1892–1896); Struthers Furnace Co. (1896–1927); Struthers Iron and Steel Co. (1927–1946); Kaiser-Frazer Parts Corporation (1946–1949); Pittsburgh Coke and Chemical Co. (1949–1961); Youngstown Sheet & Tube Co. (1961–1966)	1953	1966
Ward; Kitty; Thomas; Niles	1870, Niles	William Ward & Co. (1870–1876); Thomas Furnace Co. (1879–1899); *National Steel Co. (1899–1903); *Carnegie Steel Co. (1903–1925)	1919	1925
Warren	1870, Warren	William Richards & Sons (1870–1875)	1875	1878 (burned)
Hubbard No. 2	1872, Hubbard	Andrews & Hitchcock (1872–1892); Andrews & Hitchcock Iron Co. (1892–1916); Youngstown Sheet & Tube Co. (1916–1937)	1937	1937

TABLE A.1. Mahoning Valley Blast Furnaces, 1802–1898 *(continued)*

Name(s)	Year Built, Location	Owners (years)	Year Abandoned or Last Operated	Year Dismantled
Spiegel; Tod No. 2; Tod	1880, Youngstown	Brier Hill Iron and Coal Co. (1880–1890); Youngstown Steel Co. (1890–1912); Brier Hill Steel Co. (1912–1923); Youngstown Sheet & Tube Co. (1923–1929)	Torn down in 1887, rebuilt in 1889; abandoned in 1925	1929

*U.S. Steel subsidiary after 1901

TABLE A.2. Mahoning Valley Puddled Ironworks and Rolling Mills, 1841–1910

Name(s)	Year Built, Location	Owners (years)	Year Abandoned or Last Operated	Year Dismantled
Falcon Iron Works; Old Ward Mill	1841, Niles	James Ward & Co. (1841–1874); Ward Iron Co. (1875–1883); George Summers' Sons (1886–1887); Coleman, Shields & Co. (1887–1899); Continental Iron Co. (1899–1901)	1900	1901
Mahoning Iron Works; Brown-Bonnell Works	1846 (No. 1), 1863 (No. 2), 1879 (No. 3), 1880 (No. 4), Youngstown	Youngstown Iron Co. (1846–c.1850); Brown, Bonnell & Co. (1855–1892); Brown-Bonnell Iron Co. (1892–1899); Republic Iron and Steel Co. (1899–1909)	1902 (No. 1 Mill), 1903 (No. 2 and No. 3 Mill), 1909 (No. 4 Mill)	1902, 1903, 1909 (converted to steel finishing plant exclusively after 1909)
Enterprise Iron Works; Lower Union Mill	1863 (No. 1), 1874 (No. 2), 1890 (No. 3), Youngstown	Shedd, Clark & Co. (1863–1869); Shedd, Cartwright & Co. (1869–1871); Cartwright, McCurdy & Co. (1871–1892); Union Iron and Steel Co. (1892–1899); *American Steel Hoop Co. (1899–1903); *Carnegie Steel Co. (1903–1931)	1930 (converted to steel finishing plant in 1905)	1931
Russia Sheet Iron Mills; Russia Mill	1865, Niles	James Ward & Co. (1865–1874); L. B. Ward Co. (1875–1883); George Summers (1884–1885); Falcon Iron and Nail Co. (1886–1900); *American Sheet Steel Co. (1900–1902)	1900	1902
Niles Iron Works; Haselton Iron Works; Andrews Works	1865, Niles; moved to Youngstown in 1881	Harris, Davis & Co. (1865–1867); Harris, Blackford & Co. (1867–1870); Niles Iron Co. (1872–1883); Andrews Bros. Co. (1883–1899); Republic Iron and Steel Co. (1899–1905)	1905	1905

TABLE A.2. Mahoning Valley Puddled Ironworks and Rolling Mills, 1841–1910 (continued)

Name(s)	Year Built, Location	Owners (years)	Year Abandoned or Last Operated	Year Dismantled
Warren Rolling Mill	1866, Warren	Packard & Barnum Iron Co. (1866–1870); William Richards & Sons (1870–1875); C. Westlake & Co. (1879–1884); Alderdice, Bishop & Co. (1887); Trumbull Iron Co. (1889–1891); Youngstown Iron and Steel Co. (1891–1892); Union Iron and Steel Co. (1892–1899); *American Steel Hoop Co. (1899–1903); *Carnegie Steel Co. (1903–1905)	1905	1905
Falcon Iron and Nail; Falcon Mill	1867, Niles	James Ward & Co. (1867–1874); Falcon Iron and Nail Co. (1875–1900); *American Sheet Steel Co. (1900–1903); *American Sheet & Tin Plate Co. (1903–1904)	1904	1904
Youngstown Rolling Mill; Upper Union Mill	1871, Youngstown	Youngstown Rolling Mill Co. (1871–1891); Youngstown Iron and Steel Co. (1891–1892); Union Iron and Steel Co. (1892–1899); *American Steel Hoop Co. (1899–1903); *Carnegie Steel Co. (1903–1935); *Carnegie-Illinois Steel Corp. (1935–1949)	1949 (converted to iron and steel finishing plant in 1898)	1949
Valley Mill; Mahoning Valley Works	1871, Youngstown	Valley Iron Co. (1871–1873); Wick, Ridgeway & Co. (1873–1875); Mahoning Valley Iron Co. (1879–1899); Republic Iron and Steel Co. (1899–1913)	1913	1913–1914
Hubbard Rolling Mill	1872, Hubbard	Hubbard Iron Co. (1872–1875); Jesse Hall & Sons (1875–1886); Hubbard Iron Co. (1886–1892); Mahoning Valley Iron Co. (1892–1893); Hubbard Co-operative Iron Co. (1893–1895); Mahoning Valley Iron Co. (1895–1898)	1896	1898–1899

TABLE A.2. Mahoning Valley Puddled Ironworks and Rolling Mills, 1841–1910 *(continued)*

Name(s)	Year Built, Location	Owners (years)	Year Abandoned or Last Operated	Year Dismantled
Girard Rolling Mill; Girard Works	1873, Girard	Girard Rolling Mill Co. (1873–1878); Corns Iron Co. (1878–1883); Trumbull Iron Co. (1883–1891); Youngstown Iron and Steel Co. (1891–1892); Union Iron and Steel Co. (1892–1899); *American Steel Hoop Co. (1899–1903); *Carnegie Steel Co. (1903–1905)	1905	1905
Summers Iron Works; Struthers Works	1882, Struthers	Summers Bros. & Co. (1882–1895); Struthers Iron and Steel Co. (1895–1900); *American Sheet Steel Co. (1900–1903); *American Sheet & Tin Plate Co. (1903–1912); Youngstown Sheet & Tube Co. (1912–1913)	1910 (converted to iron and steel finishing plant in 1899)	1913
Youngstown Sheet & Tube Puddling Mill	1901–1902, East Youngstown	Youngstown Sheet & Tube Co. (1902–1930)	1930	1930–1931
Byers's Puddling Mill	1908–1909, Girard	A.M. Byers Co. (1909–1930)	1930	1930–1931

*U.S. Steel subsidiary after 1901

TABLE A.3. Steelworks in the Youngstown District, as Defined by the American Iron and Steel Institute, 1920

Mill	Year(s) Built	Type and Capacity (1920)	Owner and Location	Blast Furnaces Associated with Steelworks (1920)
Mahoning Valley				
Ohio Works	1893–1894	2 10-ton Bessemer converters	Carnegie Steel Co., Youngstown	6
	1908–1909	15 65-ton open-hearth furnaces		
Brown-Bonnell Works	1900	2 10-ton Bessemer converters	Republic Iron and Steel Co., Youngstown	0 (see *Haselton Works*)
East Youngstown Works	1905–1906	2 12-ton Bessemer converters	Youngstown Sheet & Tube Co., East Youngstown	4
	1911–1913	12 100-ton open-hearth furnaces		
Haselton Works	1911–1912	14 70-ton open-hearth furnaces	Republic Iron and Steel Co., Youngstown	5
Brier Hill Works	1912–1913	12 75-ton open-hearth furnaces	Brier Hill Steel Co., Youngstown	3
Lowellville Works	1914–1915	6 75-ton open-hearth furnaces	Sharon Steel Hoop Co., Lowellville	1
Warren Works	1917–1918	7 100-ton open-hearth furnaces	Trumbull Steel Co., Warren	1 (*Furnace built 1920–21*)
Shenango Valley				
New Castle Works*	1892	2 12-ton Bessemer converters	Carnegie Steel Co., New Castle	4
North Sharon Works**	1896–1897	6 40-ton open-hearth furnaces	Carnegie Steel Co., Sharon	1
Farrell Works	1900–1901	15 open-hearth furnaces (12 60-ton and three 75-ton)	Carnegie Steel Co., Farrell	3
Sharon Works***	1902–1904	7 30-ton open-hearth furnaces	Sharon Steel Hoop Co., Sharon	0

*Plant dismantled 1935–1937

**Steelworks dismantled 1923; furnace dismantled 1925

***Steelworks removed between 1923 and 1925

MAPS

Evolution of Iron and Steel Works in the Mahoning and Shenango Valleys, 1850–1930

Maps adapted with permission from Kenneth Warren, *The American Steel Industry, 1850–1970: A Geographical Interpretation* (Oxford: Clarendon Press, 1973), 171.

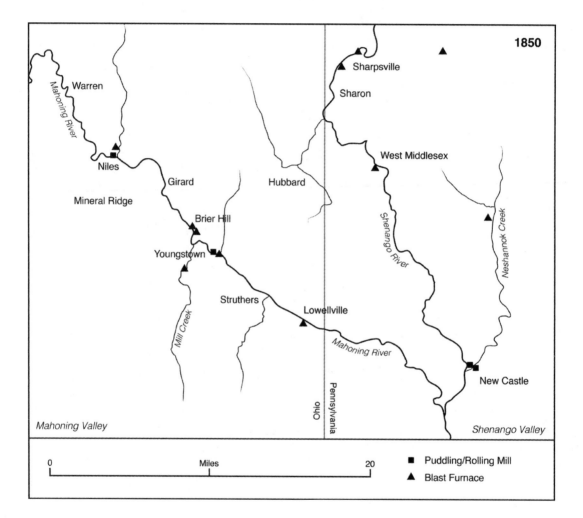

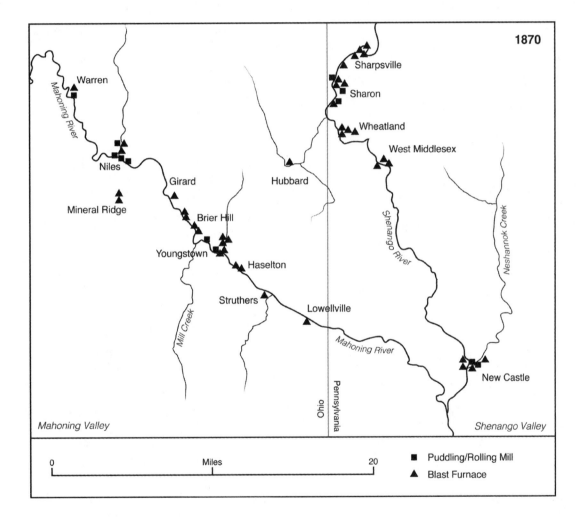

1870

Warren

Mahoning River

Niles

Girard

Mineral Ridge

Brier Hill

Youngstown

Haselton

Struthers

Mill Creek

Lowellville

Mahoning River

Sharpsville

Sharon

Wheatland

West Middlesex

Hubbard

Shenango River

Neshannok Creek

New Castle

Ohio

Pennsylvania

Mahoning Valley

Shenango Valley

0   Miles   20

■ Puddling/Rolling Mill

▲ Blast Furnace

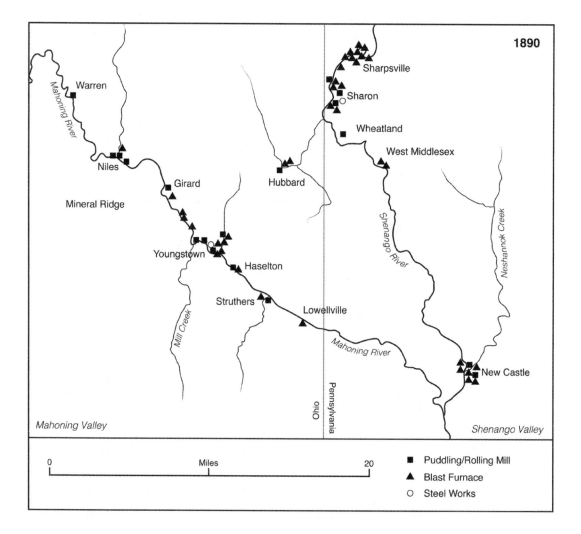

1890

Warren

Mahoning River

Niles

Girard

Mineral Ridge

Youngstown

Haselton

Struthers

Mill Creek

Hubbard

Sharpsville

Sharon

Wheatland

West Middlesex

Shenango River

Lowellville

Mahoning River

Neshannok Creek

New Castle

Ohio

Pennsylvania

Mahoning Valley

Shenango Valley

0   Miles   20

■ Puddling/Rolling Mill
▲ Blast Furnace
○ Steel Works

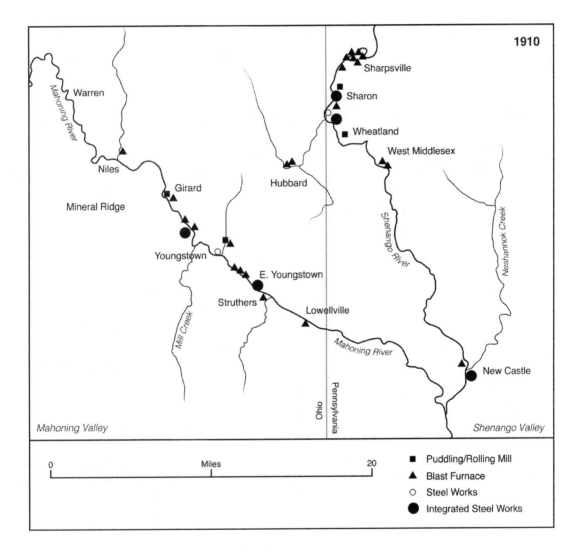

1910

Warren

Mahoning River

Niles

Girard

Mineral Ridge

Youngstown

Struthers

E. Youngstown

Lowellville

Mill Creek

Hubbard

Sharpsville

Sharon

Wheatland

West Middlesex

Shenango River

Neshannok Creek

Mahoning River

New Castle

Ohio

Pennsylvania

Mahoning Valley

Shenango Valley

0   Miles   20

■ Puddling/Rolling Mill
▲ Blast Furnace
○ Steel Works
● Integrated Steel Works

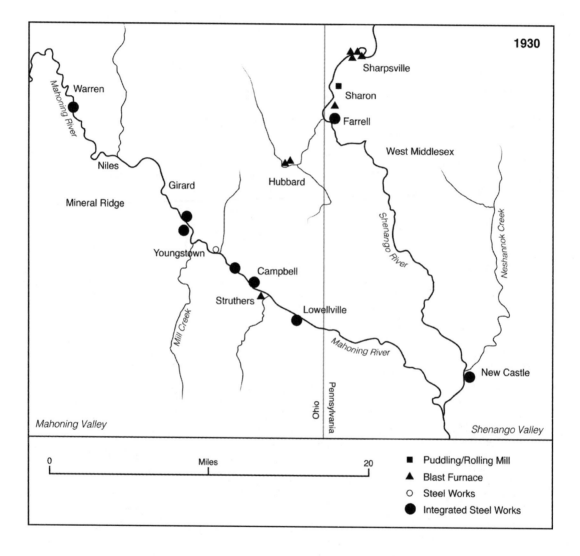

1930

Warren

Mahoning River

Niles

Girard

Mineral Ridge

Youngstown

Struthers

Campbell

Lowellville

Mill Creek

Sharpsville

Sharon

Farrell

West Middlesex

Hubbard

Shenango River

Neshannok Creek

New Castle

Mahoning River

Ohio | Pennsylvania

Mahoning Valley

Shenango Valley

0	Miles	20

■ Puddling/Rolling Mill
▲ Blast Furnace
○ Steel Works
● Integrated Steel Works

NOTES

INTRODUCTION

1. *Industrial World*, no. 52 (December 26, 1910): 1532.
2. Peter Temin, *Iron and Steel in Nineteenth Century America: An Economic Inquiry* (Cambridge, MA: The MIT Press, 1964), 275.
3. *Productivity of Labor in Merchant Blast Furnaces* (Washington, DC: Government Printing Office, 1929), iii.
4. *The Iron Age* 85 (January 27, 1910): 224; G. F. Laughlin, *Mineral Resources of the United States 1918* (Washington, DC: Government Printing Office, 1921), 561.
5. *The Iron Age* 85 (January 27, 1910): 224.
6. Ibid.
7. *Directory to the Iron and Steel Works of the United States and Canada* (New York: American Iron and Steel Institute), 1916, 363.
8. James Howard Bridge, *The Inside History of the Carnegie Steel Company: A Romance of Millions* (New York: The Aldine Book Company, 1903), 55.
9. Robert Bruno, *Steelworker Alley: How Class Works in Youngstown* (Ithaca: Cornell University Press, 1999), 36.
10. Bruno, *Steelworker Alley*, 36.
11. See John N. Ingham, *Making Iron and Steel: Independent Mills in Pittsburgh, 1820–1920* (Columbus: Ohio State University Press, 1991).
12. See David H. Wollman and Donald R. Inman, *Portraits in Steel: An Illustrated History of Jones & Laughlin Steel Corporation* (Kent, OH: Kent State University Press, 1999). Other works include numerous details regarding both Cleveland and Pittsburgh's industry, such as William T. Hogan, *Economic History of the Iron and Steel Industry in the United States*, 5 vols. (New York: Lexington Books, 1971); Kenneth Warren, *The American Steel Industry, 1850–1970: A Geographical Interpretation* (Oxford: Clarendon Press, 1973), and Thomas J. Misa, *A Nation of Steel: The Making of Modern America, 1865–1925* (Baltimore: The Johns Hopkins University Press, 1995). Primary sources and other records from the Mahoning Valley's iron companies are almost nonexistent. This is likely why nearly all historians of the iron and steel industry have neglected the Mahoning Valley.
13. *American Working People*, April 1872, 14.
14. See Samuel Di Rocco II, "In the Shadow of Steel: Leetonia, Ohio and Independent Iron Manufacturers in the Mahoning and Shenango Valleys, 1845–1920," (PhD diss., University of Toledo, 2012), for more information on Shenango Valley ironmasters and their attempts to adapt to large-scale steel production.
15. Warren, *American Steel Industry*, 168.

16. *The Bulletin of the American Iron and Steel Association* 40 (June 1, 1906): 76. In one example, editors reported, "In 1899 the [pig iron] production increased proportionately in both valleys, the Shenango Valley still maintaining its lead, and being 5,050 tons ahead of its rival. In 1900, however, the Mahoning Valley went away ahead of the Shenango Valley, leading it by 202,148 tons."

17. *The National McKinley Birthplace Memorial* (Niles, OH: The National McKinley Birthplace Memorial Association, 1918), 111.

18. Wollman and Inman, *Portraits in Steel*, 3.

CHAPTER 1

1. Frederick J. Blue, William D. Jenkins, H. William Lawson, and Joan M. Reedy, *Mahoning Memories: A History of Youngstown and Mahoning County* (Virginia Beach: The Donning Company Publishers, 1995), 13.

2. Thomas W. Sanderson, *20th Century History of Youngstown and Mahoning County, Ohio* (Chicago: Biographical Publishing Company, 1907), 27–28.

3. Howard C. Aley, *A Heritage to Share: The Bicentennial History of Youngstown and Mahoning County, Ohio* (Youngstown, OH: Youngstown Lithographing Company, 1975), 10. See also Jamie H. Eves, "'Shrunk to a Comparative Rivulet': Deforestation, Stream Flow, and Rural Milling in 19th Century Maine," *Technology and Culture* 33, no. 1 (1992): 38–65.

4. Joseph G. Butler Jr., *History of Youngstown and the Mahoning Valley*, vol. 1 (New York: American Historical Society, 1921), 602.

5. *Historical Collections of the Mahoning Valley*, vol. 1 (Youngstown, OH: Mahoning Valley Historical Society, 1876), 172–73.

6. Butler, *History of Youngstown*, vol. 1, 88.

7. Aley, *Heritage*, 28–29.

8. Butler, *History of Youngstown*, vol. 1, 653.

9. Dean Heaton, *The Heaton Families: 350 Years In America* (Baltimore: Gateway Press, 1982), 256.

10. Helen Vogt, *Westward of Ye Laurall Hills* (Brownsville, PA: published by the author, 1976), 143–234. Myron B. Sharp and William H. Thomas's *A Guide to the Old Stone Blast Furnaces in Western Pennsylvania* (Pittsburgh: Historical Society of Western Pennsylvania, 1966) ascertain that Isaac constructed only a single furnace on the western side of the Monongahela River.

11. United States Department of the Interior, National Park Service, *Colver-Rogers Farmstead, Greene County, Pennsylvania* (NPS form 10–900–1), 8–9, retrieved from https://www.dot7.state.pa.us/ce_imagery/phmc_scans/H114419NOM2.PDF.

12. *History of Trumbull and Mahoning Counties*, vol. 1 (Cleveland: H. Z. Williams & Bro., 1882), 432; Heaton, *Heaton Families*, 260.

13. Butler, *History of Youngstown and the Mahoning Valley*, vol. 1, 473–74.

14. Heaton, *Heaton Families*, 262.

15. Accounts in the 1882 *History of Trumbull and Mahoning Counties* cite the construction of the furnace at 1805 to 1806. Swank's 1884 *History of the Manufacture of Iron in All Ages* lists the furnace being in production by 1804, while Butler in his *History of Youngstown and the Mahoning Valley* contradicts himself, saying it was erected in 1802 and 1803.

16. John R. White, "The Rebirth and Demise of Ohio's Earliest Blast Furnace: An Archaeological Postmortem," *Midcontinental Journal of Archaeology* 21 (1996): 222.

17. Warren, *American Steel Industry*, 35; J. Ramsey Speer, *Chronology of Iron and Steel* (Pittsburgh: Pittsburgh Iron & Steel Foundries Company, 1920), 77.

18. *National McKinley Birthplace Memorial*, 116–17.

19. Ibid., 171.

20. See Joel Mokyr, "Technological Change, 1700–1830," in *The Economic History of Britain Since 1700*, 2nd ed., ed. Roderick Floud and Deirdre McCloskey (Cambridge: Cambridge University Press, 1994), 26.

21. Sir William Crookes, *The Chemical News and Journal of Physical Science* 79 (May 26, 1899): 249.

22. James Moore Swank, *History of the Manufacture of Iron in All Ages* (Philadelphia: The American Iron and Steel Association, 1892), 454.

23. Robert B. Gordon, *American Iron: 1607–1900* (Baltimore: Johns Hopkins University Press, 1996), 1.

24. John Struthers Stewart, *History of Northeastern Ohio,* vol. 1 (Indianapolis: Historical Publishing Co., 1935), 418.

25. J. E. Johnson Jr., *The Principles, Operations, and Products of the Blast Furnace* (New York: McGraw-Hill Book Company, 1918), 1.

26. Ralph Sweetser, *Blast Furnace Practice* (New York: McGraw-Hill Book Company, 1938), 278.

27. L. W. Spring, "Non-technical Chats on Iron and Steel and Their Application to Modern Industry," *The Valve World* 13 (January 1916): 52.

28. Ibid.

29. *The Mahoning Register,* November 26, 1874.

30. *Western Reserve Chronicle,* November 1, 1871.

31. *Mahoning County Register,* November 2, 1871; *Western Reserve Chronicle,* November 1, 1871.

32. Charles Patrick Neill, *Report on Conditions of Employment in the Iron and Steel Industry in the United States,* vol. 4 (Washington, DC: Government Printing Office, 1913), 92.

33. Ibid., 94.

34. Ibid.

35. John R. White, "Rebirth and Demise," 220; Butler, *History of Youngstown,* vol. 1, 174.

36. John R. White, "Rebirth and Demise," 239.

37. Aley, *Heritage,* 35.

38. Swank, *History of the Manufacture of Iron,* 301.

39. Ingham, *Making Iron and Steel,* 22; James Moore Swank, *The American Iron Trade in 1876* (Philadelphia: The American Iron and Steel Association, 1876), 137.

40. Swank, *American Iron Trade,* 137.

41. John R. White, "Rebirth and Demise," 240.

42. Ibid.

43. *The Vindicator,* March 27, 1938.

44. John R. White, "Rebirth and Demise," 222.

45. L. C. Hunter, "The Pittsburgh Iron Industry," *Journal of Economic and Business History* 1 (February 1929): 245, quoted from the *Pittsburgh Gazette,* August 17, 1833.

46. Temin, *Iron and Steel,* 83.

47. Swank, *Iron in All Ages,* 301.

48. Louisa Maria Edwards, *A Pioneer Home Maker, 1787–1866: A Sketch of the Life of Louisa Maria Montgomery* (n.p., 1903), 41.

49. Temin, *Iron and Steel,* 264.

50. Butler, *History of Youngstown,* vol. 1, 661.

51. Temin, *Iron and Steel,* 15.

52. Kenneth Warren, *Wealth, Waste, and Alienation: Growth and Decline in the Connellsville Coke Industry* (Pittsburgh: Pittsburgh University Press, 2001), 1.

53. Swank, *Iron in All Ages,* 41.

54. Kenneth Warren, *Bethlehem Steel: Builder and Arsenal of America* (Pittsburgh: University of Pittsburgh Press, 2008), 5.

55. Butler, *History of Youngstown,* vol. 1, 653.

56. Aley, *Heritage,* 34.

57. Swank, *Iron in All Ages,* 233.

58. *National McKinley Birthplace Memorial,* 118.

59. Butler, *History of Youngstown,* vol. 1, 473.

60. *American Biography: A New Cyclopedia,* vol. 11 (New York: The American Historical Society, 1922), 39.

61. *National McKinley Birthplace Memorial,* 120.

62. *The Vindicator,* March 27, 1938.

63. Jack Gieck, *A Photo Album of Ohio's Canal Era, 1825–1913* (Kent, OH: Kent State University Press, 1988), 200–201.

64. Swank, *Iron in All Ages,* 302–3.

65. William B. McCord, *History of Columbiana County, Ohio and Representative Citizens* (Chicago: Biographical Publishing Co., 1905), 133. In McCord's history of the Rebecca furnace, he incorrectly notes that it was the first iron furnace constructed west of the Allegheny Mountains.

66. Allan Louis Rodgers, "The Iron and Steel Industry of the Mahoning and Shenango Valleys" (PhD diss., University of Wisconsin, 1950), 170.

67. Gordon, *American Iron*, 77.

68. Joseph Nimmo Jr., *Report on the Internal Commerce of the United States* (Washington, DC: Government Printing Office, 1884), 157.

69. J. M. Peck, *New Guide for Emigrants to the West*, 2nd ed. (Boston: Gould, Kendall & Lincoln, 1837), 117.

70. William F. Durfee, "The Development of American Industries Since Columbus: Iron Mills and Puddling Furnaces," *The Popular Science Monthly* 38 (January 1891): 328.

71. Ingham, *Making Iron and Steel*, 34.

72. John A. Fitch, *The Steel Workers* (New York: Charities Publication Committee, 1910), 34.

73. James Davis, *The Iron Puddler: My Life in the Rolling Mills and What Came of It* (Indianapolis: The Bobbs-Merrill Company, 1922), 87.

74. Ibid., 85–86.

75. Gordon, *American Iron*, 147.

76. Butler, *History of Youngstown*, vol. 1, 473.

77. *Western Reserve Chronicle*, September 12, 1833.

78. Richard M. Ruffolo and Charles N. Ciampaglio, *From the Shield to the Sea: Geological Field Trips from the 2011 Joint Meeting of the GSA Northeastern and North-Central Sections* (Boulder, CO: The Geological Society of America, 2011), 162.

79. Freeman Hunt, *Merchants' Magazine and Commercial Review* 16 (1847): 630.

80. Butler, *History of Youngstown*, vol. 1, 663.

81. Heaton, *Heaton Families*, 262.

82. Butler, *History of Youngstown*, vol. 1, 475, notes that various firms leased the Maria furnace throughout the 1840s and early 1850s, but the actual ownership was still vested in the heirs of the Heaton family, though they had no hand in operating the furnace.

83. John Macgregor, *The Progress of America, from the Discovery by Columbus to the Year 1846*, vol. 2 (London: Whittaker & Co., 1847), 453. Temin, *Iron and Steel*, 233, notes that data on the volume of pig iron production in the antebellum period are "fragmentary and of varying reliability."

84. S. D. Harris, *The Ohio Cultivator* 14 (1858): 361; Temin, *Iron and Steel*, 264.

85. Swank, *Iron in All Ages*, 312.

86. Swank, *Iron in All Ages*, 305.

87. Harris, *The Ohio Cultivator* 14 (1858): 361; John D. Knox, "A Century of Ironmaking in Southern Ohio," *The Iron Trade Review* 67 (September 30, 1920): 919.

88. Gordon, *American Iron*, 78–79.

89. *The Iron Trade Review* 67 (September 30, 1920): 921.

90. Swank, *Iron in All Ages*, 305.

91. *The Bicentennial Edition, "Lake County History"* (Painsville: Painsville Publishing Company, 1976), 254–55.

92. William Richard Cutter, *Genealogical and Family History of Western New York*, vol. 2 (New York: Lewis Historical Publishing Company, 1912), 553.

93. Marc Harris, "John Wilkeson," in *Encyclopedia of American Business History and Biography: Iron and Steel in the Nineteenth Century*, edited by Paul F. Paskoff (New York: Bruccoli Clark Layman, 1989), 368.

94. *The Vindicator*, January 25, 1900.

95. *Report on the Manufacturers of the United States at the Tenth Census* (Washington, DC: Government Printing Office, 1883), 103; Marc Harris, "John Wilkeson," in *Encyclopedia of American Business History and Biography: Iron and Steel in the Nineteenth Century*, edited by Paul F. Paskoff (New York: Bruccoli Clark Layman, 1989), 368.

96. Harris, "John Wilkeson," 368.

97. *The Bulletin of the American Iron and Steel Association* (Philadelphia: American Iron and Steel Association, 1894), 75; Swank, *Iron in All Ages*, 282.

CHAPTER 2

1. *Compendium of the Enumeration of the Inhabitants and Statistics of the United States* (Washington, DC: Printed by Thomas Allen, 1841), 78.

2. Aley, *Heritage*, 34, 70.

3. *Inhabitants and Statistics of the United States*, 1841, 78.

4. James Harrison Kennedy, *A History of the City of Cleveland* (Cleveland: The Imperial Press, 1896), 257.

5. *An Accompaniment to Mitchell's Map of the World* (Philadelphia: R. L. Barnes, 1842), 182.

6. Butler, *History of Youngstown*, vol. 1, 177.

7. Gieck, *Photo Album*, 200.

8. John Kilbourn, *Public Documents Concerning the Ohio Canals* (Columbus: Published by I. N. Whiting, 1832), 383.

9. Butler, *History of Youngstown*, vol. 1, 178.

10. Aley, *Heritage*, 60.

11. Ibid., 54.

12. Temin, *Iron and Steel*, 52.

13. Larry Thomas, *Coal Geology* (West Sussex: John Wiley & Sons, 2002), 55. In England, bituminous coal was known as "steam" coal.

14. Butler, *History of Youngstown*, vol. 1, 769.

15. Ibid.

16. Aley, *Heritage*, 87.

17. *Report of the Geological Survey of Ohio*, vol. 5 (Columbus: G. J. Brand & Co., 1884), 318.

18. *Mahoning Register*, January 25, 1872.

19. *The Mahoning Valley, Condensed Statement of Its Resources, and Exhibit of Its Mines, Blast Furnaces & Rolling Mills, Its Commercial Centre, Youngstown, Ohio* (Youngstown, OH: Mahoning Valley Centennial Association, 1876), 3.

20. *Youngstown Past and Present* (Cleveland: Wiggins & McKillop, 1875), 61.

21. Joseph F. Froggett, "The Mahoning Valley as an Iron Center," *The Iron Trade Review* 44 (Cleveland: January 21, 1909): 183.

22. *The Vindicator*, November 10, 1871.

23. Di Rocco, "In the Shadow of Steel," 216.

24. James Macfarlane, *The Coal Regions of America: Their Topography, Geology, and Development* (New York: D. Appleton and Company, 1873), 208.

25. Macfarlane, *Coal Regions*, 320; Gordon, *American Iron*, 161. It is worth noting that coal in Ohio's Hocking Valley was similar in structure and composition, but it did not have the same effect on the region's iron industry. It was largely used in steamboats, locomotives, and rolling mills. By the early 1870s, ironmasters had some success in smelting iron with the coal at furnaces in Columbus and Zanesville.

26. Sanderson, *20th Century History*, 542.

27. *Thirty-Sixth General Assembly of the State of Ohio*, vol. 36 (Columbus: Samuel Medary, Printer to the State, 1838), 328.

28. Kennedy, *History*, 373; *First Annual Report of the State Inspector of Mines for the Year 1874* (Columbus: Nevins and Myers, State Printers, 1875), 20.

29. Ronald L. Lewis, *Welsh Americans: A History of Assimilation in the Coalfields* (Chapel Hill: University of North Carolina Press, 2008), 62.

30. Ibid., 63–64.

31. Kennedy, *History*, 373.

32. Maurice Joblin, *Cleveland Past and Present* (Cleveland: Fairbanks, Benedict & Co., 1869), 321.

33. Aley, *Heritage*, 86.

34. Butler, *History of Youngstown*, vol. 1, 769.

35. Andrew Roy, *A History of the Coal Miners of the United States* (Columbus: Press of J. L. Trauger Printing Company, 1907), 56.

36. Kennedy, *History*, 373.

37. *Nelson's Biographical Dictionary and Historical Reference Book of Erie County, Pennsylvania* (Erie, PA: S. B. Nelson Publisher, 1896), 583.

38. John G. White, *A Twentieth Century History of Mercer County Pennsylvania*, vol. 1 (Chicago: The Lewis Publishing Company, 1909), 365.

39. Swank, *Iron in All Ages*, 372.

40. *Report of the Geological Survey of Ohio*, 452.

41. Ibid.

42. Ibid.

43. Swank, *Iron in All Ages*, 373.

44. See chapter 1.
45. Butler, *History of Youngstown*, vol. 1, 667. Though no other sources confirm Tod's partnership in Wilkeson Wilkes Co., it is possible owing to Tod's development of the Valley's coal mines and Wilkeson's intention to use only Brier Hill and Sharon seam coal in his furnace.
46. Gordon, *American Iron*, 160.
47. Swank, *Iron in All Ages*, 373.
48. Ibid.
49. Ibid., 373–74. Joshua became a prominent blast furnace engineer in western Pennsylvania and constructed, among others, the Isabella furnaces in Pittsburgh in 1872.
50. *Mahoning Index*, September 2, 1846.
51. Ibid.
52. Freeman Hunt, *The Merchant's Magazine and Commercial Review*, vol. 16 (New York: Published at 142 Fulton Street, 1847), 630.
53. Ibid.
54. Swank, *Iron in All Ages*, 373.
55. Gordon, *American Iron*, 160.
56. James Moore Swank, *Introduction to a History of Ironmaking and Coal Mining in Pennsylvania* (Philadelphia: Published by the author, 1878), 77.
57. William F. Durfee, "The Development of American Industries Since Columbus: Iron Smelting by Modern Methods," *The Popular Science Monthly* 38 (February 1891): 458–59.
58. Henry Howe, *Historical Collections of Ohio* (Cincinnati: Published by the author by Derby, Bradley & Co., 1847), 341.
59. *The Mahoning Index*, May 20, 1846.
60. Joblin, *Cleveland Past and Present*, 324.
61. Ibid., 325.
62. Di Rocco, "In the Shadow of Steel," 132.
63. *History of Trumbull and Mahoning Counties*, 371; Di Rocco, "In the Shadow of Steel," 132.
64. *History of Trumbull and Mahoning Counties*, 371.
65. *Mahoning Free Democrat*, January 25, 1854; J. P. Lesley, *The Iron Manufacturer's Guide to the Furnaces, Forges and Rolling Mills of the United States* (New York: John Wiley, Publisher, 1859), 110.
66. Ibid.
67. George Irving Reed, *Century Cyclopedia of History and Biography of Pennsylvania*, vol. 2 (Chicago: The Century Publishing and Engraving Company, 1904), 97.
68. Erasmus Wilson, *Standard History of Pittsburgh, Pennsylvania* (Chicago: H. R. Cornell & Company, 1898), 284.
69. Warren, *American Steel*, 41.
70. Ralph D. Williams, *The Honorable Peter White: A Biographical Sketch of the Lake Superior Iron Country* (Cleveland: The Penton Publishing Co., 1907), 49–50.
71. Swank, *Iron in All Ages*, 222–23.
72. Ralph D. Williams, *The Honorable Peter White*, 50.
73. *The Vindicator*, January 25, 1900.
74. Butler, *History of Youngstown*, vol. 1, 475.
75. *History of Trumbull and Mahoning Counties*, 241.
76. *Western Reserve Chronicle*, August 4, 1858.
77. Butler, *History of Youngstown*, vol. 1, 475.
78. *Western Reserve Chronicle*, February 14, 1866.
79. *National McKinley Birthplace Memorial*, 125.
80. Ibid.
81. Butler, *History of Youngstown*, vol. 1, 603.
82. *Western Reserve Chronicle*, February 14, 1866.
83. Butler, *History of Youngstown*, vol. 1, 681–82.
84. Sanderson, *20th Century History*, 508.
85. Ibid., 671.
86. *Mahoning Free Democrat*, August 23, 1854.
87. Hogan, *Economic History*, vol. 1, 2.
88. Warren, *American Steel*, 42.

89. Swank, *Iron in All Ages*, 240, 313. Cleveland's first rolling mill was a plate mill put into operation in 1854 or 1855. It produced iron directly from iron ore using bloomeries. Cleveland Rolling Mill Co., which opened a rolling mill to reroll iron rails in 1857, built the first blast furnace in Cleveland in 1864, called the Newburg furnace.

90. George H. Thurston, *Pittsburgh as It Is; Or, Facts and Figures Exhibiting the Past and Present of Pittsburgh, its Advantages, Resources, Manufactures, and Commerce* (Pittsburgh: W. S. Haven, 1857), 111.

91. Warren, *American Steel*, 56.

92. *The Mining and Statistic Magazine: Devoted to Mines, Mining Operations, Metallurgy, &c., &c.*, vol. 10 (New York: Published at No. 1 Spruce Street, January to June, 1858), 327. After completing his furnace in 1858, Robson tried using raw coal. The furnace cracked and he consequently switched to using charcoal.

93. Aley, *Heritage*, 73; *Mahoning Free Democrat*, March 1, 1854.

94. Aley, *Heritage*, 73.

95. Butler, *History of Youngstown*, vol. 1, 302–30.

96. *The Vindicator*, February 21, 1897.

97. Cutter, *Genealogical and Family History*, 106.

98. Ralph D. Williams, *The Honorable Peter White*, 77.

99. Ibid.

100. W. J. Comley and W. D'Eggville, *Ohio: The Future Great State* (Cincinnati and Cleveland: Comley Brothers Manufacturing and Publishing Company, 1875), 158.

101. *Mahoning Free Democrat*, December 29, 1853.

102. *Mahoning Free Democrat*, August 2, 1854.

103. William J. Tenney, *The Mining Magazine*, vol. 3 (New York: Published at 98 Broadway, from July to December, 1854), 337.

104. *Mahoning Free Democrat*, August 2, 1854.

105. Ibid.

106. Hogan, *Economic History*, vol. 1, 28; Abram W. Foote, *Foote Family, Comprising the Genealogy and History of Nathaniel Foote of Wethersfield, Conn. and His Descendants*, vol. 1 (Rutland, VT: Marble City Press, 1907), 379.

107. *Mahoning Free Democrat*, January 25, 1854.

108. *Mahoning Free Democrat*, August 30, 1854.

109. Ibid.

110. *The Vindicator*, February 21, 1897; *Mahoning Register*, April 24, 1856; Di Rocco, "In the Shadow of Steel," 136.

111. *History of the Great Lakes*, vol. 1 (Chicago: J. H. Beers & Co., 1899), 561.

112. *Mahoning Register*, July 17, 1856.

113. *History of the Great Lakes*, 561.

114. Froggett, "The Mahoning Valley as an Iron Center," 184.

115. *The Vindicator*, August 3, 1905.

116. *History of the Great Lakes*, 561.

117. Terry S. Reynolds and Virginia P. Dawson, *Iron Will: Cleveland-Cliffs and the Mining of Iron Ore, 1847–2006* (Detroit: Wayne State University Press, 2011), 13.

118. Ibid., 17.

119. Ibid., 15.

120. Ralph D. Williams, *The Honorable Peter White*, 67.

121. William H. Egle, *An Illustrated History of the Commonwealth of Pennsylvania*, 2nd ed. (Philadelphia: E. M. Gardner, 1880), 936; Di Rocco, "In the Shadow of Steel," 221.

122. Ibid.

123. Egle, *Illustrated History*, 70.

124. Reynolds and Dawson, *Iron Will*, 25.

125. Ralph D. Williams, *The Honorable Peter White*, 75–76.

126. Ibid., 68.

127. Reynolds and Dawson, *Iron Will*, 25; Clayton J. Ruminski, "From Mary Furnace to Sharon Steel: Evolution and Integration of the United States' Last Manually Filled Blast Furnace, 1845–1963," *IA: The Journal of the Society for Industrial Archeology* 38, no. 2 (2012): 44.

128. Ralph D. Williams, *The Honorable Peter White*, 78.

129. *Mahoning Register*, September 11, 1856.

130. Ibid., April 17, 1859. According to Lesley's *Iron Manufacturer's Guide*, furnaces in and around Mercer County in western Pennsylvania averaged only five to ten tons of iron a day. Alexander Crawford's Tremont furnace reportedly produced five tons of iron per day over a period of twenty-one weeks in 1856. Other furnaces, such as J. B. Curtis's Sharon furnace, averaged slightly less than five tons daily in 1855.

131. *Mahoning County Register*, April 17, 1859. The Eagle furnace used thirty to forty tons of coal, 160 to 165 tons of Lake ores, and only four to five tons of limestone per day.

132. Warren, *American Steel*, 56.

133. *Mahoning Free Democrat*, August 23, 1854.

134. Ibid., September 20, 1854.

135. Sanderson, *20th Century History*, 585.

136. Ibid., 607.

137. Ibid., 585.

138. Ibid.

139. Ibid.

140. *Biographical History of Northeastern Ohio, Embracing the Counties of Ashtabula, Trumbull and Mahoning* (Chicago: The Lewis Publishing Company, 1893), 601.

141. Sanderson, *20th Century History*, 607.

142. John G. White, *Twentieth Century History*, 378.

143. Froggett, "The Mahoning Valley as an Iron Center," 184.

144. Sanderson, *20th Century History*, 607.

145. *Mahoning Register*, September 10, 1857.

146. *Western Reserve Chronicle*, March 26, 1856.

147. Joseph D. Weeks, *Report on the Statistics of Wages in Manufacturing Industries* (Washington, DC: Government Printing Office, 1886), 219–32.

148. Lesley, *Iron Manufacturer's Guide*, 106.

149. *Western Reserve Chronicle*, January 16, 1856.

150. Joseph G. Butler Jr., *Recollections of Men and Events* (New York: Putnam Publishing, 1927), 57.

151. Richard Meade, *The Coal and Iron Industries of the United Kingdom* (London: Crosby Lockwood and Co., 1882), 719.

152. *Annual Report of the Secretary of State to the Governor of the State of Ohio for the Year 1873* (Columbus: Nevins & Myers, State Printers, 1874), 284–285.

153. Ibid., 283. Before calcination, black-band iron ore contained about 27.12% iron; however, after calcination, black-band yielded about 54.4% iron.

154. Charles Whittlesey, *Report on the Mineral Ridge Mining Property Owned By Rice, French & Co., Located in Weathersfield, Trumbull Co., Ohio* (Cleveland: Harris, Fairbanks & Co., 1856), 9.

155. *National McKinley Birthplace Memorial*, 125.

156. *The Mahoning Valley, condensed statement of its Resources*, 6.

157. Frederick Overman, *The Manufacture of Iron, in All its Various Branches*, 3rd ed. (Philadelphia: Henry C. Baird, 1854), 174.

158. Warren, *Bethlehem Steel*, 8–9.

159. Ingham, *Making Iron and Steel*, 27–28. Ingham notes that in 1857, Pittsburgh's rolling mills produced 67,100 tons of bar iron; 33,488 tons of nails, spikes, and rivets; 10,850 tons of blister, plow, spring, and cast steel; 5,637 tons of sheet iron; 3,212 tons of boiler iron; and several thousand tons of other products.

160. *The Mining and Statistic Magazine*, vol. 10, 327. One furnace in Lawrence County, Ohio, used bituminous coal.

161. Warren, *American Steel*, 52.

162. David D. Van Tassel and John Vacha, *Behind Bayonets: The Civil War in Northern Ohio* (Kent, OH: Kent State University Press, 2006), 82.

163. *Mahoning Register*, May 6, 1858.

CHAPTER 3

1. George W. Knepper, *Ohio and Its People*, 3rd ed. (Kent, OH: Kent State University Press, 2003), 217; Aley, *Heritage*, 46.

2. Anne Kelly Knowles, *Mastering Iron: The Struggle to Modernize an American Industry, 1800–1868* (Chicago: University of Chicago Press, 2013), 187.

3. Sanderson, *20th Century History*, 542.
4. Delmer J. Trester, "The Political Career of David Tod," (MA thesis, Ohio State University, 1950), 73.
5. Joseph P. Smith, *History of the Republican Party in Ohio*, vol. 2 (Chicago: The Lewis Publishing Company, 1898), 425.
6. Knepper, *Ohio and Its People*, 153.
7. Ibid., 218.
8. Leland D. Baldwin, *Pittsburgh: The Story of a City, 1750–1865* (Pittsburgh: University of Pittsburgh Press, 1937), 197.
9. "The Coal and Iron Industry of Cleveland," *Magazine of Western History* 2 (May–October, 1885): 339.
10. Knepper, *Ohio and Its People*, 218.
11. *Journal of the Senate of the Commonwealth of Pennsylvania*, vol. 2 (Harrisburg, PA: J. M. G. Lescure, Printer to the State, 1848), 456–57.
12. *First Annual Report of the Directors of the Cleveland and Mahoning Railroad* (Cleveland: J. W. Gray, Plain Dealer Steam Press, 1853), 5.
13. Ibid., 9.
14. Ibid.
15. *History of Trumbull and Mahoning Counties*, 104.
16. *First Annual Report of the Directors of the Cleveland and Mahoning Railroad*, 11.
17. Ibid., 11–12.
18. Ibid., 26.
19. Butler, *History of Youngstown*, vol. 1, 760.
20. *History of Trumbull and Mahoning Counties*, 104.
21. Ibid., 324.
22. Ibid.
23. *Mahoning Register*, July 3, 1856.
24. *Western Reserve Chronicle*, July 9, 1856.
25. The original route adopted by the C&M Railroad was established by the old Ohio & Erie Railroad in the late 1820s and went from Ashtabula south to Columbiana, but was never constructed. The route for the C&M Railroad was changed to accommodate the coal and iron trade developed by the canal.
26. Aley, *Heritage*, 74.
27. *First Annual Report of the Directors of the Cleveland and Mahoning Railroad*, 29.
28. Aley, *Heritage*, 76.
29. *Mahoning Register*, November 13, 1856.
30. *American Railroad Journal* 30, no. 27 (July 4, 1857): 419.
31. Ibid.
32. *Western Reserve Chronicle*, March 3, 1858.
33. *Mahoning Register*, April 3, 1856.
34. *History of Trumbull and Mahoning Counties*, 236; *Report of the Geological Survey of Ohio*, 320.
35. *Report of the Geological Survey of Ohio*, 174.
36. Ibid.
37. *Western Reserve Chronicle*, December 19, 1855.
38. Ibid.
39. *Annals of Cleveland: Court Record Series*, vol. 3 (Cleveland: Works Project Administration, 1939), 90.
40. *Western Reserve Chronicle*, May 13, 1857.
41. Ibid.
42. *Western Reserve Chronicle*, July 15, 1857.
43. Ibid.
44. Ibid. Rice, French & Co. shipped 10,000 tons of coal between May 1, 1857 and July 15, 1857.
45. *Mahoning Register*, May 6, 1858.
46. Ibid.
47. Butler, *History of Youngstown*, vol. 1, 266.
48. *Western Reserve Chronicle*, April 28, 1858.
49. Ibid.
50. *Mahoning Register*, May 6, 1858; *Western Reserve Chronicle*, April 28, 1858.
51. Ibid.

52. *Western Reserve Chronicle,* July 23, 1856.
53. *Western Reserve Chronicle,* August 10, 1870.
54. Ronald L. Lewis, *Welsh Americans,* 155.
55. *Western Reserve Chronicle,* December 21, 1859.
56. Ibid.
57. *Western Reserve Chronicle,* July 15, 1857.
58. *Mahoning County Register,* December 8, 1859; Di Rocco, "In the Shadow of Steel," 146.
59. *Mahoning County Register,* December 8, 1859
60. *Western Reserve Chronicle,* January 27, 1858.
61. Di Rocco, "In the Shadow of Steel," 146.
62. *History of Trumbull and Mahoning Counties,* 105; John S. Newberry, *Annual Report of the Directors and Chief Engineer of the Ashtabula & New Lisbon Rail Road Co.* (Cleveland: Printed by E. Cowles & Co., 1857), 5.
63. S. H. Church, *Corporate History of the Pennsylvania Lines West of Pittsburgh,* vol. 2 (Baltimore: The Friedenwald Company, 1898), 295; *History of Trumbull and Mahoning Counties,* 105.
64. *Mahoning County Register,* September 22, 1859.
65. *Mahoning County Register,* December 11, 1862.
66. *Annual Report of the Directors and Chief Engineer of the Ashtabula & New Lisbon Rail Road Co.,* 5.
67. *New Castle News,* April 1, 2000.
68. *Western Reserve Chronicle,* February 14, 1866; *Mahoning County Register,* September 8, 1859.
69. *Western Reserve Chronicle,* March 26, 1856.
70. *Western Reserve Chronicle,* October 26, 1859. It is unclear how much working capital James Ward & Co. had following the inclusion of George C. Reis and Andrew Berger into the company.
71. Ibid.
72. Ibid.
73. Temin, *Iron and Steel,* 90.
74. Ibid.; R. N. Grosse, "Determinants of the Size of Iron and Steel Firms in the United States, 1820–1880" (PhD diss., Harvard University, 1948), 270, notes that, of the number of blast furnaces he studied that were in production in 1820, less than twenty percent were integrated with forges.
75. Temin, *Iron and Steel,* 90.
76. Warren, *American Steel,* 88.
77. Gordon, *American Iron,* 162; Swank, *Iron in All Ages,* 256.
78. John N. Boucher, *The Cambria Iron Company* (Harrisburg, PA: Meyers Printing and Publishing House, 1888), 3.
79. Sharon A. Brown, *Cambria Iron Company: Historic Resource Study* (Washington, DC: U.S. Department of the Interior, 1989), 34.
80. Aaron L. Hazen, *20th Century History of New Castle and Lawrence County* (Chicago: Richmond-Arnold Publishing Co., 1908), 118; Lesley, *Iron Manufacturer's Guide,* 108, 252; *Atlas of the County of Lawrence and the State of Pennsylvania* (Philadelphia: G. M. Hopkins & Co., 1872), 47.
81. Lesley, *Iron Manufacturer's Guide,* 108–9, 252.
82. Lesley, *Iron Manufacturer's Guide,* 252–53; Gordon, *American Iron,* 162, states that Winslow split rail was a compound rail, which "made a continuous track from two half-rails fastened together with the joints offset half a rail length."
83. Hazen, *20th Century History,* 120.
84. *Western Reserve Chronicle,* June 27, 1860.
85. *Western Reserve Chronicle,* April 28, 1858; *Atlas of the County of Lawrence and the State of Pennsylvania,* 57.
86. Ibid., 206.
87. *Weekly Telegram,* April 22, 1895.
88. Di Rocco, "In the Shadow of Steel," 134–35.
89. *Weekly Telegram,* April 22, 1895.
90. Swank, *Introduction,* 67–68.
91. *Weekly Telegram,* April 22, 1895.
92. *Western Reserve Chronicle,* June 27, 1860.
93. Ibid.

94. Ibid.
95. Warren, *Wealth, Waste, and Alienation,* 32.
96. *Western Reserve Chronicle,* June 27, 1860.
97. Ibid.
98. Ibid.
99. *Western Reserve Chronicle,* April 17, 1861.
100. *Western Reserve Chronicle,* December 18, 1861.
101. *Western Reserve Chronicle,* June 18, 1862.
102. *Mahoning Dispatch,* May 10, 1878.
103. Ibid.
104. *Western Reserve Chronicle,* June 18, 1862.
105. John G. White, *A Twentieth Century History,* 144.
106. *Second Annual Report of the Bureau of Statistics of Pennsylvania* (Harrisburg, PA: B. F. Meyers, State Printers, 1875), 262.
107. *Western Reserve Chronicle,* June 18, 1862.
108. *History of Trumbull and Mahoning Counties,* 237.
109. Temin, *Iron and Steel,* 283.
110. *Western Reserve Chronicle,* November 12, 1862.
111. *Atlas of the County of Lawrence and the State of Pennsylvania,* 47; *Western Reserve Chronicle,* March 1, 1876.
112. *History of Trumbull and Mahoning Counties,* 237; Butler, *History of Youngstown,* vol. 1, 686.
113. Di Rocco, "In the Shadow of Steel," 147.
114. *Western Reserve Chronicle,* March 11, 1863.
115. Butler, *History of Youngstown,* vol. 1, 193.
116. Ibid., 201.
117. *Youngstown Past and Present,* 76.
118. *Mahoning County Register,* June 30, 1859.
119. *Mahoning Register,* December 24, 1871; *Proceedings of the Lake Superior Mining Institute,* vol. 19 (Ishpeming, MI: Published by the Institute, 1914), 305.
120. *Proceedings of the Lake Superior Mining Institute,* vol. 19, 306; Reynolds and. Dawson, *Iron Will,* 35.
121. Ralph D. Williams, *The Honorable Peter White,* 71.
122. Joseph Wiggins, *Directory of Beaver, Shenango and Mahoning Valleys, for 1869* (Pittsburgh: Printed and Bound by Bakewell & Marthens, 71 Grant Street, 1869), 224–25.
123. Marsena R. Patrick, *Memorial of Hon. William Kelly* (Albany: Joel Munsell, Printer, 1873), 31.
124. Ibid.
125. *Mahoning County Register,* February 23, 1860.
126. *Mahoning County Register,* January 10, 1861.
127. *The Railway News* 4 (October 28, 1865): 461.
128. The Cleveland and Mahoning Railroad, along with its Hubbard Branch, was leased for ninety-nine years by the Atlantic and Great Western Railroad in October 1863.
129. J. Fletcher Brennan, *A Bibliographical Cyclopedia and Portrait Gallery of Distinguished Men* (Cincinnati: John C. Yorston & Company, 1879), 384.
130. Wiggins, *Directory,* 224.
131. J. Disturnell, *The Great Lakes, or Inland Seas of America* (New York: Charles Scribner, 1863), 36–37.
132. Van Tassel and Vacha, *Behind Bayonets,* 82.
133. *Mahoning Register,* May 6, 1858.
134. Ibid., 156.
135. *Mahoning Register,* August 19, 1869. The length of time the furnace was in operation is not clear.
136. *Annual Report of the Secretary of State, for the Year 1860* (Columbus: Richard Nevins, State Printer, 1861), 7.
137. Ibid.
138. Ibid., 672; *Western Reserve Chronicle,* August 10, 1859.
139. Wiggins, *Directory,* 157.
140. Blue, Jenkins, Lawson, and Reedy, *Mahoning Memories,* 37.
141. Ibid., 38.
142. *Mahoning Register,* April 29, 1858; *Western Reserve Chronicle,* March 1, 1876.

143. Copy of the original articles of incorporation of the Tod Iron Co. provided by Marcia Buchanan. Original located at the Trumbull County Archives and Microfilm Department, Warren, Ohio; John N. Ingham, *The Iron Barons: A Social Analysis of an American Urban Elite, 1874–1965* (Westport, CT: Greenwood Press, 1978), xvii; Di Rocco, "In the Shadow of Steel," 148; *Biographical History of Northeastern Ohio*, 470.

144. Wiggins, *Directory*, 226.

145. Warren Van Tine and Michael Pierce, *Builders of Ohio: A Biographical History* (Columbus: Ohio State University Press, 2003), 91.

146. Eugene B. Willard, Daniel W. Williams, George O. Newman, and Charles B. Taylor *A Standard History of the Hanging Rock Iron Region of Ohio*, vol. 1 (Chicago: The Lewis Publishing Company, 1916), 285.

147. Ibid.

148. *The Iron Trade Review* 67 (September 30, 1920): 921.

149. Willard, *Standard History*, 286.

150. Knepper, *Ohio and Its People*, 219–20.

151. Whitelaw Reid, *Ohio in the War: Her Statesmen, Generals and Soldiers*, vol. 2 (Cincinnati: The Robert Clarke Company, 1895), 492.

152. Butler, *History of Youngstown*, vol. 1, 818.

153. Much of the pig iron used in casting guns at Fort Pitt came from the Juniata region in Pennsylvania and the Hanging Rock region in southern Ohio.

154. *The Vindicator*, September 16, 1962. This information regarding General Morgan's intentions to destroy the Mahoning furnace is only reported in this article. No other firsthand accounts or sources confirm the information.

155. *Mahoning County Register*, August 27, 1863.

156. Homer Hamilton began his career as a machinist in Warren before coming to Youngstown and forming a partnership in the foundry of J & C Predmore. Tod and Stambaugh purchased Predmore's interest in 1861.

157. *The Vindicator*, October 1, 1892.

158. *Mahoning County Register*, August 27, 1863.

159. *Report of the Commissioner of Patents for the Year 1863*, vol. 1 (Washington, DC: Government Printing Office, 1866), 855.

160. Ibid.

161. Gordon, *American Iron*, 221.

162. Misa, *Nation of Steel*, 6.

163. Gordon, *American Iron*, 221.

164. Butler, *Recollections*, 45; *Year Book of the American Iron and Steel Institute, 1917* (New York: American Iron and Steel Institute, 1918), 287, 321. Joseph G. Butler Jr. recalls that during Kelly's visit to the Ward household, Kelly was "much exercised over the fact that he had neglected to patent his discovery," but he had high hopes that he would still benefit financially from his process.

165. Misa, *Nation of Steel*, 5.

166. *Mahoning County Register*, August 27, 1863; *The Vindicator*, October 1, 1892.

167. Ibid.

168. Temin, *Iron and Steel*, 283.

169. *Working People*, April 1872, 15.

170. *The Bulletin of the American Iron and Steel Association* 26 (September 28, 1892): 285.

171. *Mahoning County Register*, May 21, 1863.

172. Ibid.

173. *Mahoning County Register*, June 9, 1864.

174. Butler, *History of Youngstown*, vol. 1, 672.

175. *The Vindicator*, October 7, 1940.

176. *Mahoning County Register*, June 9, 1864.

177. Ibid.

178. Butler, *History of Youngstown*, vol. 1, 358.

179. *Mahoning County Register*, March 17, 1859.

180. *Mahoning County Register*, September 18, 1860.

181. Ibid. In 1862, Westerman purchased a share in the Sharon Iron Works in Sharon, Pennsylvania; by 1865, it was renamed the Westerman Iron Company.

182. Annette Blaugrund, *Dispensing Beauty in New York and Beyond: The Triumphs and Tragedies of Harriet Hubbard Ayer* (Charleston: The History Press, 2011), 23.

183. *Mahoning County Register*, September 12, 1863.

184. *The Iron Age* 90 (October 3, 1912): 763; *Mahoning County Register*, September 12, 1863.

185. *Mahoning County Register*, March 5, 1863; *Mahoning County Register*, September 18, 1860.

186. *Mahoning Register*, July 15, 1858.

187. *Statistics of the American and Foreign Iron Trades for 1902* (Philadelphia: The American Iron and Steel Association, 1903), 10.

188. Warren, *Wealth, Waste, and Alienation*, 22.

189. *Mahoning County Register*, May 25, 1865.

190. *Mahoning County Register*, October 18, 1865.

191. *The Iron Trade Review* 44 (April 22, 1909): 746.

192. *Executive Documents Printed by Order of the House of Representatives during the Second Session of the Fortieth Congress, 1867–68* (Washington, DC: Government Printing Office, 1868), 663.

193. *The Vindicator*, August 23, 1931.

194. *Pittsburgh: Its Industry & Commerce* (Pittsburgh: Barr & Myers, 145 Wood Street, 1870), 28; Warren, *American Steel*, 58; T. Sterry Hunt, *The Coal and Iron of Southern Ohio* (Salem, MA: Naturalists' Agency, 1874), 29.

195. F. A. Herwig, *The Present Management of the Reading Railroad, as it Affects the Coal Regions, the Coal Miners and Consumers* (Pottsville, PA: Standard and Chronicle Print, 1879), 15.

196. *Mahoning County Register*, June 15, 1865.

197. Warren, *American Steel*, 56.

CHAPTER 4

1. Butler , *History of Youngstown*, 201.

2. Ibid.

3. Aley, *Heritage*, 83.

4. Ibid.

5. Ibid.

6. *American Working People*, April 1872, 15.

7. *The Engineering and Mining Journal* XLI (February 27, 1886): 152.

8. Temin, *Iron and Steel*, 274; *Hearings Before the Committee on Investigation of the United States Steel Corporation*, vol. 5 (Washington DC: Government Printing Office, 1912), 3466.

9. *Western Reserve Chronicle*, February 3, 1864.

10. Butler, *History of Youngstown*, vol. 1, 827.

11. Ibid.

12. Smith, *History of the Republican Party*, 559.

13. Butler, *Recollections*, 41.

14. Ibid.

15. Ibid., 43.

16. "Sketch of Career of Joseph G. Butler, Jr.," *The Blast Furnace & Steel Plant* 10 (January, 1922): 61.

17. Butler, *Recollections*, 60.

18. Butler, *History of Youngstown*, vol. 1, 821.

19. Ibid.

20. Ibid.

21. David Bremner, *The Industries of Scotland* (Edinburgh: Adam and Charles Black, 1869), 57.

22. *Western Reserve Chronicle*, February 14, 1866.

23. *Western Reserve Chronicle*, December 7, 1864; *Western Reserve Chronicle*, August 3, 1864.

24. Ibid., 45.

25. *History of Pittsburgh and Environs*, vol. 3 (New York and Chicago: American Historical Society, 1922), 843. In 1848, William H. Brown formed a partnership with Alexander Miller and George Black, who operated the Kensington Iron Works in Pittsburgh, in order to supply the company coal.

26. *Mahoning County Register*, January 18, 1866.

27. Gordon, *American Iron*, 194.

28. *The Hancock Jeffersonian* (Findlay, Ohio), July 19, 1867.

29. Gordon, *American Iron*, 194.

30. Ibid.

31. *Niles' National Register* 75 (April 4, 1849): 216.

32. John Percy, *The Manufacture of Russian Sheet-Iron* (Philadelphia: Henry Carey Baird, 1871), 12.

33. Ibid.

34. *Cleveland Daily Leader* (Morning Edition), August 22, 1866. The paper reports that several companies in the United States sent agents to the vicinity of the Russian works to steal their secret, but all efforts were failures.

35. *The Niles Times* (Niles, Ohio), July 8, 1993.

36. *Western Reserve Chronicle*, January 31, 1866.

37. Ibid.

38. Butler, *History of Youngstown*, vol. 1, 476.

39. *Cleveland Daily Leader* (Morning Edition), August 22, 1866; *American Artisan and Patent Record* 3 (October 10, 1866): 356.

40. *The Evening Telegraph* (Philadelphia), February 28, 1867; *The Daily Phoenix* (Columbia, SC), March 19, 1867; *Evening Star* (Washington DC), March 2, 1867; *The Wheeling Daily Intelligencer* (Wheeling, WV), March 6, 1867. These papers reported the spelling of Kungonchieff's name in several different variations, including *Kouongshiff* and *Keengancheff*.

41. *Western Reserve Chronicle*, February 14, 1866.

42. J. Leander Bishop, *A History of American Manufactures from 1608–1860*, vol. 3 (Philadelphia: Edward Young & Co., 1868), 475.

43. *The Mechanics' Magazine* 83 (July 14, 1865): 24. Likewise, three furnaces built by the Acklam Iron Works in Middlesborough, England, in 1865 and 1866 were seventy feet in height with boshes twenty-two-and-a-half feet in diameter. These were reported to produce 350 tons of iron per week. John F. Frazer, *Journal of the Franklin Institute of the State of* Pennsylvania 51 (1866): 40.

44. *Western Reserve Chronicle*, February 14, 1866.

45. *Western Reserve Chronicle*, April 24, 1867; A. T. Andreas, *History of Chicago, from the Earliest Period to the Present Time*, vol. 3 (Chicago: The A. T. Andreas Company, Publishers, 1886), 506.

46. Ibid.

47. *Mahoning Dispatch*, May 10, 1878.

48. Di Rocco, "In the Shadow of Steel," 130; McCord, *History of Columbiana County*, 135.

49. Copy of the original articles of incorporation of the Mineral Ridge Iron and Coal Co. provided by Marcia Buchanan. Original located at the Trumbull County Archives and Microfilm Department, Warren, Ohio.

50. *Mahoning County Register*, September 19, 1867.

51. Copy of the original articles of incorporation of the Brown Iron Co. provided by Marcia Buchanan. Original located at the Trumbull County Archives and Microfilm Department, Warren, Ohio.

52. *History of Trumbull and Mahoning Counties*, 232.

53. *Western Reserve Chronicle*, September 28, 1870.

54. Ibid.

55. Jacob Greenebaum Jr. was a brother of Michael Greenebaum, a partner with James Ward Jr. and W. H. Brown in the Falcon Iron and Nail Co.

56. Hogan, *Economic History*, vol. 1, 14; Temin, *Iron and Steel*, 283.

57. *Pittsburgh: Its Industry & Commerce*, 6.

58. Ibid. The steel mills referenced are small-scale, crucible steel works, not Bessemer.

59. *The Ironworks of the United States* (Philadelphia: The American Iron and Steel Association, 1874), 71.

60. George H. Thurston, *Pittsburgh and Allegheny in the Centennial Year* (Pittsburgh: A. A. Anderson & Son, 1876), 176; Warren, *American Steel*, 58.

61. *American Working People*, April 1872, 15.

62. Temin, *Iron and Steel*, 97.

63. *The Repertory of Patent Inventions*, vol. 30 (London: Alexander Macintosh, Great New-Street, July–December, 1857), 177–79.

64. *American Working People*, April 1872, 15.

65. *Mahoning Register*, September 24, 1868.

66. Temin, *Iron and Steel*, 97; George Wilkie, *The Manufacture of Iron in Great Britain* (London: A. Fullarton & Co., 1857), 85; Samuel Baldwin Rogers, *An Elementary Treatise of Iron Metallurgy* (New York: H. Bailliere, Publisher and Foreign Bookseller, 1857), 195.

67. "Apparatus for Utilizing the Waste Gases from Blast Furnaces," *Scientific American* 4, no. 4 (January 26, 1861): 49–50.

68. *The Vindicator*, August 23, 1931.

69. Gordon, *American Iron*, 165–66.

70. Sanderson, *20th Century History*, 262; *Western Reserve Chronicle*, November 20, 1867.

71. *American Working People*, April 1872, 16.

72. Butler, *History of Youngstown*, vol. 1, 503.

73. *History of Trumbull and Mahoning Counties*, 436; Butler, *History of Youngstown*, vol. 1, 503.

74. Egbert Cleave, *City of Cleveland and Cuyahoga County, Taken from Cleave's Biographical Cyclopedia of the State of Ohio* (Cleveland: Fairbanks, Benedict & Co., 1875), 27.

75. *Western Reserve Chronicle*, April 30, 1873.

76. Wiggins, *Directory*, 224.

77. U. A. Swogger, "Coalburg Once Was Booming, Rip-Roaring Coal Mining Community of 1,500 People," *Youngstown Sheet & Tube Bulletin*, February 1959: 16.

78. N. J. Drohan, *History of Hubbard, Ohio: Its People, Churches, Industries and Institutions—from Early Settlement in 1798 to 1907* (Hubbard: H. W. Ulrich Print Co., 1907), 20.

79. *Western Reserve Chronicle*, April 30, 1873. Some small mines were not accounted for.

80. Joseph G. Butler Jr., "Early History of Iron and Steel Making in Mahoning Valley," *The Iron Trade Review* 67 (August 20, 1925): 426.

81. *Western Reserve Chronicle*, January 26, 1870.

82. Knowles, *Mastering Iron*, 7–8.

83. *The Vindicator*, February 26, 1876.

84. Butler, *History of Youngstown*, vol. 1, 688.

85. Ibid.

86. Knowles, *Mastering Iron*, 7. As was the case in the Mahoning Valley, Welsh immigrant ironworkers played a significant role in transforming the early American iron industry. Along with English iron artisans, Knowles notes that Welsh puddlers and blast furnace workers showed up at nearly every iron mill in the early to mid-nineteenth century.

87. Philip Jenkins, *A History of Modern Wales, 1536–1990* (New York: Routledge, 2014), 221. The Dowlais area had forty-four blast furnaces in the 1840s and employed about 7,000 workers. Almost 200,000 people depended on the iron furnaces for subsistence.

88. Butler, *History of Youngstown*, vol. 1, 686.

89. *Mahoning Register*, February 14, 1867.

90. Ibid.

91. In February 1867, the *Mahoning Register* reported that the furnace produced about twenty-five tons of "good quality" pig iron per day. They also noted that when "its working is more thoroughly understood," the furnace's yield would increase to thirty tons per day.

92. Butler, "Early History," 426.

93. Ibid.

94. Ibid.; *The Vindicator*, February 4, 1870.

95. *The Vindicator*, October 6, 1900. In the early 1870s, partners of the Girard Iron Co. included Byers and Fleming of Pittsburgh; Butler, John Tod, and Charles D. Arms of Youngstown; and Evan Morris of Girard. By 1900, Byers was the sole owner of the Girard Iron Co.

96. *The Ironworks of the United States, 1876*, 47.

97. Rodgers, "Iron and Steel Industry," 172.

98. *Western Reserve Chronicle*, October 31, 1866.

99. Butler, *History of Youngstown*, vol. 1, 434; Rodgers, "Iron and Steel Industry," 171–72.

100. *Twenty-Fifth Annual Report of the Ohio State Board of Agriculture for the Year 1870* (Columbus: Nevins & Myers, State Printers, 1871), xvi.

101. *Western Reserve Chronicle*, May 1, 1872.

102. *Western Reserve Chronicle*, June 5, 1867.

103. *Western Reserve Chronicle*, March 27, 1867.

104. Ibid.

105. Ibid.

106. *Western Reserve Chronicle*, October 20, 1869; Contract between Warren Packard and William Richards, January 18, 1870, provided by Shelley Richards.
107. *Western Reserve Chronicle*, November 11, 1873.
108. Ibid.
109. *Western Reserve Chronicle*, September 14, 1870.
110. *Western Reserve Chronicle*, October 31, 1866.
111. *The Vindicator*, February 4, 1870.
112. Hogan, *Economic History*, vol. 1, 14.
113. "Iron and Steel: The Pennsylvania Lines in the Development of Youngstown and Vicinity," *Mahoning Bank Bulletin* 3 (October, 1923): 1.
114. *The Mahoning Register*, June 9, 1870.
115. *First Annual Report of the State Inspector of Mines*, 56.
116. *Western Reserve Chronicle*, January 20, 1869.
117. *Western Reserve Chronicle*, April 30, 1873.
118. Frederick E. Saward, *The Coal Trade* (New York: Published at 111 Broadway, 1878), 50; *History of Trumbull and Mahoning Counties*, 99.
119. *The Iron Age* 11 (January 23, 1873): 5. Trumbull County produced 3,080,804 bushels of coal in 1872, whereas Mahoning County only produced 648,440 bushels.
120. Warren, *Wealth, Waste, and Alienation*, 25.
121. *The International Review* 1 (1874): 771.
122. Temin, *Iron and Steel*, 201; Warren, *American Steel*, 110; Mark Aldrich, *Safety First: Technology, Labor, and Business in the Building of American Work Safety, 1870–1939* (Baltimore: The Johns Hopkins University Press, 1997), 87.
123. Robert C. Allen, "The Peculiar Productivity History of American Blast Furnaces, 1840–1913," *The Journal of Economic History* 37 (September 1977): 615.
124. *Report of the Geological Survey of Ohio*, 535–36.
125. *The Iron Age* 104 (August 14, 1919): 432.
126. Joel Sabadasz, "The Development of Modern Blast Furnace Practice: The Monongahela Valley Furnaces of the Carnegie Steel Company, 1872–1913," *IA: The Journal of the Society for Industrial Archeology* 18 (1992): 95.
127. Ibid.
128. Bridge, *Inside History*, 56.
129. Ibid., 55.
130. *The Mahoning Register*, October 26, 1871.
131. Ibid.
132. Bridge, *Inside History*, 56; *Directory to the Iron and Steel Works of the United States* (Philadelphia: The American Iron and Steel Association, 1878), 57.
133. Joseph G. Butler Jr., *Fifty Years of Iron and Steel* (Cleveland: The Penton Press, 1922), 32.
134. Ibid.
135. Ibid.; Warren, *Wealth, Waste, and Alienation*, 39.
136. *American Working People*, April 1872, 16.
137. Franklin Ellis, *History of Fayette County, Pennsylvania, with Biographical Sketches of Many of its Pioneers and Prominent Men* (Philadelphia: L. H. Everts & Co., 1882), 519.
138. Saward, *Coal Trade*, 30; Warren, *Wealth, Waste, and Alienation*, 59.
139. *The Vindicator*, June 4, 1933.
140. Brennan, *Biographical Cyclopedia*, 384.
141. Wiggins, *Directory*, 224.
142. *History of Trumbull and Mahoning Counties*, 105.
143. Brennan, *Biographical Cyclopedia*, 384.
144. *Western Reserve Chronicle*, December 7, 1870.
145. *Statistical Report for the National Association of Iron Manufacturers for 1872* (Philadelphia: J. A. Wagenseller, 1873), 18.
146. I. Lowthian Bell, "Notes of a Visit to Coal and Iron Mines and Ironworks in the United States," *The Journal of the Iron and Steel Institute* 9 (1875): 117.
147. *The Mahoning Register*, November 26, 1874.
148. *The Mahoning Register*, December 3, 1874.
149. *The Mahoning Register*, January 28, 1875.
150. Lorett Treese, *Railroads of Pennsylvania: Fragments of the Past in the Keystone Landscape* (Mechanicsburg, PA: Stackpole Books, 2003), 216; Albert J. Churella, *The Pennsylvania*

Railroad: Building an Empire, 1846–1917, vol. 1 (Philadelphia: University of Pennsylvania Press, 2013), 496.

151. *Annual Report of the Commissioner of Railroads and Telegraphs of Ohio for the Year Ending June 30, 1880* (Columbus: J. G. Brand & Co., 1880), 1198.

152. *Bulletin of the American Iron and Steel Association* 7, no. 37 (May 14, 1873): 292.

153. Aley, *Heritage*, 59.

154. *Western Reserve Chronicle*, May 7, 1873.

155. *Western Reserve Chronicle*, April 30, 1873.

156. *Bulletin of the American Iron and Steel Association* 7, no. 27 (March 9, 1870): 209–10.

157. *Statistical Report of the National Association of Iron Manufacturers for 1872*, 18.

158. *Leading Manufacturers and Merchants of Ohio Valley* (New York: International Publishing Company, Publishers, 1887), 133.

159. *Statistics of the American and Foreign Iron Trades for 1877*, 38.

160. Temin, *Iron and Steel*, 274.

161. *Bulletin of the American Iron and Steel Association* 7, no. 37 (May 14, 1873): 292.

162. Benson J. Lossing, *History of American Industries and Arts* (Philadelphia: Porter & Coates, 1876), 175; Hogan, *Economic History*, vol. 1, 40.

163. *The Ironworks of the United States, 1874*, 92.

164. Ibid., 82, 92.

165. Christopher Peter Sandberg, "The Manufacture and Wear of Rails," *Van Nostrand's Eclectic Engineering Magazine* 7 (July–December, 1872): 93.

166. Temin, *Iron and Steel*, 121.

167. *The Engineering and Mining Journal* XLI (February 27, 1886): 152.

168. *The Successful American* IV, no. 2 (1901): 485.

169. *The Mahoning Register*, March 30, 1871.

170. *American Working People*, April 1872, 15.

171. Ibid.

172. *The Mahoning Register*, August 1, 1872.

173. Ibid.; *The Mahoning Register*, August 15, 1872.

174. *Bulletin of the American Iron and Steel Association* 7, no. 53 (September 3, 1873): 417.

175. Allan Nevins, *The Emergence of Modern America, 1865–1878* (New York: The Macmillan Company, 1927), 298.

176. *The Vindicator*, September 21, 1913.

177. *History of the Terrible Financial Panic of 1873* (Written and Compiled by a Journalist, 1873), 4.

178. J. R. Vernon, "Unemployment Rates in Postbellum America: 1869–1899," *Journal of Macroeconomics* 16, no. 4 (1994): 710.

179. M. John Lubetkin, *Jay Cooke's Gamble: The Northern Pacific Railroad, the Sioux, and the Panic of 1873* (Norman: University of Oklahoma Press, 2006), 285.

180. Weeks, *Statistics of Wages*, 121.

181. Butler, *History of Youngstown*, vol. 1, 205.

182. Aley, *Heritage*, 98.

183. *The Mahoning Register*, November 12, 1874.

184. Ingham, *Making Iron and Steel*, 48.

185. Nevins, *Emergence of Modern America*, 298.

186. Butler, *History of Youngstown*, vol. 1, 476–77.

187. See Elmus Wicker, *Banking Panics of the Gilded Age* (Cambridge: Cambridge University Press, 2000), and Charles P. Kindleberger and Robert Z. Aliber, *Manias, Panics, and Crashes: A History of Financial Crises* (New York: Palgrave Macmillan, 2011).

188. *Western Reserve Chronicle*, February 18, 1874.

189. Ibid.

190. Knowles, *Mastering Iron*, 152.

191. Bridge, *Inside History*, 22–23.

192. *The Vindicator*, March 4, 1891.

193. David Montgomery, *The Fall of the House of Labor: The Workplace, the State, and American Labor Activism, 1865–1925* (Cambridge: Cambridge University Press, 1987), 21.

194. Butler, *Recollections*, 123.

195. *The Mahoning Register*, February 29, 1872.

196. *Western Reserve Chronicle*, February 10, 1875.

197. *Western Reserve Chronicle*, September 9, 1874.

198. Sanderson, *20th Century History*, 608. William Bonnell was a partner in Brown's Chicago company before his death.
199. *Engineering News* 8 (July 2, 1881): 270.
200. *The New York Times*, April 27, 1887.
201. Weeks, *Statistics of Wages*, 122.
202. Butler, "Early History," 426.
203. Kevin Hillstrom and Laurie Collier Hillstrom, *The Industrial Revolution in America: Communications*, vol. 1 (Santa Barbara: ABC-Clio INC., 2007), 39–40.
204. Ingham, *Making Iron and Steel*, 17–18.

CHAPTER 5

1. Aley, *Heritage*, 103.
2. *Compendium of the Tenth Census, Part I* (Washington DC: Government Printing Office, 1883), 461.
3. Aley, *Heritage,* 98; Butler, *History of Youngstown*, 210–12.
4. James E. Homans, *The Cyclopaedia of American Biography*, vol. 8 (New York: The Press Association Compilers, 1918), 46.
5. Hogan, *Economic History*, 94–100.
6. *Statistics of the American and Foreign Iron Trades for 1881*, 50.
7. Ibid., 25.
8. Warren, *Bethlehem Steel*, 33.
9. *History of Trumbull and Mahoning Counties*, 99–100.
10. Ingham, *Making Iron and Steel*, 183.
11. *Western Reserve Chronicle*, October 30, 1878.
12. *The Bulletin of the American Iron and Steel Association* 17 (Philadelphia, August 1, 1883): 203.
13. It was highly uncommon for women to run an iron company. One of the few instances in the nineteenth century was in Chester County, Pennsylvania. Rebecca Lukens ran the Brandywine Iron Works after her husband, the founder of the mill, died in 1825. By her retirement in 1847, Lukens had added another plant to the mill and turned a profit; she left management in the hands of her brother-in-law and son-in-law. Madeleine B. Stern, *We the Women: Career Firsts of Nineteenth-Century America* (Lincoln: University of Nebraska Press, 1994), 244.
14. Swank, *Iron in All Ages*, 451.
15. *Statistics of the American and Foreign Iron Trades For 1888*, 40.
16. *Directory*, 1882, 138.
17. Butler, *History of Youngstown*, vol. 2, 1–2.
18. *Western Reserve Chronicle*, November 22, 1882.
19. Ibid.
20. *Directory*, 1886, 130; *Directory*, 1892, 156.
21. Butler, *History of Youngstown*, vol. 3, 805.
22. *Reports of the United States Board of Tax Appeals* 6 (Washington DC: Government Printing Office, 1927), 9. Thomas acquired Lizzie B. Ward's interests in Niles Fire Brick in 1879.
23. Joseph F. Froggett, "The Mahoning Valley as an Iron Center," *The Iron Trade Review* 44 (January 28, 1909): 225; *Directory*, 1882, 63.
24. *Report of the Select Committee on Ordnance and War Ships* (Washington: Government Printing Office, 1886), 291.
25. *Directory*, 1882, 137. According to the 1884 *Directory*, the company with the second most puddling furnaces was Pittsburgh's Allegheny, Monongahela, and Birmingham Iron Works owned by Oliver Brothers & Phillips. The mills had a combined 107 furnaces.
26. *The Vindicator*, April 29, 1893.
27. *Chicago Daily Tribune*, January 13, 1899.
28. Blaugrund, *Dispensing Beauty*, 52.
29. *The Public* 23 (February 22, 1883): 115.
30. *The Bulletin of the American Iron and Steel Association* 15 (July 20, 1881): 180.
31. See E. L. DeWitt, *Reports of Cases Argued and Determined in the Supreme Court of Ohio* 35 (Cincinnati: Robert Clarke & Co., 1880), 10.

32. *Weekly Register*, March 8, 1882.

33. *Directory*, 1882, 137–38; *Register & Tribune*, June 17, 1880.

34. See Donna M. DeBlasio and Martha I. Pallante, "Memories of Work and the Definition of Community: The Making of Italian Americans in the Mahoning Valley," *Ohio History* 121 (2014): 89–111.

35. "An English Opinion of American Iron Works," *The Bulletin of the American Iron and Steel Association* 16 (June 7, 1882): 155.

36. *Statistics of the American and Foreign Iron Trade for 1882*, 36.

37. *The Vindicator*, April 29, 1893.

38. *The Bulletin of the American Iron and Steel Association* 16 (June 7, 1882): 155.

39. David Brody, *Steelworkers in America: The Nonunion Era* (Cambridge, MA: Harvard University Press, 1960), 2–3.

40. *The Vindicator*, February 4, 1870.

41. Ibid.

42. *Youngstown Past and Present*, 53, 62. Andrews & Hitchcock's mines and furnaces in Hubbard Township experienced a six-month strike in 1873.

43. Hogan, *Economic History*, vol. 1, 86.

44. Carroll D. Wright, "The National Amalgamated Association of Iron, Steel, and Tin Workers, 1892–1901," *The Quarterly Journal of Economics* 16 (November 1901): 37–38.

45. *History of Trumbull and Mahoning Counties*, 423.

46. Jesse S. Robinson, *The Amalgamated Association of Iron, Steel and Tin Workers* (Baltimore: The Johns Hopkins Press, 1920), 20–21.

47. *Annual Report of the Secretary of Internal Affairs of the Commonwealth of Pennsylvania, Part III: Industrial Statistics*, vol. 10 (Harrisburg: Lane S. Hart, 1883), 176.

48. Wollman and Inman, *Portraits in Steel*, 30–31.

49. *Bulletin of the American Iron and Steel Association* 16 (May 31, 1882): 149.

50. John R. Commons, *History of Labour in the United States*, vol. 2 (New York: The MacMillan Company, 1921), 316.

51. *The Vindicator*, June 2, 1882.

52. *Bulletin of the American Iron and Steel Association* 16 (August 9 and 16, 1882): 221.

53. *The Vindicator*, June 9, 1882.

54. *The Vindicator*, September 22, 1882.

55. *Bulletin of the American Iron and Steel Association* 16 (September 27, 1882): 261.

56. *The Wheeling Daily Intelligencer*, September 8, 1882.

57. Ibid.

58. Blaugrund, *Dispensing Beauty*, 52.

59. *Statistics of the American and Foreign Iron Trades for 1882*, 9; Temin, *Iron and Steel*, 284.

60. *Statistics of the American and Foreign Iron Trades for 1883*, 67.

61. *The New York Times*, February 20, 1883.

62. *The New York Times*, February 19, 1883.

63. Ibid.

64. *Bulletin of the American Iron and Steel Association* 17 (May 2, 1883): 115.

65. *The New York Times*, February 19, 1883.

66. *Transactions of the American Institute of Mining Engineers* 20 (1892): 273.

67. *History of Trumbull and Mahoning Counties*, 231.

68. *Bulletin of the American Iron and Steel Association* 17 (August 1, 1883): 203.

69. *Western Reserve Chronicle*, August 1, 1883.

70. *Bulletin of the American Iron and Steel Association* 19 (August 5, 1885): 205.

71. *Western Reserve Chronicle*, November 4, 1885.

72. Ibid.

73. Ibid.

74. Ibid.

75. Ibid.

76. *Western Reserve Chronicle*, February 18, 1885.

77. Butler, *History of Youngstown*, vol. 2, 51.

78. *Western Reserve Chronicle*, September 11, 1878.

79. Sanderson, *20th Century History*, 166.

80. *The Engineering and Mining Journal* 46 (July 28, 1888): 69; Butler, *History of Youngstown*, vol. 3, 692.

81. Sanderson, *20th Century History*, 551.

82. *Biographical History of Northeastern Ohio*, 586.

83. *Register and Tribune,* February 5, 1880.

84. Butler, *History of Youngstown,* vol. 1, 722.

85. Butler, *History of Youngstown,* vol. 3, 763.

86. Ibid.

87. *The Vindicator,* December 21, 1941.

88. *The Vindicator,* April 29, 1893.

89. Abner C. Harding, "Fifty Years of Mechanical Engineering," *The Popular Science Monthly* 24 (February 1884): 536.

90. *Arguments Before the Committee of Ways and Means on the Morrison Tariff Bill* (Washington DC: Government Printing Office, 1884), 120.

91. *The Vindicator,* October 25, 1878.

92. *Directory,* 1888, 56.

93. Robert Forsythe, *The Blast Furnace and the Manufacture of Pig Iron,* 2nd edition (New York: David Williams Company, 1909), 287.

94. Butler, *History of Youngstown,* vol. 1, 686; *The Iron Age* 55 (March 28, 1895): 670.

95. *The Iron Age* 55 (March 28, 1895): 670; *The Mahoning Valley, condensed statement of its Resources* (Youngstown: Mahoning Valley Centennial Association, 1876), 6.

96. John D. Knox, "Mary Blast Furnace: Last of Hand-Filled Stacks in America," *Steel* 125 (October 1949): 140.

97. Butler, *Recollections,* 57.

98. Edward Kirk, *A Practical Treatise on Foundry Irons* (Philadelphia: Henry Carey Baird & Co., 1911), 30.

99. *The Journal of the Iron and Steel Institute* (New York: 44 Murray Street, 1886), 426.

100. *Statistics of the American and Foreign Iron Trades for 1888,* 14.

101. *American Machinist* 4 (October 15, 1881): 9.

102. *The William B. Pollock Company Presents the Seventy-Five Year History of its Contributions to the Advancement of the Art of Iron and Steelmaking* (Youngstown: William B. Pollock Co., 1939), 10–11; Butler, *History of Youngstown,* vol. 1, 723.

103. *The Foundry* 16 (March 1900): 38.

104. *Directory,* 1884, 380.

105. Chas. W. Sisson, *The ABC of Iron* (Louisville: Press of the Courier-Journal Job Printing Co., 1893), 106.

106. *Directory,* 1884, 57–58.

107. *Directory,* 1882, 61. The Licking Iron Company in Newark, Ohio, reported producing American Scotch from black-band iron ore found near their furnaces.

108. Ibid., 31.

109. *The Mahoning Register,* December 17, 1874.

110. Butler, "Early History," 426.

111. Ibid.

112. Ibid., 426–27.

113. *Statistics of the American and Foreign Iron Trades for 1880,* 22.

114. Butler, "Early History," 427.

115. *Appendix to Swineford's History of the Lake Superior Iron District* (Marquette, MI: Mining Journal Office, 1872), 5–6.

116. Temin, *Iron and Steel,* 195.

117. A. P. Swineford, *Annual Review of the Iron Mining and Other Industries of the Upper Peninsula* (Marquette, MI: The Mining Journal, 1882), 161–62.

118. Hogan, *Economic History,* vol. 1, 20.

119. Swineford, *Annual Review,* 42.

120. *The Iron Age* 47 (June 18, 1891): 1177.

121. George J. Snelus, "On the Manufacture and Use of Spiegeleisen," *The American Chemist* 6, no. 1 (July 1875–June 1876): 16.

122. Ulrich Wengenroth, *Enterprise and Technology: The German and British Steel Industries, 1865–1895* (Cambridge: Cambridge University Press, 1994), 20.

123. George J. Snelus, "On the Manufacture and Use of Spiegeleisen," *The American Chemist* 6, no. 1 (July 1875–June 1876): 16.

124. Albert Williams Jr., *Mineral Resources of the United States* (Washington DC: Government Printing Office, 1885), 564.

125. *Statistics of the American and Foreign Iron Trades for 1877,* 20–21; Williams, *Mineral Resources of the United States,* 564.

126. *Statistics of the American and Foreign Iron Trades for 1878,* 20.

127. Williams, *Mineral Resources of the United States*, 263.

128. Butler, "Early History," 427.

129. Knox, "Mary Blast Furnace," 138.

130. Butler, "Early History," 428.

131. *The Vindicator*, July 5, 1927.

132. Butler, "Early History," 428.

133. *Arguments Before the Committee of Ways and Means on the Morrison Tariff Bill*, 120.

134. Ibid.

135. *The Bulletin of the American Iron and Steel Association* 19 (January 21, 1885): 20.

136. Ibid., 16 (August 2, 1882): 211.

137. *Directory*, 1886, 131.

138. *Report of the Select Committee on Ordnance and War Ships* (Washington DC: Government Printing Office, 1886), 232–34. By the mid-1880s, Cleveland Rolling Mill Co. and Otis Iron and Steel Co. in Cleveland contained batteries of four to five open-hearth furnaces ranging from seven to fifteen tons capacity each. In 1884, these two companies had the largest open-hearth steelworks in the United States.

139. "Revival of Washed Metal for Tool Steel," *Raw Material* 3, no. 6 (December 1920): 243.

140. Henry D. Hibbard, "Process of Making Washed Metal," *The Iron Trade Review* 59 (August 10, 1916): 275.

141. *Arguments Before the Committee of Ways and Means on the Morrison Tariff Bill*, 120; E. C. Evans, *The Cambrian: A National Monthly Magazine* 9, no. 2 (February 1889): 57. The American Iron and Steel Association's 1886 *Directory* lists the Youngstown Steel Company as the only plant in the United States that produced washed metal.

142. *Report of the Select Committee on Ordnance and War*, 291, 232. Butler tried to persuade Bethlehem Iron Co. superintendent John Fritz to build a steel gun foundry in Youngstown. Fritz had issues finding cheap, high-quality iron with which to produce heavy steel ordnance, a problem that washed metal could alleviate. Butler responded to Fritz, saying, "We believe it is just the material needed for manufacturing the proposed guns."

143. *Arguments Before the Committee of Ways and Means on the Morrison Tariff Bill*, 121.

144. *Statistics of the American and Foreign Iron Trades for 1878*, 20.

145. Warren, *Wealth, Waste, and Alienation*, 47.

146. *Report of the Geological Survey of Ohio*, 461.

147. *Directory*, 1884, 58.

148. *Annual Report of the Secretary of Internal Affairs*, 59–61.

149. *Report of the Geological Survey of Ohio*, 535–36.

150. Ibid., 536; Sabadasz, "Development of Modern Blast Furnace Practice," 95.

151. *The Journal of the Iron and Steel Institute*, issue 1 (London: E. & F. N. Spon, 16, Charing Cross, 1883), 158.

152. Andrew Carnegie, *Autobiography of Andrew Carnegie* (Boston and New York: Houghton Mifflin Company, 1920), 179.

153. *The Journal of the Iron and Steel Institute*, issue 1, 158.

154. "Cowper's Hot-Blast Stoves," *Scientific American Supplement* 10, no. 235 (July 3, 1880): 3,742.

155. *Register & Tribune*, March 18, 1880.

156. *The Bulletin of the American Iron and Steel Association* 16 (December 30, 1882): 357.

157. *Transactions of the American Institute of Mining Engineers* 19 (May 1890 to February 1891): 1039.

158. *The Bulletin of the American Iron and Steel Association* 16 (December 30, 1882): 357.

159. *Statistics of the American and Foreign Iron Trades for 1895*, 38.

160. *Report of the Committee of the Senate Upon the Relations Between Labor and Capital*, vol. 1 (Washington DC: Government Printing Office, 1885), 1138; Mansel G. Blackford, *A History of Small Business in America*, 2nd ed. (Chapel Hill: The University of North Carolina Press, 2003), 82; Myron R. Stowell, *"Fort Frick," or the Siege of Homestead: A History of the Famous Struggle between the Amalgamated Association of Iron and Steel Workers and the Carnegie Steel Company (Limited) of Pittsburgh, PA* (Pittsburgh: Pittsburgh Printing Co., 1893), 24.

161. *The Iron Age* 42 (November 1, 1888): 6.

162. *The Bulletin of the American Iron and Steel Association* 26 (April 27, 1892): 117.

163. *Leading Manufacturers and Merchants of Ohio Valley* (New York: International Publishing Company, 1887), 133.

164. Warren, *American Steel*, 168–70.

165. Ibid., 169.

166. *Directory*, 1886, vii; *Directory*, 1894, vii.

167. *Statistics of the American and Foreign Iron Trades for 1895*, 44.

168. *The Bulletin of the American Iron and Steel Association* 20 (July 21, 1886): 189.

169. *Statistics of the American and Foreign Iron Trades for 1890*, 49.

170. *Directory*, 1884, 174.

171. Butler, *History of Youngstown*, vol. 1, 478.

172. *Western Reserve Chronicle*, September 21, 1887.

173. *The Iron Trade Review* 74 (April 24, 1924): 1,098.

174. *The Vindicator*, March 1, 1892.

175. *The Vindicator*, March 12, 1892.

176. Fitch, *The Steel Workers*, 141.

177. *The Vindicator*, May 30, 1892.

178. *The Vindicator*, July 9, 1892.

179. *Directory*, 1894, 154.

180. Joseph Butler, "Early History of Iron and Steel Making in Mahoning Valley," *The Iron Trade Review* 67 (August 27, 1925): 483.

181. *The Vindicator*, May 30, 1892.

182. Ibid.

183. *The Bulletin of the American Iron and Steel Association* 27 (February 15, 1893): 53; *The Vindicator*, September 14, 1892.

184. "The Shenango Valley Bessemer Plant," *The Iron Age* 50 (October 13, 1892): 676.

185. *The Vindicator*, April 29, 1893.

186. *The Vindicator*, June 4, 1892.

187. *The Iron Trade Review* 27 (August 16, 1894): 10.

188. *The Vindicator*, August 8, 1892.

189. *The Vindicator*, August 12, 1892.

190. *The Vindicator*, September 14, 1892; *The Bulletin of the American Iron and Steel Association* 32 (March 1, 1898): 35.

191. *The Bulletin of the American Iron and Steel Association* 27 (October 11, 1893): 297. The number of men employed in Youngstown's iron companies in the fall of 1892 versus the fall of 1893 was drastically different. Brown-Bonnell Iron Co. employed 1,800 in 1892 and only 50 in 1893; Union Iron and Steel Co. employed 2,500 in 1892 versus 25 in 1893; and Mahoning Valley Iron Co. employed 1,200 in 1892 versus 50 in 1893.

192. *The Iron Trade Review* 44 (February 4, 1909): 266.

193. "The Ohio Steel Company's Plant at Youngstown, Ohio," *Scientific American Supplement* 40, no. 1018 (July 6, 1895): 16,264.

194. *The Iron Trade Review* 27 (August 16, 1894): 10.

CHAPTER 6

1. Joseph F. Froggett, "The Mahoning Valley as an Iron Center," *The Iron Trade Review* 44 (February 4, 1909): 266.

2. "The Ohio Steel Company's Plant at Youngstown, Ohio," *Scientific American Supplement* 40, no. 1018 (July 6, 1895): 16,265.

3. Ibid., 16,264; *The Journal of the Iron and Steel Institute* 47 (1895): 442.

4. "A Model Steel Plant," *The Iron Trade Review* 27 (August 16, 1894): 8–10.

5. Ibid., 9.

6. *Biographical Directory of the American Iron and Steel Institute* (New York: American Iron and Steel Institute, 1911), 150.

7. *Directory*, 198.; Froggett, "The Mahoning Valley as an Iron Center," 266.

8. *The Bulletin of the American Iron and Steel Association* 30 (February 20, 1896): 45.

9. Froggett, "The Mahoning Valley as an Iron Center," 267.

10. *Statistics of the American and Foreign Iron Trades for 1897*, 29.

11. Naomi Lamoreaux, *The Great Merger Movement in American Business, 1895–1904* (Cambridge: Cambridge University Press, 1985), 14.

12. Hogan, *Economic History*, vol. 1, 237.

13. *Statistics of the American Iron Trade for 1897*, 20.

14. Ibid.

15. *The Bulletin of the American Iron and Steel Association* 27 (March 1, 1893): 69.
16. *The Iron Trade Review* 28 (March 28, 1895): 5.
17. Ibid.
18. Lamoreaux, *Great Merger Movement*, 16.
19. *Directory*, 1898, 162–63.
20. *The Iron Trade Review* 28 (April 18, 1895): 10.
21. *The Vindicator,* March 13, 1893.
22. *Directory*, 1898, 163–64.
23. "The Morgan Continuous Billet Mill," *The Iron Age* 53 (May 24, 1894): 996.
24. *The Iron Age* 58 (November 19, 1896): 967.
25. Ibid.
26. George Higley, *History of the Union Works* (Youngstown: n.p., n.d.), 31.
27. *The Engineering and Mining Journal* 64 (November 20, 1897): 618.
28. Higley, *History of the Union Works*, 31.
29. Sanderson, *20th Century History*, 257.
30. *The Iron Trade Review* 28 (March 28, 1895): 5.
31. Temin, *Iron and Steel*, 284.
32. *Directory*, 1892, 53–55; *Directory*, 1898, 46–49.
33. *The Bulletin of the American Iron and Steel Association* 31 (May 1, 1897): 101.
34. *The Iron Trade Review* 28 (March 28, 1895): 5.
35. *Directory*, 1898, 47.
36. *The Iron Trade Review* 27 (July 26, 1894): 10.
37. *The Vindicator,* July 21, 1900.
38. Ibid.
39. *Transactions of the American Institute of Mining Engineers* 20 (1892): 273.
40. *The Bulletin of the American Iron and Steel Association* 27 (March 1, 1893): 69.
41. Butler, *History of Youngstown*, vol. 2, 32.
42. E. A. Uehling, "Pig-Iron Casting and Conveying Machinery," *Cassier's Magazine* 24 (May–October 1903): 118.
43. W. David Lewis, *Sloss Furnaces and the Rise of the Birmingham District: An Industrial Epic* (Tuscaloosa: The University of Alabama Press, 1994), 205; *Cassier's Magazine* 24 (May–October 1903): 118. Iron loaders had to carry 100- to 125-pound bars of pig iron six to ten paces over "loose hot sand" and load them onto railroad cars. This process was repeated 250 to 300 times in a short period.
44. "The Age of Steel," *The Iron and Machinery World* 90 (November 30, 1901): 22.
45. "The Uehling Metal Conveying Machine," *American Manufacturer and Iron World* 63 (November 18, 1898): 727.
46. *American Manufacturer and Iron World* 63 (November 18, 1898): 727.
47. *The Stevens Indicator* 15 (1898): 302.
48. *Western Reserve Chronicle*, August 15, 1900.
49. Duncan L. Burn, *The Economic History of Steelmaking, 1867–1939* (Cambridge: Cambridge University Press, 1961), 188–90; Ruminski, "From Mary Furnace to Sharon Steel," 50.
50. *The Iron Trade Review* 32 (December 21, 1899): 13.
51. Sabadasz, "Development of Modern Blast Furnace Practice," 98.
52. Ibid., 99.
53. Froggett, "The Mahoning Valley as an Iron Center," 267.
54. *The Iron Trade Review* 32 (December 21, 1899): 13; *Directory*, 1901, 17.
55. *The Iron Trade Review* 32 (December 21, 1899): 13–16.
56. Ibid., 22.
57. G. O. Haven Jr., "Important Industrial Combinations," *The Chicago Banker*, vol. 1 (Chicago: The Chicago Banker Company, 1899), 444–45.
58. National Steel also acquired interests in the Oliver Iron Mining Company and a quarter interest in a lease of the Biwabik Mine in the Mesabi Range.
59. Haven, *The Chicago Banker*, vol. 1, 447; Aley, *Heritage*, 166.
60. *Western Reserve Chronicle*, March 15, 1899.
61. "The Industrial, Social and Educational Advantages of Warren and Niles," *Warren Daily Tribune*, June 1898.
62. Hogan, *Economic History*, vol. 1, 238.
63. Lamoreaux, *Great Merger Movement*, 120. According to Lamoreaux, by using this strategy, a consolidation "could remove all incentive for price cutting by allowing

the independent firms to sell as much output as they wished at whatever price the dominant firm set."

64. Hogan, *Economic History*, vol. 1, 237.

65. *The Iron Age* 62 (September 15, 1898): 16.

66. *The Vindicator*, January 15, 1897.

67. Ibid., January 15, 1897; "Coming Consolidation of Mahoning Valley Interests," *The Iron Age* 59 (January 28, 1897): 11.

68. *The Iron Age* 59 (January 28, 1897): 11.

69. *The Vindicator*, January 15, 1897.

70. *The Vindicator*, January 16, 1897.

71. Ibid.

72. Lamoreaux, *Great Merger Movement*, 2.

73. Hogan, *Economic History*, vol. 1, 239.

74. *The Iron Trade Review* 32 (May 11, 1899): 9.

75. Ibid.; Aley, *Heritage*, 166.

76. Hogan, *Economic History*, vol. 1, 298.

77. *Preliminary Report on Trusts and Industrial Combinations* (Washington DC: Government Printing Office, 1900), 953.

78. Ibid.

79. Ibid., 953–54.

80. Hogan, *Economic History*, vol. 1, 297.

81. *The Bulletin of the American Iron and Steel Association* 34 (July 15, 1900): 125.

82. *Preliminary Report on Trusts and Industrial Combinations*, 30.

83. *Statistics of the American Iron Trade for 1898*, 19.

84. Ibid.

85. Ibid., 61.

86. *The Iron Age* 63 (April 6, 1899): 18–19.

87. *The Vindicator*, May 27, 1899.

88. *The Iron Age* 63 (April 6, 1899): 18–19.

89. *Evening Bulletin* (Maysville, Kentucky), November 22, 1899.

90. *The Vindicator*, May 6, 1899.

91. *The Vindicator*, July 1, 1899.

92. Ibid.

93. *The Bulletin of the American Iron and Steel Association* 34 (January 22, 1900): 20.

94. Hogan, *Economic History*, vol. 2, 563.

95. *American Manufacturer and Iron World* 67 (October 4, 1900): 249.

96. Ibid.

97. *Statistics of the American Iron Trade for 1899*, 50.

98. *Directory*, 1901, 72.

99. *The Vindicator*, September 25, 1977.

100. *The Vindicator*, September 28, 1900.

101. *The Vindicator*, November 12, 1900.

102. *The Iron Trade Review* 34 (January 24, 1901): 25.

103. Temin, *Iron and Steel*, 284–85.

104. *Directory* 1904, 85.

105. *The Vindicator*, September 17, 1902.

106. Ibid.

107. "The New Republic Billet Mill," *The Iron Age* 70 (October 2, 1902): 17.

108. *The Vindicator*, March 29, 1902.

109. Ibid.

110. *The Iron and Steel Magazine* 9 (June, 1905): 565.

111. *The Iron Trade Review* 35 (January 30, 1902): 19.

112. *Statistics of the American and Foreign Iron Trades for 1910*, 91–92.

113. Hogan, *Economic History*, vol. 2, 563.

114. Ibid., 463.

115. James Douglas Rose, *Duquesne and the Rise of Steel Unionism* (Champaign: University of Illinois Press, 2001), 14.

116. Hogan, *Economic History*, vol. 2, 481–82.

117. *The Bulletin of the American Iron and Steel Association* 38 (January 25, 1904): 14. U.S. Steel only produced 1.3% of the country's forge, foundry, and other kinds of pig iron in 1902. Independent merchant furnace and steel companies made 5.1 million

tons of foundry pig iron versus 68,088 tons produced by U.S. Steel–owned blast furnaces.

118. *Directory,* 1901, 22; *Directory,* 1904, 14.
119. Higley, *History of the Union Works,* 37.
120. *The Iron Trade Review* 35 (January 30, 1902): 19.
121. "Youngstown, Ohio," *The Successful American* 4 (1901): 488.
122. Ibid.
123. Butler, *History of Youngstown,* vol. 1, 230.
124. *The Successful American* 4 (1901): 488.
125. Froggett, "The Mahoning Valley as an Iron Center," 269.
126. *The Vindicator,* November 22, 1900.
127. *The Iron Trade Review* 35 (December 18, 1902): 45; *The Metal Worker* 56 (September 14, 1901): 59.
128. Blue, Jenkins, Lawson, and Reedy, *Mahoning Memories,* 95.
129. *The Metal Worker* 56 (September 14, 1901): 59.
130. *American Manufacturer and Iron World* 70 (February 20, 1902): 216.
131. *The Iron Age* 67 (March 7, 1901): 29.
132. Kenneth Warren, *Big Steel: The First Century of the United States Steel Corporation, 1901–2001* (Pittsburgh: University of Pittsburgh Press, 2001), 28; Robert Hessen, *Steel Titan: The Life of Charles M. Schwab* (Pittsburgh: University of Pittsburgh Press, 1975), 125.
133. *The Metal Worker* 57 (June 7, 1902): 49.
134. *Directory,* 1904, 56.
135. *The Iron and Machinery World* 96 (July 16, 1904): 28.
136. *The Vindicator,* September 25, 1977.
137. *Hearings Before the Committee on Education and Labor of the United States Senate* (Washington DC: Government Printing Office, 1903), 462.
138. *The Bulletin of the American Iron and Steel Association* 35 (September 10, 1901): 134.
139. *The Vindicator,* July 21, 1900.
140. *Statistics of the American Iron Trade for 1904,* 37.
141. Ibid.
142. Ibid.
143. *The Age of Steel* 91 (April 26, 1902): 24.
144. *The Vindicator,* June 2, 1902; *The Iron Trade Review* 35 (June 5, 1902): 35; *Tenth Annual Report of the Ohio State Board of Arbitration for the Year Ending December 31, 1902* (Springfield: The Springfield Publishing Company, 1904), 44.
145. Ibid., 45.
146. *The Vindicator,* June 2, 1902.
147. Ibid.
148. *Tenth Annual Report of the Ohio State Board of Arbitration,* 45.
149. *The Vindicator,* June 2, 1902
150. *Tenth Annual Report of the Ohio State Board of Arbitration,* 45.
151. Brody, *Steelworkers in America,* 158.
152. Blue, Jenkins, Lawson, and Reedy, *Mahoning Memories,* 104.
153. *The Stark County Democrat,* July 5, 1901; *Wall Street Journal,* July 13, 1907.
154. *Directory,* 1916, 87; Sanderson, *20th Century History,* 255; *Directory,* 1908.
155. Warren, *Big Steel,* 29.
156. Ibid., 30.
157. *Executive Documents: Annual Reports for 1904 Made to the Seventy-Seventh General Assembly of the State of Ohio* (Columbus: F. J. Heer, State Printer, 1905), 213.
158. *The Vindicator,* August 22, 1904.
159. *Executive Documents: Annual Reports,* 214.
160. *The Vindicator,* August 22, 1904.
161. *The Vindicator,* August 23, 1904.
162. *The Vindicator,* August 30, 1904.
163. *The Vindicator,* September 4, 1904.
164. Ibid.
165. Warren, *Big Steel,* 29.
166. Ibid.
167. Ibid.
168. *Thirteenth Annual Report of the State Board of Arbitration, Year Ending December 31, 1905* (Columbus: P. J. Heer, State Printers, 1906), 880.

169. Warren, *Big Steel*, 29.

170. John N. Ingham, "Iron and Steel in the Pittsburgh Region: The Domain of Small Business," *Business and Economic History* 20 (1991): 111.

171. *Industrial World*, no. 36 (September 7, 1908): 1052.

172. *Directory*, 1908, 416; *The Iron Trade Review* 40 (April 18, 1907): 618.

173. *The Vindicator*, January 31, 1909.

174. "Passing of An Old Iron Mill," *The Iron Trade Review* 45 (July 29, 1909): 218.

175. *The Vindicator*, July 1, 1909.

176. *Eleventh Annual Report Republic Iron & Steel Company for the Fiscal Year Ending June 30, 1910*, 16.

177. *The Vindicator*, September 21, 1913.

178. Fitch, *Steel Workers*, 33; David T. Day, *Mineral Resources of the United States* (Washington DC: Government Printing Office, 1904), 119.

179. Charles Patrick Neill, *Report on the Conditions of Employment in the Iron and Steel Industry in the United States*, vol. 3 (Washington DC: Government Printing Office, 1913), 476.

180. *United States Investor* 32 (August 6, 1921): 90.

181. Fitch, *Steel Workers*, 33.

182. *The Vindicator*, July 1, 1909.

183. *United States Investor* 32 (August 6, 1921): 90.

184. Warren, *American Steel Industry*, 175.

EPILOGUE

1. *Directory*, 1904, 253–55. Other old mills such as the Pittsburgh Forge and Iron Company's works, established in 1864, made splice bars, draw bars, hammered car and locomotive axles, and other heavy forgings for the railroad.

2. *Directory*, 1890, 25–27, 52–54.

3. *The Commercial & Financial Chronicle* 82 (March 24, 1906): 696.

4. *The Hubbard Enterprise*, April 13, 1916; "Buys Andrews & Hitchcock Interests," *The Iron Trade Review* 58 (April 13, 1916): 803.

5. *Industrial World*, no. 4 (January 22, 1912): 92.

6. *The Iron Age* 84 (December 23, 1909): 1898.

7. *The Iron Trade Review* 46 (February 10, 1910): 294.

8. *The Iron Age* 91 (May 1, 1913): 1089; *The Mahoning Dispatch*, June 26, 1914.

9. Butler, "Early History," 484.

10. *Industrial World*, no. 4 (January 22, 1912): 92.

11. "The Brier Hill Steel Company's New Works," *The Iron Age* 93 (April 2, 1914): 841.

12. Ibid., 840.

13. Laughlin, *Mineral Resources*, 561; Temin, *Iron and Steel*, 283–85.

14. *Metallurgical and Chemical Engineering* 14 (January 1, 1916): 7.

15. *The Iron Trade Review* 61 (December 6, 1917): 1201.

16. *The Iron Age* 100 (September 6, 1917): 545.

17. During the war, many steel companies were not guaranteed deliveries on material for new mills or blast furnace construction.

18. *Moody's Manual of Railroads and Corporation Securities*, vol. 2 (New York: Poor's Publishing Company, 1921), 839.

19. *The Vindicator*, April 12, 1916.

20. *Twenty-Fourth Annual Report Republic Iron & Steel Company for the Year Ending December 31, 1925*, 13.

21. *The Iron Trade Review* 68 (April 7, 1921): 960.

22. *The Iron Trade Review* 75 (November 14, 1929): 1291. By the early 1920s, merchant furnace owners only made a profit one-third of the year on average.

23. Ibid.

24. "Mechanical Puddling," *The Iron Age* 110 (July 6, 1922): 9.

25. *Productivity of Labor in Merchant Blast Furnaces*, 64.

26. *The Iron Age* 123 (1929): 306.

27. *Iron Trade Review* 83 (February 7, 1929): 386.

28. *The Vindicator*, May 27, 1927.

29. Ibid.

BIBLIOGRAPHY

NEWSPAPERS

American Working People (Pittsburgh, PA)

Chicago Daily Tribune

Cleveland Morning Leader (Cleveland, OH)

Courier Express (Buffalo, NY)

The Evening Bulletin (Maysville, KY)

The Evening Times (Washington DC)

The Hancock Jeffersonian (Findlay, OH)

The Hubbard Enterprise (Hubbard, OH)

Mahoning Dispatch (Youngstown, OH)

Mahoning Free Democrat (Youngstown, OH)

Mahoning Index (Canfield, OH)

The Mahoning Register [*Mahoning County Register*] (Youngstown, OH)

New Castle News (New Castle, PA)

The New York Times

The Niles Times (Niles, OH)

Pittsburgh Gazette (Pittsburgh, PA)

The Stark County Democrat (Canton, OH)

The Vindicator (Youngstown, OH)

Wall Street Journal

Warren Daily Tribune (Warren, OH)

Weekly Telegram (Youngstown, OH)

Western Reserve Chronicle (Warren, OH)

The Wheeling Daily Intelligencer (Wheeling, WV)

Youngstown Register and Tribune (Youngstown, OH)

Annual Report of the Commissioner of Railroads and Telegraphs of Ohio for the Year Ending June 30, 1880. Columbus: J. G. Brand & Co., 1880.

Annual Report of the Secretary of Internal Affairs of the Commonwealth of Pennsylvania, Part III: Industrial Statistics. Vol. 10. Harrisburg: Lane S. Hart, 1883.

Annual Report of the Secretary of State, for the Year 1860. Columbus: Richard Nevins, State Printer, 1861.

Annual Report of the Secretary of State to the Governor of the State of Ohio for the Year 1873. Columbus: Nevins & Myers, State Printers, 1874.

Arguments Before the Committee of Ways and Means on the Morrison Tariff Bill. Washington DC: Government Printing Office, 1884.

Barnes, Phineas. *The Present Technical Condition of the Steel Industry of the United States.* Washington DC: Government Printing Office, 1885.

Blake, William P. *Report on Iron and Steel.* Washington DC: Government Printing Office, 1876.

Brown, Sharon A. *Cambria Iron Company: Historic Resource Study.* Washington DC: U.S. Department of the Interior, 1989.

Compendium of the Enumeration of the Inhabitants and Statistics of the United States. Washington DC: Printed by Thomas Allen, 1841.

Compendium of the Tenth Census, Part I. Washington DC: Government Printing Office, 1883.

Day, David T. *Mineral Resources of the United States.* Washington DC: Government Printing Office, 1904.

Executive Documents: Annual Reports for 1904 Made to the Seventy-Seventh General Assembly of the State of Ohio. Columbus: F. J. Heer, State Printer, 1905.

Executive Documents Printed by Order of the House of Representatives during the Second Session of the Fortieth Congress, 1867–68. Washington DC: Government Printing Office, 1868.

First Annual Report of the State Inspector of Mines for the Year 1874. Columbus: Nevins and Myers, State Printers, 1875.

Hearings Before the Committee on Education and Labor of the United States Senate. Washington DC: Government Printing Office, 1903.

Hearings Before the Committee on Investigation of the United States Steel Corporation. Vol. 5. Washington DC: Government Printing Office, 1912.

Journal of the Senate of the Commonwealth of Pennsylvania. Harrisburg, PA: J. M. G. Lescure, Printer to the State, 1848.

Kilbourn, John. *Public Documents Concerning the Ohio Canals.* Columbus: Published by I. N. Whiting, 1832.

Laughlin, G. L. *Mineral Resources of the United States 1918.* Washington DC: Government Printing Office, 1921.

Neill, Charles Patrick. *Report on Conditions of Employment in the Iron and Steel Industry in the United States.* Vol. 3–4. Washington DC: Government Printing Office, 1913.

Nimmo, Joseph Jr. *Report on the Internal Commerce of the United States.* Washington DC: Government Printing Office, 1884.

Preliminary Report on Trusts and Industrial Combinations. Washington DC: Government Printing Office, 1900.

Productivity of Labor in Merchant Blast Furnaces. Washington DC: Government Printing Office, 1929.

Report of the Commissioner of Patents for the Year 1863. Vol. 1. Washington DC: Government Printing Office, 1866.

Report of the Committee of the Senate Upon the Relations Between Labor and Capital. Vol. 1. Washington DC: Government Printing Office, 1885.

Report of the Geological Survey of Ohio. Vol. 5. Columbus: G. J. Brand & Co., 1884.

Report of the Select Committee on Ordnance and War Ships. Washington DC: Government Printing Office, 1886.

Report on the Manufacturers of the United States at the Tenth Census. Washington DC: Government Printing Office, 1883.

Reports of the United States Board of Tax Appeals. Washington DC: Government Printing Office, 1927.

Second Annual Report of the Bureau of Statistics of Pennsylvania. Harrisburg, PA: B. F. Meyers, State Printers, 1875.

Thirteenth Annual Report of the State Board of Arbitration, Year Ending December 31, 1905. Columbus: P. J. Heer, State Printers, 1906.

Thirty-Sixth General Assembly of the State of Ohio. Vol. 36. Columbus: Samuel Medary, Printer to the State, 1838.

Twenty-Fifth Annual Report of the Ohio State Board of Agriculture for the Year 1870. Columbus: Nevins & Myers, State Printers, 1871.

United States Department of the Interior, National Park Service. *Colver-Rogers Farmstead, Greene County, Pennsylvania.* NPS form 10–900–1.

Weeks, Joseph D. *Report on the Statistics of Wages in Manufacturing Industries.* Washington DC: Government Printing Office, 1886.

Williams, Albert Jr. *Mineral Resources of the United States.* Washington DC: Government Printing Office, 1885.

Wright, George B. *Annual Report of the Commissioner of Railroads and Telegraphs for the Year 1870.* Vol. 1. Columbus: Nevins & Myers, State Printers, 1870.

INDUSTRIAL TRADE JOURNALS, MAGAZINES, AND REPORTS

The Age of Steel. St. Louis. Unknown publisher.

American Artisan and Patent Record. New York: Brown, Coombs & Co.

The American Chemist. Philadelphia: Henry C. Lea.

The American Engineer. Chicago: Cowles & Weston.

American Machinist. New York: American Machinist Press.

American Manufacturer and Iron World. Pittsburgh: National Iron and Steel Publishing Co.

American Railroad Journal. New York: Published at No. 323 Pearl-Street.

American Railroad Journal and Iron Manufacturer's and Mining Gazette. Philadelphia: D. K. Minor.

Annual Review of the Iron Mining and Other Industries of the Upper Peninsula. Marquette: Mining Journal.

The Blast Furnace & Steel Plant. Pittsburgh: National Iron and Steel Publishing Co.

The Bulletin of the American Iron and Steel Association. Philadelphia: American Iron and Steel Association.

Cassier's Magazine. New York: The Cassier Magazine Co.

The Chemical News and Journal of Physical Science. London: Creed Lane & Ludgate Hall.

The Commercial & Financial Chronicle. New York: William B. Dana Company.

Engineering News. New York: Engineering News Publishing Co.

Engineering and Mining Journal. New York: McGraw-Hill Company, Inc.

Industrial World. Pittsburgh: 108 Smithfield Street.

The International Review. New York: A. S. Barnes & Co.

The Iron Age. New York: David Williams Company.

The Iron and Machinery World. Chicago and St. Louis. Unknown publisher.

The Iron and Steel Magazine. Boston: 446 Tremont Street.

The Iron Trade Review. Cleveland: The Penton Publishing Co.

Journal of the Franklin Institute of the State of Pennsylvania: Philadelphia: Published by the Franklin Institute at Their Hall.

The Journal of the Iron and Steel Institute. London: E. & F. N. Spon.

Magazine of Western History. New York: Magazine of Western History Publishing Co.

The Mechanics' Magazine. London: Robertson, Brooman, and Co.

Merchants' Magazine and Commercial Review. New York: Published at 142 Fulton Street.

The Metal Worker. New York and Chicago. Unknown Publisher.

Metallurgical and Chemical Engineering. New York: McGraw-Hill Company.

The Mining and Statistic Magazine: Devoted to Mines, Mining Operations, Metallurgy, &c., &c. New York: Published at No. 1 Spruce Street.

The New Technical Educator. London, Paris, and Melbourne: Cassell and Company.

Niles' National Register. Baltimore: H. Niles, editor.

The Ohio Cultivator. Columbus: S. D. Harris.

The Popular Science Monthly. New York: D. Appleton and Company.

The Public. New York: The Financier Association.

Public Opinion. New York: The Public Opinion Company, Publishers.

The Railway News. London: Published at the Office of the "Railway News."

Raw Material. New York: The Gage Publishing Co.

Scientific American. New York: Munn & Co.

Scientific American Supplement. New York: Munn & Co.

Steel. Cleveland: The Penton Publishing Co.

Stevens Indicator. Hoboken: Stevens Institute of Technology.

The Successful American. New York: The Writers' Press Association.

Transactions of the American Institute of Mining Engineers. New York: Published by the Institute.

United States Investor. Boston and New York. Unknown publisher.

The Valve World. Chicago: Crane Co.

Van Nostrand's Eclectic Engineering Magazine. New York: D. Van Nostrand, Publisher.

Wiley's American Iron Trade Journal. New York: John Wiley & Son.

Youngstown Sheet & Tube Bulletin. Youngstown: Youngstown Sheet & Tube Co.

BOOKS, ARTICLES, AND UNPUBLISHED DISSERTATIONS

Adams, Charles Kendall. *The Universal Cyclopedia.* Vol. 2. New York: D. Appleton and Company, 1900.

Aldrich, Mark. *Safety First: Technology, Labor, and Business in the Building of American Work Safety, 1870–1939.* Baltimore: The Johns Hopkins University Press, 1997.

Aley, Howard C. *A Heritage to Share: The Bicentennial History of Youngstown and Mahoning County, Ohio.* Youngstown, OH: Youngstown Lithographing Company, 1975.

Allen, Robert C. "The Peculiar Productivity History of American Blast Furnaces, 1840–1913." *The Journal of Economic History* 37 (September 1977): 605–33.

American Biography: A New Cyclopedia. Vol. 11. New York: The American Historical Society, 1922.

An Accompaniment to Mitchell's Map of the World. Philadelphia: R. L. Barnes, 1842.

Andreas, A. T. *History of Chicago, from the Earliest Period to the Present Time.* Vol. 3. Chicago: The A. T. Andreas Company, Publishers, 1886.

Annals of Cleveland: Court Record Series. Vol. 3. Cleveland: Works Project Administration, 1939.

Appleton's Annual Cyclopedia and Register of Important Events of the Year 1898. New York: D. Appleton and Company, 1899.

Atlas of the County of Lawrence and the State of Pennsylvania. Philadelphia: G. M. Hopkins & Co., 1872.

Austin, Richard Cartwright. *East of Cleveland: Moral Imagination in Industrial Culture, 1820–1940*. Dungannon, VA: Creekside Press, 2004.

Baldwin, Leland D. *Pittsburgh: The Story of a City, 1750–1865*. Pittsburgh: University of Pittsburgh Press, 1937.

Benjamin, Park. *Appleton's Cyclopedia of Applied Mechanics*. Vol. 2. New York: D. Appleton and Company, 1888.

The Bicentennial Edition: "Lake County History." Painesville, OH: Painesville Publishing Company, 1976.

Biographical Directory of the American Iron and Steel Institute. New York: American Iron and Steel Institute, 1911.

Biographical History of Northeastern Ohio, Embracing the Counties of Ashtabula, Trumbull and Mahoning. Chicago: The Lewis Publishing Company, 1893.

Bishop, J. Leander. *A History of American Manufactures from 1608–1860*. Vol. 3. Philadelphia: Edward Young & Co., 1868.

Blackford, Mansel G. *A History of Small Business in America*. 2nd ed. Chapel Hill: The University of North Carolina Press, 2003.

Blaugrund, Annette. *Dispensing Beauty in New York and Beyond: The Triumphs and Tragedies of Harriet Hubbard Ayer*. Charleston: The History Press, 2011.

Blue, Frederick J., William D. Jenkins, H. William Lawson, and Joan M. Reedy. *Mahoning Memories: A History of Youngstown and Mahoning County*. Virginia Beach: The Donning Company Publishers, 1995.

Bluestone, Berry, and Bennett Harrison. *The Deindustrialization of America: Plant Closings, Community Abandonment, and the Dismantling of Basic Industry*. New York: Basic Books, 1982.

Boucher, John N. *The Cambria Iron Company*. Harrisburg, PA: Meyers Printing and Publishing House, 1888.

Brackert, A. O. *The ABC of Iron and Steel*. Cleveland: The Penton Publishing Co., 1915.

Bremner, David. *The Industries of Scotland*. Edinburgh: Adam and Charles Black, 1869.

Brennan, J. Fletcher. *A Bibliographical Cyclopedia and Portrait Gallery of Distinguished Men*. Cincinnati: John C. Yorston & Company, 1879.

Bridge, James Howard. *The Inside History of the Carnegie Steel Company: A Romance of Millions*. New York: The Aldine Book Company, 1903.

Brody, David. *Steelworkers in America: The Nonunion Era*. Cambridge, MA: Harvard University Press, 1960.

Bruno, Robert. *Steelworker Alley: How Class Works in Youngstown*. Ithaca: Cornell University Press, 1999.

Burn, Duncan L. *The Economic History of Steelmaking, 1867–1939*. Cambridge: Cambridge University Press, 1961.

Butler, Joseph G. Jr. "Early History of Iron and Steel Making in Mahoning Valley." *The Iron Trade Review* 67 (August 20, 1925): 425–28, 438.

———. "Early History of Iron and Steel Making in Mahoning Valley." *The Iron Trade Review* 67 (August 27, 1925): 481–84.

———. *Fifty Years of Iron and Steel*. Cleveland: The Penton Press, 1922.

———. *History of Youngstown and the Mahoning Valley*. 3 vols. New York: American Historical Society, 1921.

———. *Recollections of Men and Events*. New York: Putnam Publishing, 1927.

Carnegie, Andrew. *Autobiography of Andrew Carnegie*. Boston and New York: Houghton Mifflin Company, 1920.

Century Cyclopedia of History and Biography of Pennsylvania. Vol. 2. Chicago: The Century Publishing and Engraving Company, 1904.

Church, S. H. *Corporate History of the Pennsylvania Lines West of Pittsburgh*. Vol. 2. Baltimore: The Friedenwald Company, 1898.

Churella, Albert J. *The Pennsylvania Railroad: Building an Empire, 1846–1917*. Vol. 1. Philadelphia: University of Pennsylvania Press, 2013.

Cleave, Egbert. *City of Cleveland and Cuyahoga County, Taken from Cleave's Biographical Cyclopedia of the State of Ohio*. Cleveland: Fairbanks, Benedict & Co., 1875.

Comley, W. J., and W. D'Eggville. *Ohio: The Future Great State*. Cincinnati and Cleveland: Comley Brothers Manufacturing and Publishing Company, 1875.

Commons, John R. *History of Labour in the United States*. Vol. 2. New York: The MacMillan Company, 1921.

Cutter, William Richard. *Genealogical and Family History of Western New York*. Vol. 2. New York: Lewis Historical Publishing Company, 1912.

Davis, James. *The Iron Puddler: My Life in the Rolling Mills and What Came of It*. Indianapolis: The Bobbs-Merrill Company, 1922.

DeBlasio, Donna M., and Martha I. Pallante. "Memories of Work and the Definition of Community: The Making of Italian Americans in the Mahoning Valley." *Ohio History* 121 (2014): 89–111.

Deby, Julien. *Report on the Progress of the Iron and Steel Industries in Foreign Countries*. New Castle-Upon-Tyne: M. & M. W. Lambert, 1878.

DeWitt, E. L. *Reports of Cases Argued and Determined in the Supreme Court of Ohio*. Vol. 35. Cincinnati: Robert Clarke & Co., 1880.

Di Rocco, Samuel II. "In the Shadow of Steel: Leetonia, Ohio and Independent Iron Manufacturers in the Mahoning and Shenango Valleys, 1845–1920." PhD diss., University of Toledo, 2012.

Directory to the Iron and Steel Works of the United States. Philadelphia: The American Iron and Steel Association, 1878, 1882, 1884, 1886, 1888, 1890, 1892, 1894, 1896, 1898, 1901, 1904, 1908.

Directory of the Iron and Steel Works of the United States and Canada. New York: American Iron and Steel Institute, 1916.

Disturnell, J. *The Great Lakes, or Inland Seas of America*. New York: Charles Scribner, 1863.

Doran, David, and Bob Cather. *Construction Materials Reference Book*. New York: Routledge, 2013.

Drohan, N. J. *History of Hubbard, Ohio: Its People, Churches, Industries and Institutions—from Early Settlement in 1798 to 1907*. Hubbard: H. W. Ulrich Print Co., 1907.

Edwards, Louisa Maria. *A Pioneer Home Maker, 1787–1866: A Sketch of the Life of Louisa Maria Montgomery*. n.p., 1903.

Egle, William H. *An Illustrated History of the Commonwealth of Pennsylvania*. 2nd ed. Philadelphia: E. M. Gardner, 1880.

Eleventh Annual Report Republic Iron & Steel Company for the Fiscal Year Ending June 30, 1910. Pittsburgh: Republic Iron and Steel Co., 1910.

Ellis, Franklin. *History of Fayette County, Pennsylvania, with Biographical Sketches of Many of its Pioneers and Prominent Men*. Philadelphia: L. H. Everts & Co., 1882.

Eves, Jamie H. "'Shrunk to a Comparative Rivulet': Deforestation, Stream Flow, and Rural Milling in 19th Century Maine." *Technology and Culture* 33, no. 1 (1992): 38–65.

First Annual Report of the Directors of the Cleveland and Mahoning Railroad. Cleveland: J. W. Gray, Plain Dealer Steam Press, 1853.

Fitch, John A. *The Steel Workers*. New York: Charities Publication Committee, 1910.

Floud, Roderick, and Deirdre McCloskey. *The Economic History of Great Britain Since 1700*. 2nd ed. Cambridge: Cambridge University Press, 1994.

Foote, Abram W. *Foote Family, Comprising the Genealogy and History of Nathaniel Foote of Wethersfield, Conn. and His Descendants*. Vol. 1. Rutland, VT: Marble City Press, 1907.

Forsythe, Robert. *The Blast Furnace and the Manufacture of Pig Iron*. 2nd ed. New York: David Williams Company, 1909.

Froggett, Joseph F. "The Mahoning Valley as an Iron Center." *The Iron Trade Review* 44 (January 21, 1909): 181–85.

———. "The Mahoning Valley as an Iron Center." *The Iron Trade Review* 44 (January 28, 1909): 225–29.

————. "The Mahoning Valley as an Iron Center." *The Iron Trade Review* 44 (February 4, 1909): 265–70.

Gale, W. K. V. *The Iron and Steel Industry: A Dictionary of Terms*. New York: Drake Publishers, 1971.

Gieck, Jack. *A Photo Album of Ohio's Canal Era, 1825–1913*. Kent, OH: Kent State University Press, 1988.

Gordon, Robert B. *American Iron: 1607–1900*. Baltimore: The Johns Hopkins University Press, 1996.

Grosse, R. N. "Determinants of the Size of Iron and Steel Firms in the United States, 1820–1880." PhD diss., Harvard University, 1948.

Haven, Go. O. Jr. "Important Industrial Combinations." *The Chicago Banker* 1 (1899): 443–47.

Haydinger, Earl. "The Pennsylvania and Ohio Canal." *Towpaths: Quarterly Journal of the Canal Society of Ohio* 5 (April 1967): 15–24.

————. "The Pennsylvania and Ohio Canal, Part II." *Towpaths: Quarterly Journal of the Canal Society of Ohio* 5 (July 1967): 25–34.

Hazen, Aaron L. *20th Century History of New Castle and Lawrence County*. Chicago: Richmond-Arnold Publishing Co., 1908.

Heaton, Dean. *The Heaton Families: 350 Years in America*. Baltimore: Gateway Press, 1982.

Herwig, F. A. *The Present Management of the Reading Railroad, as it Affects the Coal Regions, the Coal Miners and Consumers*. Pottsville, PA: Standard and Chronicle Print, 1879.

Hessen, Robert. *Steel Titan: The Life of Charles M. Schwab*. Pittsburgh: University of Pittsburgh Press, 1975.

Higley, George. *History of the Union Works*. Youngstown: n.p., n.d.

Hillstrom, Kevin, and Laurie Collier Hillstrom. *The Industrial Revolution in America: Communications*. Vol. 1. Santa Barbara: ABC-Clio, 2007.

Historical Collections of the Mahoning Valley. Vol. 1. Youngstown, OH: Mahoning Valley Historical Society, 1876.

History of the Great Lakes. Vol. 1. Chicago: J. H. Beers & Co., 1899.

History of Pittsburgh and Environs. Vol. 3. New York and Chicago: American Historical Society, 1922.

History of the Terrible Financial Panic of 1873. Written and Compiled by a Journalist, 1873.

History of Trumbull and Mahoning Counties. Vol. 1. Cleveland: H. Z. Williams & Bro., 1882.

Hoerr, John P. *And the Wolf Finally Came: The Decline of the American Steel Industry*. Pittsburgh: University of Pittsburgh Press, 1988.

Hogan, William T. *Economic History of the Iron and Steel Industry in the United States*. 5 vols. New York: Lexington Books, 1971.

Holley, A. L. *The Albany and Rensselaer Iron and Steel Works*. London: Offices of "Engineering," 35 and 36, Bedford Street, 1881.

Homans, James E. *The Cyclopaedia of American Biography*. Vol. 8. New York: The Press Association Compilers, 1918.

Howe, Henry. *Historical Collections of Ohio*. Cincinnati: Published by the author by Derby, Bradley & Co., 1847.

Hunt, T. Sterry. *The Coal and Iron of Southern Ohio*. Salem, MA: Naturalists' Agency, 1874.

Hunter, L. C. "The Pittsburgh Iron Industry." *Journal of Economic and Business History* 1 (February 1929): 245.

Ingham, John N. "Iron and Steel in the Pittsburgh Region: The Domain of Small Business." *Business and Economic History* 20 (1991): 107–16.

————. *The Iron Barons: A Social Analysis of an American Urban Elite, 1874–1965*. Westport, CT: Greenwood Press, 1978.

————. *Making Iron and Steel: Independent Mills in Pittsburgh, 1820–1920.* Columbus: Ohio State University Press, 1991.

The Ironworks of the United States. Philadelphia: The American Iron and Steel Association, 1874.

The Ironworks of the United States. Philadelphia: The American Iron and Steel Association, 1876.

Jenkins, Phillip. *A History of Modern Wales, 1536–1990.* New York: Routledge, 2014.

Joblin, Maurice. *Cleveland Past and Present.* Cleveland: Fairbanks, Benedict & Co., 1869.

Johnson, J. E. Jr. *The Principles, Operations, and Products of the Blast Furnace.* New York: McGraw-Hill Book Company, 1918.

Kennedy, James Harrison. *A History of the City of Cleveland.* Cleveland: The Imperial Press, 1896.

Kindleberger, Charles P., and Robert Z. Aliber. *Manias, Panics, and Crashes: A History of Financial Crises.* New York: Palgrave Macmillan, 2011.

Kirk, Edward. *A Practical Treatise on Foundry Irons.* Philadelphia: Henry Carey Baird & Co., 1911.

Knepper, George W. *Ohio and Its People.* 3rd ed. Kent, OH: Kent State University Press, 2003.

Knowles, Anne Kelly. *Mastering Iron: The Struggle to Modernize an American Industry, 1800–1868.* Chicago: University of Chicago Press, 2013.

Knox, John D. "Mary Blast Furnace: Last of Hand-Filled Stacks in America." *Steel* 125 (October 1949): 136–44.

Lake Erie and Ohio River Ship Canal. Baltimore: The Friedenwald Company, 1897.

Lamoreaux, Naomi. *The Great Merger Movement in American Business, 1895–1904.* Cambridge: Cambridge University Press, 1985.

Leading Manufacturers and Merchants of Ohio Valley. New York: International Publishing Company, Publishers, 1887.

Lesley, J. P. *The Iron Manufacturer's Guide to the Furnaces, Forges and Rolling Mills of the United States.* New York: John Wiley, Publisher, 1859.

Lewis, Ronald L. *Welsh Americans: A History of Assimilation in the Coalfields.* Chapel Hill: University of North Carolina Press, 2008.

Lewis, W. David. *Sloss Furnaces and the Rise of the Birmingham District: An Industrial Epic.* Tuscaloosa: The University of Alabama Press, 1994.

Linkon, Sherry, and John Russo. *Steeltown U. S. A.: Work and Memory in Youngstown.* Lawrence: University Press of Kansas, 2002.

Lossing, Benson J. *History of American Industries and Arts.* Philadelphia: Porter & Coates, 1876.

Lubetkin, M. John. *Jay Cooke's Gamble: The Northern Pacific Railroad, the Sioux, and the Panic of 1873.* Norman: University of Oklahoma Press, 2006.

Macfarlane, James. *The Coal Regions of America: Their Topography, Geology, and Development.* New York: D. Appleton and Company, 1873.

Macgregor, John. *The Progress of America, from the Discovery by Columbus to the Year 1846.* Vol. 2. London: Whittaker & Co., 1847.

The Mahoning Valley, Condensed Statement of Its Resources, and Exhibit of Its Mines, Blast Furnaces & Rolling Mills, Its Commercial Centre, Youngstown, Ohio. Youngstown, OH: Mahoning Valley Centennial Association, 1876.

McCord, William B. *History of Columbiana County, Ohio and Representative Citizens.* Chicago: Biographical Publishing Co., 1905.

Meade, Richard. *The Coal and Iron Industries of the United Kingdom.* London: Crosby Lockwood and Co., 1882.

Memorial Record of the County of Cuyahoga and the City of Cleveland, Ohio. Chicago: The Lewis Publishing Company, 1894.

Misa, Thomas J. *A Nation of Steel: The Making of Modern America, 1865–1925.* Baltimore: The Johns Hopkins University Press, 1995.

Montgomery, David. *The Fall of the House of Labor: The Workplace, the State, and American Labor Activism, 1865–1925*. Cambridge: Cambridge University Press, 1987.

Moody's Manual of Railroads and Corporation Securities. Vol. 2. New York: Poor's Publishing Company, 1921.

The National Cyclopedia of American Biography. Vol. 5. New York: James T. White & Company, 1897.

The National McKinley Birthplace Memorial. Niles, OH: The National McKinley Birthplace Association, 1918.

Nelson's Biographical Dictionary and Historical Reference Book of Erie County, Pennsylvania. Erie, PA: S. B. Nelson Publisher, 1896.

Nevins, Allan. *The Emergence of Modern America, 1865–1878*. New York: The Macmillan Company, 1927.

Newberry, John S. *Annual Report of the Directors and Chief Engineer of the Ashtabula & New Lisbon Rail Road Co.* Cleveland: Printed by E. Cowles & Co., 1857.

Overman, Frederick. *The Manufacture of Iron, in All Its Various Branches*. 3rd ed. Philadelphia: Henry C. Baird, 1854.

Paskoff, Paul F. *Encyclopedia of American Business History and Biography: Iron and Steel in the Nineteenth Century*. New York: Bruccoli Clark Layman, 1989.

Patrick, Marsena R. *Memorial of Hon. William Kelly*. Albany: Joel Munsell, Printer, 1873.

Peck, J. M. *New Guide for Emigrants to the West*. 2nd ed. Boston: Gould, Kendall & Lincoln, 1837.

Percy, John. *The Manufacture of Russian Sheet-Iron*. Philadelphia: Henry Carey Baird, 1871.

Pittsburgh: Its Industry & Commerce. Pittsburgh: Barr & Myers, 145 Wood Street, 1870.

Proceedings of the Lake Superior Mining Institute. Vol. 19. Ishpeming, MI: Published by the Institute, 1914.

Reed, George Irving. *Century Cyclopedia of History and Biography of Pennsylvania*. Vol. 2. Chicago: The Century Publishing and Engraving Company, 1904.

Reid, Whitelaw. *Ohio in the War: Her Statesmen, Generals and Soldiers*. Vol. 2. Cincinnati: The Robert Clarke Company, 1895.

The Repertory of Patent Inventions. Vol. 30. London: Alexander Macintosh, Great New-Street, July–December 1857.

Reynolds, Terry S., and Virginia P. Dawson. *Iron Will: Cleveland-Cliffs and the Mining of Iron Ore, 1847–2006*. Detroit: Wayne State University Press, 2011.

Robinson, Jesse S. *The Amalgamated Association of Iron, Steel and Tin Workers*. Baltimore: The Johns Hopkins Press, 1920.

Rodgers, Allan Louis. "The Iron and Steel Industry of the Mahoning and Shenango Valleys." PhD diss., University of Wisconsin, 1950.

Rogers, Samuel Baldwin. *An Elementary Treatise of Iron Metallurgy*. New York: H. Bailliere, Publisher and Foreign Bookseller, 1857.

Rose, James Douglas. *Duquesne and Rise of Steel Unionism*. Champaign: University of Illinois Press, 2001.

Roy, Andrew. *A History of the Coal Miners of the United States*. Columbus: Press of J. L. Trauger Printing Company, 1907.

Ruffolo, Richard M., and Charles N. Ciampaglio. *From the Shield to the Sea: Geological Field Trips from the 2011 Joint Meeting of the GSA Northeastern and North-Central Sections*. Boulder, CO: The Geological Society of America, 2011.

Ruminski, Clayton J. "From Mary Furnace to Sharon Steel: Evolution and Integration of the United States' Last Manually Filled Blast Furnace, 1845–1963." *IA: The Journal of the Society for Industrial Archeology* 38, no. 2 (2012): 38–60.

Sabadasz, Joel. "The Development of Modern Blast Furnace Practice: The Monongahela Valley Furnaces of the Carnegie Steel Company, 1872–1913." *IA: The Journal of the Society for Industrial Archeology* 18 (1992): 94–105.

Sanderson, Thomas W. *20th Century History of Youngstown and Mahoning County, Ohio*. Chicago: Biographical Publishing Company, 1907.

Saward, Frederick E. *The Coal Trade*. New York: Published at 111 Broadway, 1878.

Sharp, Myron B., and William H. Thomas. *A Guide to the Old Stone Blast Furnaces in Western Pennsylvania*. Pittsburgh: Historical Society of Western Pennsylvania, 1966.

Sisson, Chas. W. *The ABC of Iron*. Louisville, KY: Press of the Courier-Journal Job Printing Co., 1893.

Smith, Joseph P. *History of the Republican Party in Ohio*. Vol. 2. Chicago: The Lewis Publishing Company, 1898.

Speer, J. Ramsey. *Chronology of Iron and Steel*. Pittsburgh: Pittsburgh Iron & Steel Foundries Company, 1920.

Statistical Report for the National Association of Iron Manufacturers for 1872. Philadelphia: J. A. Wagenseller, 1873.

Statistics of the American and Foreign Iron Trades for 1877. Philadelphia: The American Iron and Steel Association, 1878.

Statistics of the American and Foreign Iron Trades for 1878. Philadelphia: The American Iron and Steel Association, 1879.

Statistics of the American and Foreign Iron Trades for 1880. Philadelphia: The American Iron and Steel Association, 1881.

Statistics of the American and Foreign Iron Trades for 1881. Philadelphia: The American Iron and Steel Association, 1882.

Statistics of the American and Foreign Iron Trades for 1882. Philadelphia: The American Iron and Steel Association, 1883.

Statistics of the American and Foreign Iron Trades for 1883. Philadelphia: The American Iron and Steel Association, 1884.

Statistics of the American and Foreign Iron Trades for 1888. Philadelphia: The American Iron and Steel Association, 1889.

Statistics of the American and Foreign Iron Trades for 1890. Philadelphia: The American Iron and Steel Association, 1891.

Statistics of the American and Foreign Iron Trades for 1895. Philadelphia: The American Iron and Steel Association, 1896.

Statistics of the American and Foreign Iron Trades for 1897. Philadelphia: The American Iron and Steel Association, 1898.

Statistics of the American and Foreign Iron Trades for 1902. Philadelphia: The American Iron and Steel Association, 1903.

Statistics of the American and Foreign Iron Trades for 1910. Philadelphia: The American Iron and Steel Association, 1911.

Stern, Madeleine B. *We the Women: Career Firsts of Nineteenth-Century America*. Lincoln: University of Nebraska Press, 1994.

Stewart, John Struthers. *History of Northeast Ohio*. Vol. 1. Indianapolis: Historical Publishing Co., 1935.

Stowell, Myron R. *"Fort Frick," or the Siege of Homestead: A History of the Famous Struggle between the Amalgamated Association of Iron and Steel Workers and the Carnegie Steel Company (Limited) of Pittsburgh, PA*. Pittsburgh: Pittsburgh Printing Co., 1893.

Swank, James Moore. *The American Iron Trade in 1876*. Philadelphia: The American Iron and Steel Association, 1876.

———. *History of the Manufacture of Iron in All Ages*. Philadelphia: American Iron and Steel Association, 1892.

———. *Introduction to a History of Ironmaking and Coal Mining in Pennsylvania*. Philadelphia: Published by the author, 1878.

Sweetser, Ralph. *Blast Furnace Practice*. New York: McGraw-Hill Book Company, 1938.

Swineford, A. P. *Annual Review of the Iron Mining and Other Industries of the Upper Peninsula*. Marquette, MI: The Mining Journal, 1882.

Temin, Peter. *Iron and Steel in Nineteenth Century America: An Economic Inquiry*. Cambridge, MA: The MIT Press, 1964.

Tenth Annual Report of the Ohio State Board of Arbitration for the Year Ending December 31, 1902. Springfield, OH: The Springfield Publishing Company, 1904.

Thomas, Larry. *Coal Geology.* West Sussex: John Wiley & Sons, 2002.

Thurston, George H. *Pittsburgh and Allegheny in the Centennial Year.* Pittsburgh: A. A. Anderson & Son, 1876.

———. *Pittsburgh as It Is: Or, Facts and Figures Exhibiting the Past and Present of Pittsburgh, its Advantages, Resources, Manufactures, and Commerce.* Pittsburgh: W. S. Haven, 1857.

Treese, Lorett. *Railroads of Pennsylvania: Fragments of the Past in the Keystone Landscape.* Mechanicsburg, PA: Stackpole Books, 2003.

Trester, Delmer J. "The Political Career of David Tod." MA thesis, Ohio State University, 1950.

Twenty-Fourth Annual Report Republic Iron & Steel Company for the Year Ending December 31, 1925. Pittsburgh: Republic Iron and Steel Co., 1926.

Van Tassel, David D., and John Vacha. *Behind Bayonets: The Civil War in Northern Ohio.* Kent, OH: Kent State University Press, 2006.

Van Tine, Warren, and Michael Pierce. *Builders of Ohio: A Biographical History.* Columbus: Ohio State University Press, 2003.

Vernon, J. R. "Unemployment Rates in Postbellum America: 1869–1899." *Journal of Macroeconomics* 16, no. 4 (1994): 701–14.

Vogt, Helen. *Westward of Ye Laurall Hills.* Brownsville, PA: Published by the author, 1976.

Warren, Kenneth. *The American Steel Industry, 1850–1970: A Geographical Interpretation.* Oxford: Clarendon Press, 1973.

———. *Bethlehem Steel: Builder and Arsenal of America.* Pittsburgh: University of Pittsburgh Press, 2008.

———. *Big Steel: The First Century of the United States Steel Corporation, 1901–2001.* Pittsburgh: University of Pittsburgh Press, 2001.

———. *Wealth, Waste, and Alienation: Growth and Decline in the Connellsville Coke Industry.* Pittsburgh: University of Pittsburgh Press, 2001.

Wengenroth, Ulrich. *Enterprise and Technology: The German and British Steel Industries, 1865–1895.* Cambridge: Cambridge University Press, 1994.

White, C. M. *Blast Furnace Blowing Engines: Past, Present and Future.* Princeton, NJ: Princeton University Press, 1947.

White, John G. *A Twentieth Century History of Mercer County Pennsylvania.* Vol. 1. Chicago: The Lewis Publishing Company, 1909.

White, John R. "The Rebirth and Demise of Ohio's Earliest Blast Furnace: An Archaeological Postmortem." *Midcontinental Journal of Archaeology* 21 (1996): 217–45.

Whittlesey, Charles. *Report on the Mineral Ridge Mining Property Owned By Rice, French & Co., Located in Weathersfield, Trumbull Co., Ohio.* Cleveland: Harris, Fairbanks & Co., 1856.

Wicker, Elmus. *Banking Panics of the Gilded Age.* Cambridge: Cambridge University Press, 2000.

Wiggins, Joseph. *Directory of Beaver, Shenango and Mahoning Valleys, for 1869.* Pittsburgh: Printed and Bound by Bakewell & Marthens, 71 Grant Street, 1869.

Wilkie, George. *The Manufacture of Iron in Great Britain.* London: A. Fullarton & Co., 1857.

Willard, Eugene B., Daniel W. Williams, George O. Newman, and Charles B. Taylor. *A Standard History of the Hanging Rock Iron Region of Ohio.* Vol. 1. Chicago: The Lewis Publishing Company, 1916.

The William B. Pollock Company Presents the Seventy-Five Year History of Its Contributions to the Advancement of the Art of Iron and Steelmaking. Youngstown, OH: William B. Pollock Co., 1939.

Williams, Ralph D. *The Honorable Peter White: A Biographical Sketch of the Lake Superior Iron Country.* Cleveland: The Penton Publishing Co., 1907.

Williams, W. Mattieu. *The Chemistry of Iron & Steel Making.* London: Chatto & Windus, 1890.

Wilson, Dreck Spurlock. *African American Architects: A Biographical Dictionary, 1865–1945.* New York: Routledge, 2004.

Wilson, Erasmus. *Standard History of Pittsburgh, Pennsylvania.* Chicago: H. R. Cornell & Company, 1898.

Wollman, David H., and Donald R. Inman. *Portraits in Steel: An Illustrated History of Jones & Laughlin Steel Corporation.* Kent, OH: Kent State University Press, 1999.

Wright, Carroll D. "The National Amalgamated Association of Iron, Steel, and Tin Workers, 1892–1901." *The Quarterly Journal of Economics* 16 (November 1901): 37–68.

Year Book of the American Iron and Steel Institute, 1917. New York: American Iron and Steel Institute, 1918.

Youngstown Past and Present. Cleveland: Wiggins & McKillop, 1875.

INDEX